Rural China in Transition

Studies on Contemporary China

The Contemporary China Institute, based at the School of Oriental and African Studies (University of London), was established in 1968. Throughout its existence, it has been a national focus for research and publications on twentieth-century China. This new series, which is edited at the Institute, seeks to maintain and extend that tradition by making available the best work of scholars throughout the world. Future volumes will embrace a wide variety of subjects, including China's social, political, and economic change, intellectual and cultural developments, and foreign relations.

Editorial Board:

Dr Robert F. Ash
Mr B. G. Hook
Professor C. B. Howe
Dr David Shambaugh
Mr David Steeds

Rural China in Transition

Non-agricultural Development in

Rural Jiangsu, 1978–1990

SAMUEL P. S. HO

CLARENDON PRESS · OXFORD

1994

Oxford University Press, Walton Street, Oxford OX2 6DP
Oxford New York Toronto
Delhi Bombay Calcutta Madras Karachi
Kuala Lumpur Singapore Hong Kong Tokyo
Nairobi Dar es Salaam Cape Town
Melbourne Auckland Madrid
and associated companies in
Berlin Ibadan

Oxford is a trade mark of Oxford University Press

Published in the United States
by Oxford University Press Inc., New York

British Library Cataloguing in Publication Data
Data available

Library of Congress Cataloging in Publication Data
Ho, Sam P. S.
Rural non-agricultural development in Post-Mao China: the Jiangsu
experience / Samuel P. S. Ho.
p. cm.—(Studies in contemporary China)
Includes bibliographical references.
1. Rural industries—China—Kiangsu Province. 2. Kiangsu Province (China)—Economic
policy. 3. Kiangsu Province (China)—Economic conditions. I. Title. II. Series.
HC428.K5H6 1994 93-48891
338.0951'136—dc20
ISBN 0–19–828823–9

1 3 5 7 9 10 8 6 4 2

Typeset by Best-set Typesetter Ltd., Hong Kong
Printed in Great Britain
on acid-free paper by
Biddles Ltd., Guildford & King's Lynn

To my wife Sharon

Preface

INTERNATIONAL and historical experience suggest a strong relationship between the shift in the sectoral composition of employment away from agriculture to non-agricultural activities and that in population from rural to urban areas. Despite this historical fact, Chinese leaders, for various economic, social, and political reasons, have long resisted changing the government's restrictive policy towards migration, particularly the permanent movement of rural residents to urban areas, adopted in the 1950s. They point to the fact that China's large cities are already congested, with grossly inadequate infrastructure and housing, that funds for urban housing and infrastructure are lacking, that employment opportunities in cities are limited, and that the environmental and social costs of rapid urbanization are potentially very high. Undoubtedly, they also worry about the difficulty of controlling a large freely mobile population.

The removal of cities as places of permanent residence and employment means that alternative ways had to be found to absorb the growing number of underemployed in the countryside and to relieve the pressure on the rural population to migrate to cities. In the late 1970s, China announced its alternative to rural–urban migration. As part of rural reform, peasants would be encouraged hereafter to leave the land but not the countryside—i.e. they would remain in the countryside but be permitted to pursue non-agricultural activities. The purpose of this volume is to describe and to analyse the consequences of this decision.

The idea of a study of the process by which peasants leave the land but not the countryside first came up in a discussion the author had with the Jiangsu Provincial Academy of Social Sciences in the summer of 1985. Shortly after that discussion we decided to launch a collaborative research project to study rural non-agricultural development in Jiangsu. The project was initiated in 1986 under the umbrella of a collaborative research arrangement between the University of British Columbia and the Jiangsu Provincial Academy of Social Sciences, and fieldwork was conducted in Jiangsu between 1986 and 1988. The findings of the field research were disseminated at a conference in Nanjing in 1989, and the conference papers were subsequently published in Chinese (see Samuel P. S. Ho, Gu Jirui, Yan Yinglong, and Bao Zongshun, *Jiangsu nongcun fei nonghua fazhan yanjiu* (*Research on Rural Non-agricultural Development in Jiangsu*) (Shanghai: Shanghai renmin chubanshe, 1991).

In preparing this volume, I have accumulated a long list of individuals and institutions to whom I am deeply indebted. Most of all I am grateful to my colleagues at the Jiangsu Provincial Academy of Social Sciences, Gu Jirui, Yan Yinglong, Bao Zongshun, and Deng Jianjun, who participated in the field research and taught me much about life in China. I also wish to express my appreciation to the many government officials, local cadres, and rural households in Ganyu, Yixing, and Zhangjiagang who participated in the surveys and interviews and patiently answered our questions. I should also state the obvious. The responsibility for the expressed views and any errors in analysis or interpretation in this volume is exclusively mine.

The field research was carried out with the aid of a grant from the International Development Research Centre, Ottawa, Canada. At various times, I also received financial support from the University of British Columbia Humanities and Social Sciences Grants Committee. A first draft of this volume was written during 1989–91, when I was in part supported by a UBC Izaak Walton Killam Memorial Senior Fellowship. To these organizations, I would like to express my appreciation for their generous support.

I am grateful to Denise Hare, Y. Y. Kueh, Bruce L. Reynolds, Edgar Wickberg, and two anonymous referees who read all or parts of the first draft of the book and made valuable comments upon it. The map in Fig. 3.1 was prepared by Jean François Proulx, and this book benefited from initial editing by Ryan Dunch.

Chinese names and phrases have been romanized for the most part according to the *pinyin* system. The only conscious exceptions are the names of Hong Kong, Macao, and authors residing outside of China which have been romanized in the traditional way. The surnames of Chinese authors are given first except when authors have Western first names or reside outside of China.

S. P. S. H.

Contents

List of Tables

List of Figures

Abbreviations

BR	*Beijing Review*
CBE	commune-brigade enterprise
CCP	Chinese Communist Party
CCPCC	Chinese Communist Party Central Committee
CQ	*China Quarterly*
FTC	foreign trade company
GLF	Great Leap Forward
GVAIO	gross value of agricultural and industrial output
GVAO	gross value of agricultural output
GVIO	gross value of industrial output
JD	*Jingji diaocha* (*Economic Survey*)
JJN	*Jiangsu jingji nianjian* (*Economic Yearbook of Jiangsu*)
JNJSFTZ	*Jiangsu sheng nongcun jingji shouru fenpei tongji ziliao, 1986* (*Statistical Material on the Distribution of Rural Income in Jiangsu Province, 1986*)
JNZXTZ	*1985 nian Jiangsu sheng nongcun zhuanyehu xin jingji lianheti tongji ziliao* (*Statistical Material on Specialized Households and New Economic Associations in Rural Jiangsu, 1985*)
JPASS	Jiangsu Provincial Academy of Social Sciences (Jiangsu sheng shehui kexueyuan)
JPSB	Jiangsu Provincial Statistical Bureau (Jiangsu sheng tongjiju)
JQ	*Jingji quanli* (*Economic Management*)
JSN	*Jiangsu sishi nian* (*Forty Years of Jiangsu*)
JTN	*Jiangsu tongji nianjian* (*Statistical Yearbook of Jiangsu*)
JY	*Jingji Yanjiu* (*Economic Research*)
JXQTZ	*1986 nian Jiangsu sheng xiangzhen qiye tongji ziliao* (*Statistical Material on Rural Enterprises, Jiangsu Province, 1986*)
NCR	north coastal region
NDR	north delta region
NMR	north mountainous region
NJYS	*Nongcun jingji yu shehui* (*Rural Economy and Society*)
NJW	*Nongye jingji wenti* (*Problems in Agricultural Economics*)
PA	*Pacific Affairs*
PE	private enterprise

QBLQDZ	*Quanguo beicun laodongli qingkuang diaocha ziliaoji* (*Collection of Material from the National Survey of Labor Conditions in 'One Hundred' Villages*)
RGVNAO	rural gross value of non-agricultural output
RGVIO	rural gross value of industrial output
RMB	*renminbi*
RP	rural population
SC	State Council (Guowuyuan)
SDR	south delta region
SE	state enterprise
SEZ	special economic zone
SMR	south mountainous region
SSB	State Statistical Bureau (Guojia tongjiju)
SSTC	State Science and Technology Commission (Guojia kexue jishu weiyuanhui)
TVE	township-village enterprise
TE	township enterprise
TG	township government
TVG	township-village government
VE	village enterprise
VG	village government
XQZFX	*Xiangzhen qiye zhengce fagui xuanbian* (*Selected Laws and Regulations Concerning Township-Village Enterprise*)
YTM	Yixing Telecommunication Material
ZJN	*Zhongguo jingji nianjian* (*Economic Yearbook of China*)
ZNN	*Zhongguo nongye nianjian* (*Agricultural Yearbook of China*)
ZNTN	*Zhongguo nongcun tongji nianjian* (*Statistical Yearbook of Rural China*)
ZSC	Zhangjiagang Steel Cable
ZTN	*Zhongguo tongji nianjian* (*Statistical Yearbook of China*)

1

Introduction

IN 1978, after three decades of development and industrial growth, three-quarters of China's population of one billion still lived in the countryside and earned its living from the land, the consequence of a long-standing government policy, that prohibited rural–urban migration, and of development policies that were frequently dominated by political rather than economic considerations. With little prospect for expanding cultivated area and a large and growing rural population, the Chinese government became persuaded that if China were to become a middle-level, developing country, a substantial share of its rural labour force would have to move from farming to more productive non-agricultural activities. However, the government remained convinced that this occupational shift had to be achieved without significantly enlarging the population of its urban centres through rural–urban migration. Accordingly, since the late 1970s, rural China has actively pursued the policy of leave the land but not the countryside, enter the factory but not the city (*li tu bu li xiang, jin chang bu jin cheng*).

The Phenomenon of Rural Non-agricultural Development

Peasants responded to the call to leave the land but not the countryside with great enthusiasm, and the rural non-agricultural sector emerged in the 1980s as the most dynamic component of the Chinese economy. For example, the share of gross value of industrial output (GVIO) produced in rural areas increased sharply, rising from below 10 per cent in 1978 to nearly 27 per cent in 1989.[1] And this has come about with little or no investment from the state. Private enterprises in rural areas and community-owned (collective) enterprises at or below the township level have become important players in all aspects of China's rural economic life. (Hereafter 'rural enterprises' refer to both private and collective enterprises in rural areas and 'township-village enterprises (TVEs)' refer to collective enterprises at the township and village levels.)[2] Rural enterprises have generated huge amounts of tax revenues for the government, given critical support to agriculture, and developed into a significant source of income for rural residents as well as an indispensable source of revenue for local community development.

The development of the rural non-agricultural sector also has important implications for the long-term modernization and reform of the economy. The sector is almost completely market orientated and thus helps to enliven the economy, providing it with a much needed degree of flexibility. Its very existence gives the rural population greater choice and provides a competitive stimulus to the state sector. The spirit of entrepreneurship and the increased competition and commercialization that have accompanied rural non-agricultural development have undoubtedly also introduced new values to the countryside and helped modify traditional values and behaviour. At the same time, rural non-agricultural development has created and will continue to create new problems—e.g. reduced ability of the central government to control aggregate demand (particularly investment), a spatial distribution of enterprises that is not always economically efficient, wasteful duplications in investments, tension between rural and state enterprises, and misuse of farm land.

Given the growing importance of the rural non-agricultural sector and the potential it holds for helping China to modernize its vast countryside, it is imperative that we understand its growth process, its present status, and its problems. Specifically, we need to understand better how the sector operates and the forces that are driving its development; to determine the extent, the composition, and the spatial distribution of rural non-agricultural activities; to investigate the effects of rural non-agricultural development on agriculture and farm households, the major problems encountered in its development, and the conditions under which rural non-agricultural development is likely to succeed in China. The purpose of this study is to shed light on these issues.

Rural Non-agricultural Development in Jiangsu

Given China's size and diversity and the inaccessibility of its countryside, to conduct fieldwork in different parts of China presents formidable problems. The alternative, and the one adopted in this study, is to restrict the in-depth study to one province, and to use the experience of that province along with scattered reports from other provinces to shed light on the entire sector. That Jiangsu has been at the forefront of China's rural non-agricultural development, has displayed different patterns of rural non-agricultural development in its various regions, and has been

TABLE 1.1. *Selected Economic Indicators, 1988, China and Jiangsu*

	China	Jiangsu	Jiangsu's rank among the 30 provincial administrative units[a]
Population (million)	1,096.1	64.4	4[b]
Cultivated area (million *mu*[c])	1,435.8	68.5	6[d]
Social product[e] (RMB billion)	2,984.7	308.9	1
National income[f] (RMB billion)	1,177.0	99.3	1
Per capita national income (RMB)	1,081	1,553	4[g]
Per capita farm household income (RMB)	545	797	6[h]
Average cultivated area[i] per person in rural household (*mu*)	2.06	1.41	19
Net agricultural product (RMB billion)	381.8	30.5	3[j]
Grain production (million tons)	394.1	32.2	2[k]
Net industrial product (RMB billion)	543.2	50.8	1
Workers employed by TVEs (million)	48.9	7.4	1
Gross revenue of TVEs (RMB billion)	423.2	82.1	1
% of GVIO			
Light industry	49.3	53.5	
Heavy industry	50.7	46.5	
% of GVIO produced by			
State enterprises	56.8	34.7	
Collective enterprises	36.2	59.4	
Individual (private) enterprises	4.3	3.2	
Other enterprises[l]	2.7	2.7	

[a] There are 27 provinces and 3 provincial-level municipalities.
[b] Behind Sichuan (105.8), Henan (80.9), and Shandong (80.6).
[c] 1 *mu* = 0.0667 ha.
[d] Behind Henan (104.3), Shandong (103.4), Hebei (98.5), Sichuan (94.7), and Nei Mongol (73.1).
[e] The gross output of the material products sectors.
[f] *Guomin shouru*, roughly equivalent to net domestic material product.
[g] Behind Shanghai, Beijing, and Tianjin.
[h] Behind Shanghai (1,301), Beijing (1,963), Zhejiang (902), Tianjin (891), and Guangdong (809).
[i] Responsibility land plus private plots.
[j] Behind Shandong (32.8) and Sichuan (31.6).
[k] Behind Sichuan (37.9).
[l] Various types of joint ventures involving state enterprises, collective enterprises, and private enterprises, and joint equity ventures involving foreign investments.

Sources: ZTN (1989), 29, 44, 56, 89, 172, 190, 202, 270, and 747; and *JTN* (1989), 33, 37, and 41.

held up as a model for other provinces to emulate makes it an obvious candidate for in-depth study.[3]

Located on the coast in the lower reaches of the Yangzi River, bordered in the south by the great port city of Shanghai and the province of Zhejiang, in the west by Anhui, and in the north by Shandong (see Fig. 3.1), Jiangsu is one of China's most developed provinces. To gauge its relative importance in the national economy, Table 1.1 compares selected economic indicators for Jiangsu with those for the country as a whole.

With an area of over 100,000 square kilometres (slightly larger than South Korea or about the same size as the states of Maine and New Hampshire combined) and a population of 64 million in 1988, Jiangsu has one of the highest population densities (640 persons per square kilometre) in the world. The province is in a leading agricultural region and is one of China's principal grain producers. Traditionally Jiangsu has been the country's foremost centre of light industry, notably textiles, processed food, and paper. However, with increasing emphasis on the development of such industries as motor vehicles, machinery and equipment, electronics, and chemicals since 1949, heavy industry has become more important, producing about 45 per cent of Jiangsu's GVIO since the mid-1980s. In 1988, 30.8 per cent of Jiangsu's national income (*guomin shouru*) (roughly equivalent to net domestic material product) originated in agriculture, 51.2 per cent in industry, 5.7 per cent in construction, 4.0 per cent in transportation, and 8.3 per cent in trade.[4] In the same year, the distribution of the national economy by sector was: agriculture, 32.4 per cent; industry, 46.2 per cent; construction, 6.7 per cent; transportation, 3.7 per cent; and trade, 11.0 per cent.[5] Thus, compared to the national economy, Jiangsu was more industrialized.

In 1988, Jiangsu ranked first in the nation by a wide margin in both total social product and national income. Jiangsu's per capita national income was the highest in China after the three provincial-level municipalities. One reason for Jiangsu's current economic prominence is its highly developed rural non-agricultural sector. Indeed, nowhere in China is the rural non-agricultural sector more prominent than in Jiangsu. Stimulated by the national policy of leave the land but not the countryside and supported by other rural reform policies, the rural non-agricultural sector in Jiangsu has grown extremely rapidly since 1978, and is largely responsible for the dramatic increases in the province's rural income and output.[6] Almost overnight, a huge number of new industrial and other non-agricultural activities have emerged and thousands of privately and collectively owned enterprises have been established in Jiangsu's countryside. Of particular importance are TVEs, enterprises collectively owned and operated by townships or villages. In 1988, Jiangsu's TVE sector, in terms both of employment and of revenue generated, ranked first in the country, and its prominence is also suggested by the fact that nearly 60 per cent of Jiangsu's GVIO in 1988 was produced by the collective sector as compared to the national average of only 36 per cent.

TVEs are active in all components of the non-agricultural sector: commerce, construction, transportation, and industry. But it is in industry that their importance is most prominent. In the late 1980s, over 45 per cent of Jiangsu's GVIO was produced by rural enterprises, and most of it by TVEs. This is all the more impressive in view of the fact that Jiangsu was ranked first in the nation in industrial production. Within industry, TVEs are most active in the following subsectors (by order of importance): machinery, textiles, construction materials and chemicals, which together account for about three-quarters of the GVIO produced by TVEs in Jiangsu. These are the same industries that are important for TVEs in China as a whole, although the relative ranking at the national level is slightly different.[7]

The proliferation of rural enterprises in Jiangsu's countryside has had far-reaching consequences. Millions of rural workers have shifted from farming to non-agricultural activities and in the process have transformed the structure of Jiangsu's rural economy. In 1988, Jiangsu had over 119,800 TVEs, and they employed nearly 7.4 million rural workers, or about 27 per cent of the rural labour force.[8] In addition there were more than 600,000 individual and joint-household enterprises. Without rural non-agricultural development, and in particular the development of the TVEs, the commercialization of the rural economy and the transfer of workers from farming to more productive activities in Jiangsu would have been impossible.

Given its size and importance, a study of Jiangsu's rural non-agricultural development is of interest in itself. However, the hope is that this study will also shed light on rural non-agricultural development in China as a whole. Is this realistic? How representative is Jiangsu's experience? The fact that Jiangsu is one of China's most developed and industrialized provinces and is at the forefront of China's rural non-agricultural development suggests that it is not an average province and its experience may not be representative.

Nevertheless, Jiangsu's experience can reveal much about rural non-agricultural development. For one thing, rural non-agricultural development in Jiangsu has been fuelled by the same economic and political forces that are present in other parts of China. Furthermore, latecomers will face many of the same difficulties that confronted Jiangsu in its effort at rural non-agricultural development, so the Jiangsu experience—the successes and the hard lessons—is also of relevance to them. Regions in Jiangsu, as in China, are not equally endowed or equally developed. Indeed, there are parts of north Jiangsu that are still underdeveloped,

even by national standards. Thus parts of Jiangsu have experienced the
same difficulties in developing the rural non-agricultural sector as other
poor regions in China. In other words, the story of rural non-agricultural
development in Jiangsu is not one of unblemished success. And both
Jiangsu's successes and its failures are likely to shed light on the process
of rural non-agricultural development at the national level. Nevertheless
Jiangsu is not China, and the diversity that is China suggests that we must
interpret Jiangsu's experience with care. In a country as large and diverse
as China, there is great danger in accepting uncritically the experience of
one region as representing that of the country as a whole.

Terminology

In his seminal work on rural industrialization in China, Jon Sigurdson
makes the useful distinction between 'industrial decentralization' and
'rural industrialization'.[9] Industrial decentralization is the process by
which industrial development is moved away from the metropolitan
centres to other communities, usually to secondary urban centres but
sometimes also to the countryside. Rural industrialization is the development of industry in rural communities, i.e. villages and market towns
(small urban settlements that are economically and socially integrated
with the surrounding rural areas). In the past four decades China has
promoted industrial decentralization,[10] and at times has also encouraged
rural industrialization. The current policy of leave the land but not the
countryside can be seen as basically a rural industrialization policy,
adopted to bring income and job opportunities to the countryside.
However, unlike previous attempts at rural industrialization, the objective of the current policy is to promote not only the development of
industry in rural areas but other types of non-agricultural activities as
well.

A few words are necessary about the coverage of the term 'rural non-
agricultural sector' as used in this study. Any definition of 'rural' inevitably involves a degree of arbitrariness since it imposes a single dividing
line on what is in fact a continuous spectrum of changing conditions. In
most less developed countries, where statistics are tabulated by administrative units, 'rural' is defined as those units not designated as municipalities or secondary cities. This practice is also followed in studies of
China. For example, in the Western literature on the Chinese economy,
'rural' usually refers to those administrative units at or below the county
level, i.e. counties (*xian*), townships (*xiang*/*zhen*), and villages (*cun*).

However, the way statistics are tabulated in China does not permit the
inclusion in 'rural industry' of all industrial enterprises located in 'rural
areas'. Thus, in past studies, 'rural industry' in China refers only to those

industrial units owned and operated by counties, townships (rural communes), and villages (brigades).[11] Excluded are state enterprises located in 'rural' areas but under the jurisdiction of an administrative unit above the county level. In other words, 'rural', as used in these studies, is not strictly a spatial definition, but a combined spatial–ownership-level concept. Because the very large state enterprises are normally under an administrative unit above the county level, 'rural industry' is also frequently referred to as 'rural small-scale industry'. However, it should be noted that county-owned state enterprises are often neither small nor rural. Most are located in county cities,[12] and many employ 400–500 or more workers.[13]

In this study, the rural non-agricultural sector is also a combined 'spatial–ownership concept. But it differs from 'rural industry' as used by other students of China in two important ways. First, 'rural' refers to only townships (*xiang/zhen*) and villages (*cun*), i.e. county cities (*xian cheng*) are excluded. Secondly, in our usage, the rural non-agricultural sector includes not only rural industry but also most other non-agricultural economic activities with ownership at the township level or below. Specifically, it includes individual activities and collective activities of teams, villages, and townships in the following sectors: industry, construction, transportation, and commerce. In other words, all 'rural' state enterprises are excluded as are rural non-agricultural activities in the non-productive service sectors (e.g. government administration, education, health).

As indicated in the above discussion, this study follows the Chinese practice and defines agriculture broadly to include cultivation, animal husbandry, fishery, forestry, and sidelines. However, the policy of leave the land but not the countryside encourages peasants not only to pursue non-agricultural activities but also to specialize and to take advantage of their natural environment. Thus in some of the less developed regions, the policy has produced not so much rural non-agricultural development as the development of fishery, forestry, or sideline activities. In these cases, it is more appropriate to describe the process as rural non-farm development. However, the focus of the policy is clearly on rural non-agricultural development, and, to avoid confusion, we will continue to use the term even though the process may in some instances be slightly broader.

Objectives and Organization of the Book

The general objective of this study is to document and to explain the process of rural non-agricultural development in Jiangsu from 1978 to 1990, and to use the Jiangsu experience to shed light on the process of rural non-agricultural development in China as a whole. More specifically, the study seeks to document the growth, ownership pattern, and regional variations of rural non-agricultural activities; to determine the extent to which rural non-agricultural development has altered the organization and structure of the rural economy, to identify the economic forces that determine the presence or absence of rural industry; to understand the institutional and regulatory environment within which rural enterprises operate; to ascertain the economic impact of rural non-agricultural development on rural households; to examine the role local governments have played in rural non-agricultural development; and to discuss some of the major issues raised by the rapid development of the rural non-agricultural sector.

The remainder of the book is organized as follows. Rural non-agricultural development has had a long history in China, and its recent resurgence did not occur in a policy vacuum. To put the Jiangsu experience in proper perspective, Chapter 2 reviews government policies towards rural non-agricultural development since the founding of the People's Republic, but particularly policy changes implemented after 1977, and discusses the pace and pattern of rural non-agricultural development in the country as a whole from 1978 to 1990.

Chapter 3 provides a statistical overview of rural non-agricultural development in Jiangsu and discusses four aspects of rural non-agricultural development that are important in Jiangsu as well as in other parts of China. The first is the spatial distribution of rural non-agricultural activities. The second is the effect rapid rural non-agricultural development has had on farm production. The third is the relationship between China's decision to turn to the outside world for trade and technology and the pace and pattern of rural industrialization. The fourth and last issue is the impact rural industrialization has had on the environment.

As in other provinces, Jiangsu's rural non-agricultural sector is composed of two components: collective activities owned and operated by rural communities (i.e. townships, villages, and teams), and private activities owned and operated by individuals or groups of individuals. Because the development of collective activities and that of private activities represent two different approaches to rural non-agricultural development, they are examined separately: the collective component in Chapters 4 and 5, and the private component in Chapter 6.

Chapter 4 examines three general features of the collective component: its ownership composition, its industrial structure, and the contract responsibility system used by rural communities to manage collective non-agricultural activities. How township-village enterprises managed their development in the 1980s in the partially reformed economy is the subject of Chapter 5. In particular, the chapter examines how TVEs mobilized the capital, the skills, and the technology needed for rural industrialization, how they coped with marketing and supply difficulties, and how they recruited and motivated workers.

Chapter 6 discusses the development of private activities in rural Jiangsu, particularly that of 'specialized households' and 'new economic associations', and analyses why it is believed that, unless reforms change China's political environment in ways unforeseen at the moment, the private sector is not likely to become much larger or more dynamic.

The final three substantive chapters return to more general issues. In China, township-village cadres have the primary responsibility for implementing the central government's rural policies and are the principal decision-makers and entrepreneurs in the countryside. Indeed, rural non-agricultural development in China has come about, to a large extent, as the result of efforts of rural cadres working at the grass-roots level with little material assistance from senior governments. Chapter 7 explores the priorities and motivations of rural cadres, and examines the relationships between TVGs, TVEs, and rural development.

One, if not the primary, objective of the policy of leave the land but not the countryside is to increase and improve income-earning opportunities in rural areas. To what extent has rural non-agricultural development contributed to the rise in rural income? Has rural non-agricultural development made rural income distribution more or less equal? Chapter 8 attempts to answer these questions. Chapter 9 summarizes the main findings and considers the future prospects for rural non-agricultural development in China.

Because this study relies heavily on interviews and surveys, how they were conducted and the quality of the data collected are discussed in the Appendix. Those uninterested in how the field research was designed and conducted may nevertheless want to read the section of the Appendix on the surveyed sites, which provides useful background materials for those sections and chapters that use the surveyed regions as illustrations.

NOTES

1. *ZTN* (1990), 49 and 333, and *ZNTN* (1985), 11. Reports of villages and townships being transformed overnight abound. For example, see the collection of such reports in *Su Zhe Yue xiangzhen qiye chenggong zhi lu (The Successful Path of Rural Enterprises in Jiangsu, Zhejiang, and Guangdong)* (Guangzhou: Guangdong renmin chubanshe, 1985).
2. The Chinese have called 'rural enterprises' by different names. For a discussion of the different terms, see Ch. 2.
3. For a description of economic conditions in Jiangsu in the early 1980s, see Jing Wei, 'Economic Newsletter', a 3-part series on economic development in

Jiangsu, *BR* (14 Nov. 1983), 17–22; (28 Nov. 1983), 17–22; and (12 Dec. 1983), 20–3.

4. Calculated from data in *JTN* (1989), 37.

5. *ZTN* (1989), 32.

6. Despite the rapid growth of rural income in Jiangsu, rural income in other parts of China may have grown even more rapidly. In 1988, Jiangsu ranked sixth in per capita rural household income. It is understandable that Jiangsu would rank behind the three provincial-level municipalities (Shanghai, Beijing, and Tianjin), where farm household income has traditionally been higher and where agriculture has become essentially a sideline activity with most farm households earning the bulk of their income from non-agricultural activities. However, Jiangsu also ranked behind Zhejiang and Guangdong, two provinces that were less developed than Jiangsu, and this requires a few words of explanation. Like Jiangsu, Guangdong is in a rich agricultural region and has a rapidly growing and robust rural non-agricultural sector (in 1988, Guangdong's TVE sector was second only to Jiangsu in size). But compared to Jiangsu's, Guangdong's rural labour-market was much tighter and therefore its rural wages were much higher than in Jiangsu (e.g. the average TVE employee in Guangdong received 60 per cent more in wages and bonuses in 1988 than one in Jiangsu). This difference helps to explain the higher per capita farm household income in Guangdong even though its rural non-agricultural sector was probably less developed than Jiangsu's. In Zhejiang the explanation is somewhat different. Zhejiang's rural non-agricultural sector is less developed than Jiangsu's, but rural individual (private) activities are much more important. Unlike collective enterprises, private enterprises distribute most of their earnings as income, and thus the impact of rural non-agricultural development on household income is much greater in Zhejiang than in Jiangsu.

7. For China as a whole, textiles is ranked behind construction materials. Another difference is that food processing is considerably less important in Jiangsu than in China as a whole.

8. *JTN* (1989), 112 and 154.

9. Jon Sigurdson, *Rural Industrialization in China* (Cambridge, Mass.: Council on East Asian Studies, Harvard University, 1977), 13–22.

10. For discussions of China's attempts at industrial decentralization see Charles R. Roll and K. C. Yeh, 'Balance in Coastal and Inland Industrial Development', in US Congress, Joint Economic Committee, *China: A Reassessment of the Economy* (Washington, DC: US Government Printing Office, 1975), 81–93, and Barry Naughton, 'The Third Front: Defence Industrialization in the Chinese Interior, *CQ* 115 (Sept. 1988), 351–86.

11. This is the coverage used in Sigurdson, *Rural Industrialization*, in American Rural Small-Scale Industry Delegation, *Rural Small-Scale Industry in the People's Republic of China* (Berkeley, Calif.: University of California Press, 1977), and in Christine P. W. Wong, 'Rural Industrialization in the People's Republic of China: Lessons from the Cultural Revolution Decade', in US Congress, Joint Economic Committee, *China Under The Four Modernizations* (Washington, DC: US Government Printing Office, 1982).

12. In China, 'small urban settlements' are classified into four categories: county cities (*xian cheng*), county towns (*xian shu zhen*), towns (*xiang zhen*), and small market towns (*jizhen*). Standards for county cities and county towns are set by the government, and before a 'small urban settlement' is designated a county city or a county town, it must have government approval. The county city is the political and economic centre of the county, and in the late 1980s many county cities in Jiangsu had a population in the range 50,000–100,000. Township governments are located in 'towns', and the more developed ones are designated 'county towns'.

13. e.g. the average number of workers employed by the eighteen county enterprises visited by the American Rural Small-Scale Industry Delegation in 1975 was 379.

2

Regulations, Economic Policies, and Rural Non-agricultural Development in Post-1949 China

THE level of rural non-agricultural activities in post-1949 China has been strongly influenced by government regulations and policies. Indeed the immediate cause of the recent explosive growth in rural non-agricultural activities in Jiangsu and other provinces was a series of changes in government regulations and policies that began in the late 1970s. Thus to understand better the process of rural non-agricultural development in China, it is necessary to examine the government's attitude towards rural non-agricultural activities and how it has changed in the past four decades. This chapter begins with a brief discussion of rural non-agricultural development in China from 1949 to the mid-1960s. Development during the Cultural Revolution decade (1966–76) is discussed in the second section. The series of changes in economic policy that made possible the explosive growth in rural non-agricultural activities after 1978 are reviewed in the third section. The chapter concludes by examining the post-1978 growth and structural changes in China's rural non-agricultural sector.

Rural Non-agricultural Development Before 1966

The precise level of rural non-agricultural activity at the time of Liberation is uncertain, although undoubtedly many rural residents were involved in commerce, craft production, and services.[1] However, we do know that, in some parts of pre-Liberation China, rural non-agricultural activities were prevalent. For example, a 1928 survey identified forty-four different household handicraft industry groups in Hebei Province alone, and, out of the 129 counties in Hebei, handicraft production existed in 127.[2] In some of these counties, upwards of two-thirds of the rural households were involved in craft production. In Ding *xian*, for example, 'every village had some form of home industry. Fifty-two per cent of the families in the *hsien* [*xian*] were engaged in some form of home industry. Fifty-seven per cent of the industrial output was produced by home industry.'[3]

Given the conditions in pre-Liberation China—low rural income, regional political instability, a primitive transportation system, a shortage of physical and human capital, and fragmented markets—rural non-agricultural activities were probably not as developed in China as a whole as they were in Ding *xian* but one suspects they were probably at a level not too different from those found in other low-income economies where rural residents engaged principally in non-agricultural activities usually account for between 20 and 30 per cent of the gainfully occupied rural population.[4] In fact a survey of 38,256 rural families in 101 representative localities selected from twenty-two provinces in China between 1929 and 1933 indicated that of the male population 7 years of age or older in the sample, 45 per cent worked only in agriculture, 27 per cent worked part-time in agriculture and part-time in other industries, and 20 per cent were engaged in non-agricultural pursuits only.[5] Jiangsu, being one of China's most developed provinces and a centre of traditional artisan and handicraft activities, undoubtedly had a level of rural non-agricultural activities in 1949 that was higher than the national average.[6]

In the early 1950s, perhaps as many as 20 million Chinese were engaged in small-scale (fewer than ten workers) non-mechanized industrial production either on a full-time basis or as a sideline. Of this number 12 million were located in rural areas—about 10 million were peasants with commercialized non-agricultural sidelines and the remaining 2 million were rural workers involved in full-time, small-scale industrial production.[7]

According to the statistical categories used in China in the early 1950s, the industrial sector was composed of modern factories, handicraft factories, handicraft co-operatives, and individual handicraft operators not engaged in agriculture.[8] Assuming that individual handicraft operators were mainly rural and all other establishments mainly urban, Sinha estimated that rural industries accounted for about one-fifth of the gross value of industrial output in the late 1940s and the early 1950s (GVIO, in 1957 constant prices, was RMB 12.6 billion in 1949 and RMB 30.8 billion in 1952).[9] This, of course, underestimates rural industrial activity, since a great deal of it occurred as sidelines in farmers' homes. Agricultural sideline activities (including home industry) produced RMB 1.16 billion in 1949 and RMB 1.83 billion in 1952 (both in 1957 constant prices).[10] Thus perhaps 25 per cent or more of China's industrial output in 1949 was produced in the countryside. In the same year, 48.2 per cent of Jiangsu's GVIO was produced by its handicraft industry, and much of this was probably produced in the countryside.[11]

In the 1950s, the government launched several major drives to restructure the rural economy. Rural workers were organized in stages, first into small groups ('mutual aid teams' for farmers and 'supply and marketing small groups' for craftsmen and small handicraft producers) and ultimately into agricultural or handicraft co-operatives. Some rural craftsmen remained an integral part of the reorganized agricultural sector, often as 'iron-wood-bamboo agricultural implements production groups (*tie mu zhu nongju shengchan zu*)'.[12] But most were organized into 'producers' co-operatives (*shengchan hezuoshe*)', relocated in towns, and administered separately from agriculture by the county.

There are reasons to believe that the rural non-agricultural sector grew slowly during the 1950s.[13] The emphasis given to grain production and to the development of large-scale urban industrial enterprises, particularly those in heavy industry,[14] meant that resources were diverted from agricultural sideline production and from rural small-scale industries, particularly those not in direct support of agricultural production.[15] The introduction of unified purchase of grain and other agricultural products forced many small establishments (e.g. the 'four mills') to close.[16] Collectivization and the socialization of rural trade disrupted sideline activities (e.g. 'cotton spinning and weaving, the traditional household sidelines, virtually disappeared'[17]), caused rural commerce and small towns to decline, and reduced opportunities for individual initiative.[18]

To accelerate industrialization, the Great Leap Forward (GLF) was initiated in 1958, and a central theme of the movement in the countryside was 'larger in size and having a higher degree of public ownership (*yida ergong*)'. One consequence was the elevation of many handicraft co-operatives to state-owned enterprises or large urban collectives. In Jiangsu, more than 88 per cent of the handicraft co-operatives in rural and urban areas were elevated.[19] With this change in ownership status, the elevated co-operatives were formally removed from the rural sector, and many rural areas were left without the capacity to produce small consumer products for daily use or the capacity to produce and service agricultural tools and equipment.[20] The major change in the countryside, however, was the reorganization of agriculture into communes. In Jiangsu, roughly each township was turned into a commune, and the township government became a part of that commune. Henceforth all aspects of rural life were placed under the unified management of communes.

A new rural industrialization policy also emerged during the GLF. As part of the 'walking-on-two-legs' strategy,[21] the government called on the newly formed communes to establish commune-brigade enterprises (CBEs) to produce industrial goods needed by agriculture. CBEs were developed according to the principles of 'one self, two mains, three locals (*jiudi*), and four serves (*yizi, erzhu, sanjiu,* and *siwei*)'.[22] The 'one self' is self-reliance, i.e. communes should rely primarily on their own resources to develop CBEs. The 'two mains' refers to the fact that CBEs should be mainly small and use mainly indigenous (*tufa*) technology. The 'three locals' is the principle that CBEs should use local raw materials, process them locally, and distribute the end-products locally. And the 'four serves' identifies the four functions of industrial production at and below the commune level—i.e. 'it should, first of all, serve the development of agriculture and the mechanization and electrification of farming; at the same time, it should serve to meet the demands of commune members for daily necessities, and serve the great industries of the country [i.e. subcontracting and doing processing work for urban factories] and the socialist market [i.e. export]'.[23] However, state policy clearly considered serving agriculture and serving the everyday needs of commune members the most important of the 'four serves' and the main purposes of CBE development.[24]

Responding to the call, local cadres, relying on primitive or intermediate technology, established a huge number of CBEs, frequently with labour and equipment confiscated from co-operatives and households.[25] In the first nine months of 1958 alone 6 million new enterprises and workshops were established by communes and brigades.[26] Presumably, some of the new CBEs established were to replace handicraft co-operatives that had been elevated to county ownership. By the end of 1958, commune-brigade industrial enterprises employed 18 million workers, and in 1959 CBEs produced RMB 10 billion in non-agricultural gross output, or over one-quarter of the total income earned by communes.[27]

Although the GLF may have produced some positive gains, they were more than offset by the serious problems it created.[28] One problem was that some of the industrial activities promoted initially in rural areas (e.g. the 'backyard furnace' or *tugaolu*) were inappropriate in the sense that the small-scale, labour-intensive technology adopted was technologically inefficient and produced goods of inferior (often unusable) quality. In other words, a large number of inefficient plants were established in the countryside, and thus huge amounts of human and material resources were wasted. Another was that during the GLF too many rural resources, particularly labour, were allocated from farming to non-agricultural activities in rural and urban areas, and this in turn caused considerable disruption to agricultural production.[29]

The communization drive demoralized the rural population and the excesses of the Leap precipitated a major agricultural crisis.[30] Food shortages began to appear in 1959. In 1960, the government, responding to the crisis, called for a nation-wide retrenchment and adjustment of rural industry. Restrictions were introduced to limit the number of workers that could be shifted from agriculture to other rural activities, many CBEs were ordered closed, and workers who had moved to cities during the GLF were returned to the countryside. In consequence the industrial gross output produced by CBEs declined from about RMB 6 billion in 1958 to RMB 1.96 billion in 1961.[31] After the government announced in 1962 that 'communes and brigades should in general not operate enterprises', additional CBEs closed and the gross output produced by commune-brigade industrial enterprises fell to a low of RMB 410 million in 1963.[32]

In Jiangsu, communes were told to allocate at least 75 to 80 per cent of their work-force to agriculture (broadly defined to include cultivation, forestry, animal husbandry, sidelines, and basic agricultural construction),[33] and the massive closure of CBEs reduced their industrial output value from a peak of RMB 462 million in 1960 to RMB 204 million in 1961 and RMB 34 million in 1963.[34] Given economic conditions in Jiangsu at the time, it is arguable that the retrenchment was unnecessarily severe. Although agricultural production declined in Jiangsu as it did in other parts of China, the crisis was less severe. Furthermore, Jiangsu's rural industry was among the most advanced, its agriculture among the most productive, and its population density among the highest in China. In other words, conditions in rural Jiangsu, particularly in the south, were ripe for rapid industrialization. One suspects that many of the CBEs created in Jiangsu were economically viable and did not merit closure.[35] It is also questionable whether Jiangsu's agriculture really required or even could have productively absorbed 75 to 80 per cent of the rural work-force. In other words, even though economic conditions in Jiangsu were favourable to rural industrialization, it was set back by a retrenchment policy that treated all regions alike.

The 1950s and the early 1960s were not stellar years for China's rural non-agricultural sector. In this period, except for a few years during the GLF, government policy on rural non-agricultural activities became increasingly restrictive. In the early 1950s selected rural non-agricultural activities were relocated from the countryside to urban areas, and in the mid-1950s, with the adoption of unified purchase and the collectivization of agriculture, rural trade was suppressed and many rural establishments closed for lack of raw materials. In 1958, as part of its GLF policy, the government called for the rapid development of rural industries, and a vast number of CBEs were established. The excesses of the GLF years created chaos in agriculture, and in the early 1960s, responding to the chaos, the government suppressed all rural industries whether or not they were appropriate or economically viable.

Rural Non-agricultural Development During the Cultural Revolution Decade

In the mid-1960s, after the economy had stabilized and recovered from the GLF, the Party and the Central Government once again advocated that 'communes operate some small collective factories'[36] and that local-ities utilize local resources to produce industrial goods in support of agriculture. Some CBEs that closed down in the early 1960s reopened. At least in Jiangsu, CBEs re-emerged in the mid-1960s partly by using urban workers who were sent back to their native villages during the 'three difficult years' (1959–61), and primarily in the form of brigade enter-prises.[37] In 1965, brigade enterprises accounted for about 76 per cent of the GVIO produced by Jiangsu's CBEs.[38] When commune enterprises were shut down in the early 1960s, many of their assets and much of their equipment were returned or distributed to brigades, and these became the primary assets of the newly formed brigade enterprises. Unfortunately, no national statistics of CBE development exist for the Cultural Revolution period. In Jiangsu, the GVIO (in current prices) produced by CBEs increased from a low of RMB 34 million in 1963 to RMB 414 million in 1966. Then, for several years in the late 1960s, it fluctuated around RMB 360 million before climbing rapidly after 1968, when commune enterprises began to expand once again, to reach RMB 719 million in 1970. Real industrial output in rural Jiangsu increased at an average annual rate of 38 per cent in 1969–70.

Throughout this period, the government promoted the development of the 'five small industries (*wu xiao gongye*)', i.e. locally operated small- and medium-scale enterprises that use intermediate technology to produce iron and steel, cement, energy (coal and hydroelectricity), chemical fer-tilizers, and agricultural machinery.[39] For our purpose, it is important to note that the loosely defined five small industries included not only some CBEs but also some county-level state enterprises,[40] many of which were located not in rural areas but in county cities. This effort at rural indus-trialization was similar to the earlier attempt during the GLF in that the emphasis was on supporting agriculture, but it was also different in that rural industry was now developed more cautiously and with greater attention to standards, scales of operation, product quality, and demand.

In the early 1970s the political environment for rural non-agricultural development further improved.[41] The 1970 annual economic plan com-mitted the country to quicken the pace of agricultural modernization and to develop small local industries so that every county would have its own farm machinery factory. The call to increase agricultural capital construc-tion and to accelerate agricultural mechanization also went out at the North China Agricultural Conference held in the autumn of 1970, and the government announced special measures to promote the five small in-dustries. A special fund to RMB 8 billion was earmarked for use by the provinces over a five-year period to support the development of the five small industries. In addition county enterprises in the five small industries were permitted to retain 60 per cent of their profits for two to three years and those that were unprofitable were given temporary subsidies and had their taxes exempted or reduced. The policies adopted in the late 1960s and the early 1970s had considerable impact. Some 1,100 small-scale, coal-based ammonia plants were built between 1966 and 1974 in contrast to only ninety plants built between 1958 and 1966.[42] The number of small cement plants in China increased from about 200 in 1965 to 2,800 in 1973 when they produced about half of China's cement.[43]

However, rural non-agricultural development was not promoted across the board. Indeed, constraints on individual activities intensified and peasants were bound tightly to the land. During the Cultural Revolution decade, farming, particularly grain production, had priority over rural non-agricultural development. For fear that they might divert resources from farming, non-agricultural activities not in direct support of agriculture were discouraged if not prohibited. Individual activities, and sometimes also CBEs, were attacked as 'the tail of capitalism' and 'taking money as the key link', and there was considerable pressure to reduce if not eliminate their presence in the countryside.[44] Some county governments in Jiangsu successfully resisted the pressure to close down CBEs to concentrate on agricultural development by arguing that since agriculture was unable to generate sufficient savings to modernize itself, it was necessary to 'develop industry around agriculture, and then use industry to support agriculture (*weirao nongye ban gongye, ban hao gongye zhi nongye)*'.[45] But even in these localities, the establishment of new CBEs in areas other than the five small industries carried considerable political risk so they were frequently established secretly, without notifying and obtaining approval from senior units.[46]

During the Cultural Revolution decade, rural industrialization was viewed not primarily as a way of changing the rural economy and its structure, or of absorbing rural labour, or of directly raising rural income in poor areas. Rather, its primary purpose was to support agricultural production by providing local self-sufficiency in the material inputs needed in the 'four transformations (*sihua)*', i.e. irrigation, agricultural mechanization, electrification, and greater chemical fertilizer application.[47] In other words, rural industrialization was seen as a way of channelling resources to farming rather than as a process of reallocating resources from cultivation to other activities.

Despite the repression of individual activities and of some CBEs, the strong political and financial support given to collective non-agricultural production in the five small industries and the general shortages of industrial goods in the countryside, caused in part by work disruptions in urban enterprises during the final years of the Cultural Revolution, provided a propitious environment for rural non-agricultural development. Accordingly, commune and brigade enterprises made a remarkable recovery in the late 1960s and the 1970s.

Nationally, between 1971 and 1975, the gross output in 1980 constant prices produced by collective industrial enterprises at or below the commune level increased from RMB 8.25 billion to RMB 17.9 billion, or at an average annual rate of 21.6 per cent.[48] In Jiangsu, the rate of increase was even more rapid. Between 1971 and 1975, the real GVIO produced by Jiangsu's CBEs increased at an average annual rate of 27 per cent to reach RMB 2.28 billion in 1980 constant prices in 1975.[49] The GVIO produced by CBEs as a share of total GVIO produced in Jiangsu increased from 3.3 per cent in 1970 to 9.3 per cent in 1975.[50] The strong performance by CBEs contributed to Jiangsu's above average industrial performance in the 1970s despite the fact that it received less central investment than most other provinces.[51]

After the downfall of the ultra-leftist Gang of Four in 1976, the dam burst, and CBEs expanded even more rapidly. During 1976 and 1977, real gross output produced by industrial enterprises at or below the commune level increased at an average annual rate of 38 per cent in China as a whole and 51 per cent in Jiangsu.[52] An unknown but probably small portion of this increase may be statistical since in 1977 some handicraft enterprises that had been transferred up to the county level were returned to communes.[53]

By 1978, the size of China's rural non-agricultural sector was substantial. Christine Wong has estimated that the 600,000+ enterprises in the five small industries produced nearly all of China's farm tools and small and medium farm machines, more than half of its chemical fertilizers (by weight), two-thirds of its cement, 45 per cent of its coal, one-third of its iron and steel, and about one-quarter of its hydroelectric output.[54] Removing the county enterprises would reduce these figures somewhat (particularly in iron and steel and chemical fertilizers, where the contribution of county enterprises was very substantial), but the proportions would still be impressive. There were more than one million CBEs in China in 1978 involved in non-agricultural activities, and they employed over 22 million workers and produced a gross output in excess of RMB 39 billion.[55] The most important were the 794,000 commune-brigade industrial enterprises; together they employed over 17 million workers and produced nearly RMB 33 billion in gross output.

Commune-brigade industrial enterprises were particularly well developed in Jiangsu. In 1978, they numbered 56,496 and employed nearly 2.5 million workers.[56] In that year, commune-brigade industrial enterprises supplied Jiangsu with 71 per cent of its small and medium-size agricultural tools, 70 per cent of its semi-mechanized agricultural machines, 77 per cent of its agricultural machinery parts, 73 per cent of its bricks, 50 per cent of its tiles, 470,000 electric motors, and 160,000 transformers.[57] In terms of gross output, they produced 20.5 per cent of the province's light industrial products and 16.3 per cent of its heavy industrial products. The gross output produced by commune-brigade industrial enterprises equalled or exceeded RMB 10 million in nearly 100 of Jiangsu's 1,885 communes.[58]

On balance, the political environment for rural non-agricultural development improved after the mid-1960s. During the Cultural Revolution decade, when much importance was attached to self-reliance and self-sufficiency, the government promoted the development of the five small industries. However, rural non-agricultural development was not permitted across the board, as the government continued to suppress activities outside the five small industries partly to keep resources from leaving agriculture (particularly grain production) and partly to prevent profit motive and capitalism from reappearing. In consequence, while rural non-agricultural growth during this period was rapid, it was also inefficient in the sense that much expansion was in the producer goods industries where the rural sector did not have a long-term comparative advantage.

Post-1978 Government Policy Towards Rural Non-agricultural Development

With the arrest of the Gang of Four in late 1976, the rural development strategy of the previous decade came under scrutiny and was severely criticized. It was noted that the policy of stressing grain production, even in areas unsuited to grain, reduced the production of cash crops and discouraged agricultural sideline activities, caused economic hardship in low-income areas, and eliminated opportunities for the development of forward linkage industries. The treatment of private production and commerce as 'the tail of capitalism' during the Cultural Revolution decade also came under criticism. These activities were viewed now as subsidiary (*fushu*) and supplemental (*buchong*) to the collective and state-owned sectors.

Chinese economists also questioned whether the considerable industrial capacity created by the promotion of the five small industries was in fact an economic asset from the viewpoint of the national economy. While the five small industries introduced some new products and technologies to rural China, many of the enterprises were apparently economically (perhaps also technically) inefficient and probably should never have been established.[59] For example, small synthetic ammonia plants had unit production costs that averaged 3.6 times that of large ammonia plants and, not surprisingly, 70 per cent of the 1,539 small synthetic ammonia plants in operation in 1979 sustained losses.[60] Indeed, enterprises in the five small industries accounted for 53 per cent of the total losses sustained by all state industrial enterprises in 1978.[61] Another problem was excess capacity caused by the duplication of facilities. For example, because of overinvestment, there were too many county enterprises and CBEs in the machinery and machine tools industries, and many of them went hungry. In 1980, Jiangsu alone boasted 13,354 machinery enterprises at the commune and brigade levels, and they employed 784,723 workers.[62]

The indiscriminate expansion of the five small industries during the Cultural Revolution decade—when self-reliance and self-sufficiency were considered more important than comparative advantage, availability of local resources, and scale economies—is partly to blame for the problems faced by the five small industries in the late 1970s. But high prices for producer goods, a cumbersome and inefficient allocation system, and the principle that 'whoever builds and manages a plant has the use of its products' also encouraged the proliferation of inefficient plants in dispersed locations and permitted their continued existence.[63]

Beginning in early 1978, China took steps to consolidate its five small industries and thousands of inefficient and unprofitable enterprises (particularly those in iron and steel, chemical fertilizers, and machinery) were asked to close down, suspend operation, merge, or switch to other lines of production (*quan, ting, bing, zhuan*). County enterprises in the five small industries were severely affected but so were many CBEs, particularly those engaged in the production of machinery and machine tools. For example, in Jiangsu, many CBEs that had produced machine tools and machinery stopped production or switched to produce consumer durables (e.g. electric fans) and components for large enterprises.[64]

A dramatically different rural development strategy, one that gave rural non-agricultural activities a more prominent and flexible role, began to emerge in 1978. The major turning-point was in December 1978 when the Resolutions on Some Questions Concerning the Acceleration of Agricultural Development were adopted at the Third Plenary Session of the Eleventh Central Committee of the CCP.[65] Three of the decisions were particularly important for rural non-agricultural development: those (i) to encourage 'the simultaneous development of farming, forestry, animal husbandry, sideline occupations, and fishery';[66] (ii) to allow the individual economy greater flexibility to develop in rural areas; and (iii) to promote the vigorous and systematic development of CBEs.

The all-round development of agriculture is important to rural non-agricultural development because such an approach is bound to produce a more rational and diversified agricultural output mix, and greater diversity tends to strengthen agriculture's forward linkages. For example, the development of animal husbandry creates opportunities for the processing of animal and dairy products. Increased production of cash crops and the development of fish culture also creates employment and income in processing, storage, and transportation. Thus the decision to diversify agriculture, when combined with the increases in the purchase prices of agricultural products that were also announced, greatly improved the environment for rural non-agricultural development in rural China.

The scope and the flexibility of the individual economy in rural China were increased in three main ways by the decisions made at the Third Plenum. First, it was decided that up to 15 per cent of the total cultivated area could be allocated to private plots, and, if local circumstances permitted, peasants could devote all their time to sideline occupations. Second, the scope of sideline occupations was widened. Regulations that restricted the type of crops that could be raised on private plots were abolished, and peasants were encouraged to engage in traditional home industries (e.g. sewing, embroidering, various handicrafts using bamboo, grass, and straw). Individuals were also permitted to operate all industrial, handicraft, commercial, food and beverage, services, repairs, transport, and house renovation undertakings as long they had the permission of the brigade (now the village). Third, new marketing channels were opened. Products from sideline activities could now be sold directly to end-users. In effect, with these changes, the government gave official sanction to the return of individual entrepreneurship and of private production and commerce to the countryside.

The decision at the Third Plenum to develop CBEs vigorously so that their contribution to total rural income would rise significantly was of immense importance to rural non-agricultural development. Furthermore, the decision stated explicitly that:

> as long as it is consistent with the principle of economic rationality, the processing of all farm and sideline products should be taken over gradually by CBEs. Urban factories should, in a planned way, transfer to CBEs the production of those products or components that can be produced in rural areas, and help them to acquire the necessary equipment and technology.[67]

Finally, the documents adopted at the Third Plenum also called for tax holidays for CBEs, and subsequently several important tax concessions were enacted.[68] In the 1970s, because CBEs paid a flat 20 per cent income tax rather than the eight-grade progressive income tax faced by other collective industrial and transport enterprises, they were already taxed less heavily.[69] Of the new tax incentives, the three most important were: (i) CBEs directly serving agriculture or the people's livelihood could apply to have specific products and services exempted from income taxes; (ii) small iron mines, coal-mines, power stations, and cement plants were exempted from industrial-commercial taxes and from income taxes for three years starting in 1978; and (iii) new CBEs experiencing start-up difficulties could apply for exemption from industrial-commercial taxes and from income tax for two to three years.

Other documents issued by the Central Committee and the State Council in 1979 and the 1980s gave further encouragement and protection to the development of rural non-agricultural activities. In contrast to earlier years, when rural enterprises were operated according to the principle of the 'three locals (*san jiudi*)', CBEs were now encouraged not only to help improve agricultural production and the people's livelihood but also to operate construction teams in urban areas, to serve the needs of large-scale industry, and to export.[70] The promulgation of the Provisional Regulations Concerning the Development and Protection of Socialist Competition in October 1980 made it easier for CBEs to compete with state enterprises by outlawing the blockades and partitioning of markets previously practised by some government departments.[71] Because the temptation to elevate commune enterprises to a higher degree of public ownership is always present, the No. 1 Document of 1983 issued a strong warning to local governments not to weaken or disperse CBEs in the process of carrying out economic reform.[72] Localities were also told to promote lateral linkages between CBEs and county enterprises and between CBEs and team enterprises, and to encourage different forms of ownership (e.g. team enterprises, joint ownership by county and commune or by several communes,[73] joint household (*lianhu*) enterprises, and individually owned enterprises).

Rural non-agricultural development relies on rural trade and labour mobility, so it is significant that the State Council and the Party Central Committee issued new regulations in 1984 liberalizing rural trade and labour mobility.[74] Rural residents were now permitted to engage in the transport and sale of selected goods between town and countryside, to purchase and own motor vehicles and boats for the purpose of transporting goods and passengers, and, in selected areas, to establish household registration in small market towns (*jizhen*) to engage in commerce, service, or industry as long as they took full responsibility for their own grain rations.

In 1984, the central government adopted two other changes that made CBEs more like other collective enterprises. First, it told government departments responsible for planning, material allocation, public finance, banking, and transportation to include CBEs in their plans and to provide them with guidance and support.[75] This announcement gave CBEs additional legitimacy and security. Second, beginning in 1984, the government also required CBEs to pay the eight-grade progressive income tax instead of the flat 20 per cent income tax, thus removing one of the advantages

CBEs had over other collective enterprises.[76] However, because profits from CBEs were used not only for reinvestment but also to support agriculture and rural development in general, to reduce the negative impact the higher tax might have on rural development, the government decentralized the power to grant tax relief to the county.[77]

Together these policies formed a new rural development strategy that was summarized by the slogan 'five wheels turning at the same time, and ten industries moving upwards simultaneously (*wuge lunzi yiqi zhuan, shige hangye yiqi shang*)'.[78] The simultaneous development of all rural activities had wide support in the countryside since it was recognized as the only way rural China could absorb the huge number of surplus labour being released from collective agriculture by the widespread implementation of the household responsibility system.[79] In other words, rural non-agricultural development was now viewed not so much as a way to serve and support agriculture but as a way of generating rural employment opportunities so that surplus labour could be reallocated from cultivation to more productive activities.

The explosive expansion of CBEs that resulted from the new strategy also revealed numerous problems, e.g. misuse of agricultural land, excess capacity in some industries, pollution, too many financially weak and inefficient enterprises, abuse of tax incentives, and competition between state enterprises and CBEs for raw materials. Thus the promotion of CBE growth was accompanied by measures to consolidate the rural non-agricultural sector and partially to redirect its focus.

The main problem was identified as 'the blind development of CBEs, expanding without sufficient attention to efficiency or resources'.[80] The transfer of farm land to non-agricultural use was put under tighter control. Communes and brigades were prohibited from establishing new plants or expanding production capacity in industries where excess capacity existed in the economy. CBEs producing cigarettes and cotton textile products (two areas where the competition for agricultural raw materials between state enterprises and CBEs was particularly serious) were closed down or switched to other products, and CBEs with substandard technology or excess capacity (e.g. machine tools) were also closed down or shifted to other production. Warnings went out to communes and brigades to give careful consideration to resources, energy, transportation, and technology before establishing factories. To avoid duplication of investment in energy, transport, water supply, and pollution control and to capture scale economies in production, the government recommended that rural industry be concentrated in market towns.

The government also moved to tighten its control of CBE development. Efforts were made to strengthen enterprise management and financial control. To arrest the 'blind expansion' of CBEs, the government removed some of the tax incentives granted to CBEs only a few years earlier. For example, in January 1981, new tax regulations for CBEs were issued that rescinded some of the tax concessions granted to CBEs in 1979.[81] Another reason for the removal of the tax incentives was that in some areas they were being misused. For example, under the tax regulations only new CBEs experiencing start-up difficulties were supposed to receive tax concessions, but in many counties in Jiangsu they were apparently granted indiscriminately.[82]

In recognition of the changes under way in the countryside—e.g. the replacement of the commune by the township (*xiang*) and the emergence of new ownership forms such as joint households/associations (*lianhu/ lianheti*) and joint ownership by different administrative units—and anticipating that rural industrial enterprises would gradually be concentrated in rural towns (*zhen*), the government replaced the term 'CBE' with 'township-town enterprise' (*xiangzhen qiye*) at the end of 1983.[83] For the purpose of clarity, we shall hereafter use 'township-village enterprises' (TVEs) to denote enterprises owned collectively by townships (formerly communes) and villages (formerly brigades),[84] 'private enterprises' (PEs) to denote associations and enterprises owned by individuals, and 'rural enterprises' to denote all collectively and privately owned enterprises at or below the township level.

In the mid- and the late 1980s, new economic policies were adopted in two areas, which, if they remain unchanged, have the potential of modernizing China's rural industry and making it an important force in the country's effort to reform and develop the economy. The first was the development of a new science and technology policy that encouraged urban–rural technology transfer, and the second was the emergence of an export-orientated development strategy that assigned a prominent role to rural enterprises.

In a major policy directive issued on 1 January 1985, the government removed several long-standing restrictions that had obstructed the transfer of technology to rural areas. Specifically, it announced that hereafter technical personnel in urban areas may take leaves of absence from their work units to work in rural areas (without risking their job or residence), and that technical institutes and urban enterprises may accept contract research from, and provide consulting services to, rural enterprises for compensation.[85] This was followed by an announcement in May 1985 that the State Science and Technology Commission (SSTC) was launching the 'spark plan (*xinghuo jihua*)' to accelerate the modernization of the rural sector through technological change. The plan's main objectives were to disseminate technology to the countryside, to train rural entrepreneurs, managers, and skilled workers, and to develop appropriate technology, equipment, and management methods for rural enterprises.[86]

To achieve these goals, SSTC encouraged rural enterprises and science and technology units to co-operate in production and mobilized the Ministry of Agriculture, the Ministry of Education, and various mass organizations to provide short-term training to rural cadres and workers.[87] In addition, SSTC prodded government units at all levels to develop technology improvement projects involving TVEs using as incentive the prospect of bank loans for those proposals selected as spark projects.[88] Between 1986 and 1989, 10,345 spark projects were completed, largely financed by rural enterprises themselves and bank loans.[89] Although the government did not fund the spark policy generously, the policy was nevertheless important for it gave official sanction and support to urban–rural technology transfer. This is important because, in a partially re-formed command economy, the flow of information and technology is extremely difficult if it does not have official approval.

In 1987, China adopted an export-oriented coastal development strategy that linked rural production closely to the export sector.[90] In brief, the strategy called for the use of cheap rural labour to produce labour-intensive goods for export, and then use the export earnings to import foreign equipment and technology needed to modernize its industry. Chinese leaders found this strategy attractive because it would simultaneously generate foreign exchange earnings and create rural employment, and thus help solve two of China's most intractable problems.

The adoption of a labour-intensive, export-oriented development strategy that links rural industry with the export sector has tremendous importance for the long-term growth of China's rural non-agricultural sector. China's rural industry has a comparative advantage in the production of simple, low-tech, labour-intensive products. Thus the strategy plays into rural China's strength. The potential impact of this strategy on rural China is suggested by the fact that the early industrialization of Taiwan and South Korea also relied on the production and the export of labour-intensive products using labour that was drawn primarily from rural areas. Furthermore, by permitting and encouraging rural enterprises to interact with international markets, the export-oriented strategy opens to them a new source of raw materials, capital, and technology, factors that rural enterprises have had difficulty finding domestically. Thus the export-oriented strategy may also help relax some of the constraints on rural industrial growth.

In summary, the climate for rural non-agricultural activities improved dramatically in the 1980s. Although the state gave little direct stimulus to rural non-agricultural development by way of tax concessions, subsidies, or investments, it helped indirectly by adopting reform measures that significantly improved the environment for rural non-agricultural activities. The rural reform measures adopted in the late 1970s and the early 1980s may be described as the deregulation of rural China. The household responsibility system, which transferred many of the agricultural production and investment decisions from the communes to individual households, reduced government control of agriculture. Other reforms relaxed or dismantled the controls and regulations that had previously prohibited or severely restricted rural non-agricultural activities by either individuals or collectives. Then, in the mid- and late 1980s, the rural sector, particularly rural industry, was induced to turn increasingly to the outside world for markets and inputs. The deregulation of rural activities and the subsequent opening of the rural sector to foreign trade and investment released a tremendous amount of productive and entrepreneurial energy in rural areas that had previously been constrained by central planning and tight government control. More than any other factor, it was this release of energy at the grass-roots level that was the underlying cause behind the impressive growth and structural changes experienced by the rural non-agricultural sector in the 1980s and discussed in the following section.

Growth and Structural Change in the Rural Non-agricultural Sector, 1978–1989

China's rural non-agricultural sector expanded rapidly after 1978, and its growth was particularly strong in the mid-1980s. Statistics presented in Table 2.1 show that the output value (in 1980 prices) of the four major rural non-agricultural material products sectors—industry (except for family sidelines), construction, transportation, and commerce—increased from about RMB 66 billion in 1978 to RMB 517 billion in 1989, or at an annual average rate of over 20 per cent. The growth of RGVIO (rural GVIO) was particularly rapid. Between 1980 and 1989, RGVIO in 1980 prices increased at an average rate of over 26 per cent per year, rising from about RMB 55 billion in 1978 to RMB 420 billion in 1989. The growth of net non-agricultural output, particularly that of rural industry, is somewhat lower than that of gross output since there is evidence that the share of raw material in gross output has increased since 1978.[91] Even with this qualification, however, the pace of rural non-agricultural development in China has been impressive.

Increased non-agricultural production in the countryside was accompanied by increased rural commercial activity. Rural retail sales increased from RMB 81.04 billion in 1978 to RMB 456.7 billion in 1989.[92] As a share of total retail sales, rural retail sales rose from 52 per cent in the late 1970s to about 57 per cent in the late 1980s. Another indication of

Regulations and Economic Policies

TABLE 2.1. *Growth and Structure of China's Rural Non-agricultural Sector,*
1978–1989

Average annual growth (%)	1978–80	1981–3	1984–6	1987–9
Real RGVNAO[a]	14.6[d,e]	16.8[e]	29.8	20.2
Real RGVIO[b]	17.3	12.8	41.8	25.2
Rural non-agricultural labour force[c]	−1.6[d]	11.6	21.9	4.2
Rural industrial labour force	5.6[d]	1.2	16.1	1.2
Structural composition	**1978**	**1983**	**1986**	**1989**
Rural social product (% of total)				
Agriculture	68.5	66.7	53.1	45.1
(Cultivation)	(52.6)	(47.1)	(33.1)	(25.4)
Industry	19.4	20.0	31.5	40.6
Construction	6.6	7.8	7.8	6.3
Transportation	1.7	2.0	3.2	3.6
Commerce	3.7	3.5	4.3	4.3
Rural labour force (% of total)[f]				
Agriculture	89.7	87.9	80.2	79.2
(Cultivation)	(83.8)	(81.5)	(66.8)	(n.a.)
Industry[g]	5.7	5.8	8.3	8.0
Construction	0.7	1.4	3.4	3.7
Transportation	0.3	0.5	1.3	1.5
Commerce[h]	0.2	0.4	1.4	1.6
Other	3.5	4.0	5.4	6.0

[a] GVNAO (gross output of industry except for household sidelines, construction, transport, and commerce) in rural China in current prices deflated by the implicit price deflator for total GVNAO.
[b] 1978–86 based on linked indexes of GVIO in rural China in constant prices. 1987–9 based on GVIO in rural China in current prices deflated by the implicit price deflator for total GVIO.
[c] Labour force at or below the township (commune) level (*xiangcun laodongli*) with main occupation outside of agriculture.
[d] 1979–80.
[e] Annual compound rate calculated by using only the year-end values.
[f] Labour force at or below the township (commune) level (*xiangcun laodongli*) distributed by main occupation.
[g] Includes workers in village industrial enterprises.
[h] Includes trade, food, and drink establishments.

Sources: Calculated from data in *ZNTN* (1985, 1986, 1987, 1988); *ZTN* (1987, 1988, 1990); and *Zhongguo gongye jingji tongji ziliao, 1949–1984* (*China Industrial Economic Statistical Material, 1949–1984*).

the new vitality in rural commerce is the rising volume of agricultural products sold directly by farmers to urban residents, increasing from RMB 3.1 billion in 1978 to RMB 69.8 billion in 1989 (nearly 9 per cent of that year's total retail volume).[93]

The rural non-agricultural sector has also absorbed an increasing number of rural workers. Between 1978 and 1987, rural non-agricultural employment increased by 50.3 million, or at an average rate of 15.5 per cent per annum.[94] Of this increase, 26 per cent was absorbed by rural industry, 23 per cent by construction, 9 per cent by transportation, 11 per cent by commerce (trade, food and drink, and related services), and 30 per cent by other services.

Rural non-agricultural growth has been uneven over time. Relative slow growth in the early 1980s was followed by a sharp rise in growth after 1983 and a severe decline in growth in 1989–90. The principal underlying cause of rapid economic growth in the 1980s was economic reform, which released a tremendous amount of productive and entrepreneurial energy that had previously been constrained by central planning and rigid government control. But another consequence of reform was inflation.[95] Because inflation was an ever present threat, a distinctive feature of government policy in the 1980s was its stop-go characteristic. A period of economic stimulation and reform would be followed by a halt in reform and a period of credit control and tight restrictions on investments and expenditures.

The relatively slow pace of rural non-agricultural development in the early 1980s reflected in part the tighter government control on expenditures introduced to cool an overheated economy, and in part the adjustment and consolidation then under way to eliminate some of the country's most energy wasteful and unprofitable enterprises, among them many TVEs.[96] A lingering fear among many in rural areas that the government's rural policy might change also contributed to the slower growth in this period. Once the threat of inflation had passed, the control on expenditure relaxed and reform accelerated. Thus, in early 1984, the government affirmed its strong support of rural non-agricultural activities (private and collective), directed government departments (including banks) to support rural enterprises, gave county governments the authority to grant tax relief to those TVEs experiencing difficulties, and urged local cadres actively to promote rural non-agricultural development. These measures brought about a rapid expansion of rural non-agricultural activities. The annual rate of growth of real GVIO in rural areas was 41 per cent in 1984 and a spectacular 53 per cent in 1985.[97] The economy as a whole also grew rapidly during these years, and, with rapid growth, inflation returned. Control on local expenditure and rural credit was reimposed briefly, and brought the annual growth rate of RGVIO down to 31 per cent in 1986.

In 1987 and 1988, the economy expanded rapidly under the steady
stimulation of easy credit, and by 1988 China was in the grip of double-
digit inflation.[98] When the government announced in mid-1988 that, as
part of its ongoing reform, price controls would be lifted on more goods
in 1989, panic buying ensued. To preserve economic order and to bring
the inflation under control, the government responded in September 1988
by imposing its usual package of austerity measures. Credit was severely
tightened, interest rates raised, enterprises and government units told to
cut their expenditures drastically and halt investment projects, and price
controls reimposed. While these measures were effective in reducing
inflation,[99] they also brought economic growth to a halt, and in 1989–90
China suffered its most severe economic recession in recent decades.[100]
National output, which had been rising at an annual rate of 11 per cent in
1987–8, rose by only 1 per cent in 1989.[101]

In 1989–90, as part of the government's package of austerity measures,
bank credits to TVEs were drastically reduced and some 20,000 rural
development projects worth about RMB 10 billion were cancelled.[102] The
rural non-agricultural sector, which had become increasingly dependent
on bank credit for investment and working capital, was severely affected
by these measures.[103] The average annual growth rate of real RGVIO
dropped from 33 per cent in 1987–8 to 10 per cent in 1989. A large
number of TVEs closed down, and most of those that remained open
operated at levels significantly below capacity.[104] The number of workers
employed by TVEs declined from 48.9 million in 1988 to 47.2 million in
1989 and 45.9 million in 1990.[105] However, the number of rural workers
who lost non-agricultural employment during these years was substantially greater since the figures above do not include those who were laid
off by team and private enterprises.[106]

Rapid growth in the 1980s was accompanied by dramatic structural
changes in the rural economy. Table 2.1 presents two measures of the
economic structure of rural China: (i) the distribution of rural social
product (i.e. the gross value of output of the material products sectors) by
sector, and (ii) the distribution of all rural workers by sector. The second
of the two measures is, of course, the more comprehensive since it
includes activities outside the material products sectors. Both measures
show that rural non-agricultural development in the past decade has
dramatically altered China's countryside.

In one decade, the share of agriculture in China's rural gross output
declined from 68 per cent to 45 per cent, and that of cultivation from 53
to 25 per cent. Approximately 2 per cent of China's rural gross output is
composed of household crafts, and if this is included in non-agricultural
activity, then the share of rural non-agricultural activity in the gross value
of material products produced in rural China increased from 33 per cent

in 1978 to nearly 57 per cent in 1989. The changes in rural labour deployment show an equally dramatic shift. Between 1978 and 1989, even as the total rural labour force expanded by more than 100 million, the share employed in non-agricultural activities increased from 10 per cent to over 20 per cent.[107] Within agriculture, there was also a significant shift away from cultivation. In 1978 only slightly less than 6 per cent of China's rural labour force was involved in agricultural activities other than cultivation; by 1986 it had risen to over 13 per cent.

By the late 1980s, rural enterprises, which accounted for nearly all of the products and services produced by the rural non-agricultural (material products) sector, were of considerable importance in China. In 1987, they absorbed nearly 68 million workers (47 million by collective enterprises and 21 million by private enterprises) as compared to the 131 million employed by state and urban collective enterprises, and rural construction teams accounted for 44 per cent of China's construction workers.[108] Of the RMB 2,202 billion GVIO produced in China in 1989, 25 per cent was accounted for by rural enterprises (20 per cent by collective enterprises and 5 per cent by private enterprises).[109]

Nowhere in China is the rural non-agricultural sector more prominent than in Jiangsu. Table 2.2 shows that in 1978 rural production already accounted for nearly 19 per cent of Jiangsu's non-agricultural material products. After a decade of rapid rural non-agricultural growth, the rural share of total provincial non-agricultural material products had increased to 46 per cent. In 1989, rural enterprises accounted for 45 per cent of the gross value produced by Jiangsu's industry, 66 per cent of that produced by its construction, 52 per cent of that produced by its transportation, and 38 per cent of that produced by its commerce. The performance of Jiangsu's rural industry is truly remarkable considering that in the late 1980s Jiangsu ranked first in GVIO in China, producing about 11 per cent of the country's total GVIO. The importance of Jiangsu's rural enter-

TABLE 2.2. *Rural Production as a Percentage of Total Production in Jiangsu by Major Sector, Selected Years*

	1978	1983	1987	1989
Non-agricultural material products	18.8	29.7	44.7	46.0
Industry	18.5	26.8	43.9	44.5
Construction	14.1	55.9	54.7	66.0
Transportation	18.4	30.8	52.5	51.8
Commerce	32.6	22.6	33.8	38.3

Source: Calculated from data in *JTN* (1990), 67 and 144.

prises can be measured in another way: in the late 1980s, more than one-fifth of the GVIO produced in rural China was produced in rural Jiangsu.[110] We turn now to a more careful examination of Jiangsu's rural non-agricultural sector.

NOTES

1. Until very recently the Chinese statistical system did not systematically report on the rural non-agricultural sector, so documenting its size and its historical development before the late 1970s is difficult.
2. Lin Juhong, 'Xin jiu shengchan guanxi jiaoti zhi ji de nongcun fuye (Rural sidelines at the transition between new and old production relations)', *NJYS* 3 (1988), 51–2.
3. Sidney D. Gamble, *Ting Hsien: A North China Rural Community* (New York: Institute of Pacific Relations, 1954), 287. For evidence of industrial sideline activities in pre-1949 rural Jiangsu, see Philip C. C. Huang, *The Peasant Family and Rural Development in the Yangzi Delta, 1350–1988* (Stanford, Calif.: Stanford University Press, 1990), 127–9. Huang argues that pre-1949 rural industrialization in the Yangzi Delta was 'growth without development' because income per workday declined even though output per labourer rose. However he offers little or no evidence to support the assertion that income per workday declined.
4. The incidence of rural non-agricultural activities in a traditional agrarian economy is, of course, sensitive to the definition of 'rural' and the coverage of 'non-agricultural activities'. Here, 'rural' is defined to include rural market towns and 'non-agricultural activity' is defined to include all activities other than cultivation, animal husbandry, hunting, and gathering. For a discussion of the available evidence, see Samuel P. S. Ho, 'Rural Nonagricultural Development in Asia: Experiences and Issues', in Yang-Boo Choe and Fu-Chen Lo (eds.), *Rural Industrialization and Non-Farm Activities of Asian Farmers* (Seoul: Korea Rural Economics Institute/Asian and Pacific Development Centre, 1986); Enyinna Chuta and Carl Liedholm, 'Rural Small-Scale Industry: Empirical Evidence and Policy Issues', in Carl Eicher and John M. Staatz (eds.), *Agricultural Development in the Third World* (Baltimore: Johns Hopkins University Press, 1984); Rizwanul Islam (ed.), *Rural Industrialisation and Employment in Asia* (New Delhi: International Labour Organisation, Asian Employment Programme, 1987); R. T. Shand (ed.), *Off-Farm Employment in the Development of Rural Asia* (Canberra: National Centre for Development Studies, ANU, 1986); and Swapna Mukhopadhyay and Chee Peng Lim (eds.), *Development and Diversification of Rural Industries in Asia* (Kuala Lumpur: Asian and Pacific Development Centre, 1985).
5. John L. Buck, *Land Utilization in China* (Chicago: University of Chicago Press, 1937), 371–2.
6. Penelope B. Prime, 'The Impact of Self-Sufficiency on Regional Industrial Growth and Productivity in Post-1949 China: The Case of Jiangsu Province', Ph.D. diss., University of Michigan, 1987, 29–30.

7. Ma Hong and Sun Shangqing (eds.), *Zhongguo jingji jiegou wenti yanjiu* (*Research on the Problem of China's Economic Structure*) (Beijing: Renmin chubanshe, 1981), 172–3.

8. In the early 1950s, at the time of the land reform, as part of its effort to promote the division of labour in society the Chinese Government made the decision to separate handicraft activities from agricultural activities. Henceforth, Chinese statistics combined rural handicraft workers who did not participate in farming with urban handicraft workers and counted them as part of industry. Those who farmed and produced handicrafts were considered agricultural workers, and the handicrafts they produced were included in agricultural sidelines. See Mo Yuanren (ed.), *Jiangsu xiangzhen gongye fazhan shi* (*History of the Development of Rural Industry in Jiangsu*) (Nanjing: Nanjing gongxueyuan chubanshe, 1987), 70–1.

9. Radha Sinha, 'Rural Industrialization in China', in E. Chuta and S. V. Sethuraman (eds.), *Rural Small-Scale Industries and Employment in Africa and Asia* (Geneva: International Labour Office, 1984), 124–5.

10. *Zhongguo nongye de guanghui chengjiu, 1949–1984* (*The Glorious Achievements of China's Agriculture, 1949–1984*) (Beijing: Zhongguo tongji chubanshe, 1984), 105.

11. Mo Yuanren (ed.), *Jiangsu xiangzhen gongye fazhan shi*, 81.

12. In less developed areas, these groups often did not even have a regular production site, but, instead, their members travelled to different parts of the collective and did repair work. This was the case with one of the township enterprises we surveyed that began as an iron–wood producers' co-operative (*tie mu hezuoshe*) in 1954.

13. The slow growth of rural non-agricultural activities is suggested by the fact that agricultural sideline production as a share of GVAO remained nearly constant at around 20 per cent throughout the 1950s.

14. In Guangdong Province, for example, more than two-thirds of its total capital investment went to large-scale industries. Carl Riskin, 'Small Industry and the Chinese Model of Development', *CQ* 46 (Apr./June 1971), 251.

15. For what happened in one village in southern Jiangsu, see Fei Xiaotong, 'Kaixian'gong revisited', reproduced in his *Rural Development in China: Prospect and Retrospect* (Chicago: University of Chicago Press, 1989).

16. *ZJN* (1981), 4. 55. The 'four mills (*sifang*)' are grinding mill, flour mill, oil mill, and bean curd mill.

17. Huang, *Peasant Family*, 207–8.

18. For an interesting discussion of the ups and downs of rural industry and of small towns in south Jiangsu, see Fei Xiaotong, 'Xiao chengzhen da wenti (Small towns, big issues)', in Jiangsu sheng xiao chengzhen yanjiu ketizu (ed.), *Xiao chengzhen da wenti* (*Small Towns, Big Issues*) (Jiangsu renmin chubanshe, 1984), 14–19.

19. Mo Yuanren (ed.), *Jiangsu xiangzhen gongye fazhan shi*, 86.

20. e.g. this happened in Haitou, one of the counties we surveyed, when its iron–wood producers' co-operative was elevated to county ownership during the GLF.

21. Put simply, the 'walking-on-two-legs' strategy advocated the simultaneous development of agriculture and industry, of light industry and heavy in-

dustry, of large enterprises and small enterprises, of state enterprises and local enterprises, and of enterprises at the high end and enterprises at the lower end of the technology spectrum (technological dualism). For more details, see Jon Sigurdson, *Rural Industrialization in China* (Cambridge, Mass.: Council on East Asian Studies, Harvard University, 1977), 8–13, and Carl Riskin, *China's Political Economy* (Oxford: Oxford University Press, 1987), 117–19.

22. This paragraph draws heavily from Mo Yuanren (ed.), *Jiangsu xiangzhen gongye fazhan shi*, 91–2.

23. Resolution on Some Questions Concerning the People's Communes, adopted by the CCPCC on 10 Dec. 1958, in *The People's Republic of China, 1949–1979: A Documentary Survey* (Wilmington, NC: Scholarly Resources Inc., 1980), 725.

24. In Jiangsu, between 80 and 90 per cent of the GVIO produced by commune enterprises during 1958–60 was comprised of goods that served either agriculture or commune members (see Mo Yuanren (ed.), *Jiangsu xiangzhen gongye fazhan shi*, 92).

25. e.g. some 30,000 handicraft co-operatives, accounting for one-third of the national total, were annexed by communes and became commune enterprises (see Ma Hong (ed.), *Xiandai Zhongguo jingji shidian* (*A Record of Contemporary Chinese Economic Affairs*) (Beijing: Zhongguo shehui kexue chubanshe, 1982), 213).

26. Mark Selden, *The People's Republic of China: A Documentary History of Revolutionary Change* (New York: Monthly Review Press, 1979), 80. Jiangsu alone announced its intention to establish 40,000 new enterprises in 1958, and most of these were CBEs; see Mo Yuanren (ed.), *Jiangsu xiangzhen gongye fazhan shi*, 87. By the end of 1958, there were 600,000+ indigenous blast furnaces and some 100,000 small coal pits in the countryside; see Liu Suinian and Wu Qungan (eds.), *China's Socialist Economy: An Outline History, 1949–1984* (Beijing: Beijing Review, 1986), 237–8.

27. S. Lee Travers, 'Peasant Nonagricultural Production in the People's Republic of China', in US Congress, Joint Economic Committee, *China's Economy Looks Toward the Year 2000*, i (Washington, DC: US Government Printing Office, 1986), 377. Jiangsu experienced a similar pattern of expansion. The GVIO (in current prices) produced by CBEs in Jiangsu was RMB 298 million in 1958 and rose to RMB 462 million in 1960 (see *JSN* 137).

28. Probably the most important positive consequence was the build-up of physical and human capital in rural areas. A sizeable amount of rural infrastructure was created (we saw evidence of this in several regions we visited in Jiangsu), and there was some 'learning-by-doing' from rural industrialization.

29. A large number of rural workers were absorbed by CBEs or moved to cities as temporary workers in the state sector. Between 1957 and 1958, the agricultural labour force declined by some 41 million or about 21 per cent (see Riskin, *China's Political Economy*, 142).

30. For an assessment of the GLF and a discussion of the retrenchment that followed and the new development strategy that emerged in the early 1960s,

see Riskin, *China's Political Economy*, chs. 6 and 7, particularly 133–45 and 149–58.

31. Ma Hong (ed.), *Xiandai Zhongguo jingji shidian*, 213.
32. *ZJN* (1981), 4. 55.
33. Mo Yuanren (ed.), *Jiangsu xiangzhen gongye fazhan shi*, 96.
34. *JSN* 127.
35. The number of commune-brigade industrial enterprises in Wuxi County declined from a peak of 514 in 1959 to 20 in 1964 (see Mo Yuanren (ed.), *Jiangsu xiangzhen gongye fazhan shi*, 97). It is hard to believe that, in one of China's most industrialized counties, 96 per cent of the CBEs were economically unviable and merited closure. During the Cultural Revolution the closure of rural enterprises in the early 1960s was criticized as 'dragging the advanced backward', and it was said that the motive behind the closure was 'to reduce local control and mass initiative' (see Riskin, *China's Political Economy*, 151, and his article, 'China's Rural Industries: Self-Reliant Systems or Independent Kingdoms', *CQ* 73 (Mar. 1978), 78–83).
36. *ZJN* (1981), 4. 55.
37. Mo Yuanren (ed.), *Jiangsu xiangzhen gongye fazhan shi*, 97–9. During the GLF, urban factories recruited a large number of new workers from the countryside, and many of them as well as some recruited earlier were sent back to the countryside during the 'three difficult years'. In total, perhaps some 20 million workers nation-wide were returned to the countryside.
38. This and the other figures in the paragraph are from *JSN* 137.
39. The five small industries have been the focus of several studies of China's 'rural small-scale industry'. See Christine P. W. Wong, 'Rural Industrialization in the People's Republic of China: Lessons from the Cultural Revolution Decade', in US Congress, Joint Economic Committee, *China Under the Four Modernizations*, part 1 (Washington, DC: US Government Printing Office, 1982), 394–418; American Rural Small-Scale Industry Delegation, *Rural Small-Scale Industry in the People's Republic of China* (Berkeley, Calif.: University of California Press, 1977), and Sigurdson, *Rural Industrialization*.
40. Strictly speaking some were 'large urban collective enterprises', but they were operated as state enterprises.
41. Unless noted otherwise, the information in this paragraph is from Liu Suinian and Wu Qungan (eds.), *China's Socialist Economy*, 360–3.
42. American Rural Small-Scale Industry Delegation, *Rural Small-Scale Industry*, 156.
43. Riskin, *China's Political Economy*, 213.
44. See Mo Yuanren (ed.), *Jiangsu xiangzhen gongye fazhan shi*, 144–6, for a brief discussion of the situation in Jiangsu during this period. The priority given to grain production, the absence of opportunity for individual initiative, the restrictions placed on craftsmen interested in working outside the village, and the difficulties faced by some CBEs during the 1970s were repeatedly emphasized by all those we interviewed at the village and township levels.

45. *Lun shedui qiye* (*On Commune-Brigade Enterprises*) (Beijing: Zhongguo shedui qiye baoshe and Nongye bu renmin gongshe qiye guanli zongju, 1982), 5.
46. Township-Village Interview Notes and Enterprise Interview Notes. Also see Fei Xiaotong, 'Xiao chengzhen zai tansuo (A further inquiry into small towns)', in Jiangsu sheng xiao chengzhen yanjiu ketizu (ed.), *Xiao chengzhen*, 45.
47. Much has been written about why China relied on local small-scale industry to modernize agriculture. For example, see Sigurdson, *Rural Industrialization*, chs. 1 and 6; American Rural Small-Scale Industry Delegation, *Rural Small-Scale Industry*, chs. 1, 4, and 9; and Riskin, *China's Political Economy*, 213–18. Among the rationales most frequently mentioned are the following: (1) small-scale industry requires less capital than large-scale industry and therefore can be developed with less outside assistance; (2) rural small-scale industry utilizes scattered, low-quality resources that would not otherwise be used, in other words, the opportunity cost of the resources consumed by the five small industries is, if not zero, very close to zero; (3) the decentralization of industry to the countryside was seen as a way of overcoming China's inadequate rural transport system and its deficient commercial system; (4) finally, rural small-scale enterprises, by being closer to the end-users of their products, are in a better position to understand and to respond to local needs.
48. *ZTN* (1987), 157 and 259; and *Zhongguo gongye jingji tongji ziliao, 1949– 1984* (*China Industrial Statistical Material, 1949–1984*) (Beijing: Zhongguo tongji chubanshe, 1985), 31–2.
49. *JJN* (1986), 3. 34.
50. In some industries, e.g. construction materials, edible oil processing, and tailoring, the share reached as high as 30 per cent. See Mo Yuanren (ed.), *Jiangsu xiangzhen gongye fazhan shi*, 141.
51. For an analysis of state investment and industrial growth in Jiangsu in the 1970s, see Prime, 'Impact of Self-Sufficiency', chs. 3 and 4.
52. *JTN* (1991), 127.
53. See 'Report Concerning the Return of Rural Handicraft Industrial Enterprises to the Leadership and Management of People's Communes', in *XQZFX* 4–6.
54. Wong, 'Rural Industrialization', table 1.
55. Some of the excluded agricultural CBEs may have been involved in non-agricultural activity. These and the following data are from *Zhongguo nongye de guanghui chengjiu, 1949–1984*, 125–6.
56. Data provided by JPASS.
57. *Lun shedui qiye* (*On Commune-Brigade Enterprises*), 5.
58. Ibid. 10.
59. A production process is technically inefficient if there exists another feasible process that uses no more of any input and less of some input to produce a given output. A technically efficient production process may be economically (or allocatively) inefficient if there exists another feasible process that produces that same output at lower cost.

60. Ma Hong and Sun Shangqing (eds.), *Zhongguo jingji jiegou wenti yanjiu*, 360.

61. Wong, 'Rural Industrialization', 395. In fact, in the twelve years between 1968 and 1979, local small- and medium-scale iron and steel works alone lost a total of RMB 6.0 billion (see 'Woguo difang zhongxiao gangtie jingji xiaoyi fenxi (Analysis of the economic effectiveness of small-medium iron and steel works in our country', in Jingji diaocha bianjizu, *Jingji Diaocha (Economic Investigation)*, 2 (1983), 18.

62. *Lun shedui qiye (On Commune-Brigade Enterprises)*, 134. In fact, in 1978, commune-brigade industrial enterprises in Jiangsu owned almost 60,000 metal-cutting machines.

63. See Wong, 'Rural Industrialization', for an analysis of these problems.

64. See Mo Yuanren (ed.), *Jiangsu xiangzhen gongye fazhan shi*, 197–8, and *Lun shedui qiye (On Commune-Brigade Enterprises)*, 7, and Zhonggong Wuxi shiwei zhengze yanjiushi, 'Wuxi xian fazhan, xiangzhen gongye de jiben jingyan (The basic experience of rural industry in the development of Wuxi County)', *NJW* 11 (1984), 44.

65. This document was formally approved at the Fourth Plenary Session in Sept. 1979. The document can be found in *ZJN* (1981), 2. 100–7.

66. *BR* (29 Dec. 1978), 12.

67. *ZJN* (1981), 2. 104.

68. See article 11 of SC, 'Guanyu fazhan shedui qiye ruogan wenti de guiding (shixing caoan) (Regulations on some questions concerning the development of commune-brigade enterprises (draft for trial use)), 3 July 1979' in *ZJN* (1981), 2. 98.

69. Until the income tax law was revised in 1985, the marginal tax rate started at 7 per cent for income below RMB 300 but rose quickly to a maximum of 55 per cent for income exceeding RMB 80,000. The marginal rate for income between RMB 600 and RMB 1,000 was 20 per cent. The 1985 reform changed the rates so that the marginal tax rate started at 10 per cent for income below RMB 1,000 and rose to a maximum of 55 per cent for income exceeding 200,000. See *XQZFX* 205–6. For a discussion of the taxes paid by TVEs before 1984 see Xu Shanda, 'Guanyu xiangzhen qiye shuishou zhengce he shuifu wenti de sikao (Some thoughts on tax collection policy and tax burden of rural enterprises), *NJW* 9 (1987), 46–8.

70. SC, 'Guanyu fazhan shedui qiye ruogan wenti de guiding (shixing caoan)', 2. 96.

71. This document can be found in *XQZFX* 29–33.

72. See CCPCC and SC, 'Dangqian nongcun jingji zhengce de ruogan wenti (Some questions concerning current rural economic policies)', 2 Jan. 1983. The warning was obviously needed. In 1983, on the basis of the 1983 No. 1 Document, the Jiangsu Provincial Government ordered Nantong City to return to their original owners 210 commune enterprises that were elevated to 'large urban collectives' earlier in the year. See Mo Yuanren (ed.), *Jiangsu xiangzhen gongye fazhan shi*, 219–20.

73. e.g. the 'Report Concerning the Start of a New Phase in CBEs', issued by the CCPCC and SC in Dec. 1983 states that 'as long as they make economic

sense and are voluntary, all types of joint enterprises are permitted, without restriction as to industry or administrative division' (see *XQZFX* 120–1).

74. CCPCC, 'Zhonggong zhongyang quanyu 1984 nian nongcun gongzuo de tongzhi, zhongfa yihao (Circular of the Central Committee of the Chinese Communist Party on rural work during 1984, document no. 1)', in *Nongcun gongzuo tungxun (Rural Work Communications)*, 7 (1984), 3–8.

75. For the first time, government departments assigned CBEs their own accounts (*hutou*). See *XQZFX* 124.

76. One reason for making the change was to control tax abuse. Before the change, some counties were assigning urban collective enterprises to commune jurisdiction so that they would escape the eight-grade progressive income tax (see Y. Y. Kueh, 'Economic Reform in China at the "xian" Level', *CQ* 96 (Dec. 1983), 684.

77. *XQZFX* 126. For example, county governments have the authority to reduce by one-third to two-thirds that portion of the income tax above the 20 per cent rate (for details, see Gu Jirui (ed.), *Xiangzhen qiye shouce (Handbook for Rural Enterprises)* (Beijing: Zhongguo qingnian chubanshe, 1985), 406–7).

78. The five wheels refers to the five levels of enterprise ownership: township (*xiang*, formerly the commune), village (*cun*, formerly the brigade), group (*zu*, formerly the production team), joint-household (*lianhu*, a partnership involving several households), and individual (*geti*). The ten industries are cultivation, forestry, animal husbandry, household sidelines, fishery, industry, commerce, construction, transportation, and services.

79. The household responsibility system that was eventually adopted throughout Jiangsu and in most parts of China is the *baogan daohu* (or *da baogan*) system. Under this system, land continues to be owned collectively but is distributed to households for management. Distribution is usually according to population but sometimes according to labour. After meeting certain fixed obligations to the team, households may keep the remaining output for consumption or disposal on the market.

80. SC, 'Guanyu shedui qiye guanche guomin jingji tiaozheng fangzhen de ruogan guiding (Concerning various stipulations to be implemented by commune-brigade enterprises according to the direction of adjustment of the national economy)', *ZJN* (1982), 3. 13.

81. SC, 'Guowuyuan guanyu tiaozheng nongcun shedui qiye gongshang shuishou fudan de ruogan guiding (State Council regulations regarding the adjustment of the burden of industrial-commercial taxes on CBEs)', *ZJN* (1982), 3. 55.

82. In 1980, in Wuxi county alone, more than 1,000 CBEs received tax exemptions (amounting to about RMB 35 million), and most of them should not have qualified for the tax exemption. See *Lun shedui qiye (On Commune-Brigade Enterprises)*, 137.

83. *XQZFX* 127.

84. In Jiangsu, many small groups (formerly production teams) also own enterprises. We shall use the term 'rural collective enterprises' to denote enterprises owned by townships, villages, and small groups.

85. See SC, 'Guanyu jin jibu huoyue nongcun jingji de shi xiang zhengce (Ten policies to further enliven the rural economy)' (1 Jan. 1985). This document is reproduced in Wang Jiye and Zhu Yuanzhen (eds.), *Jingji tizhi gaige shouce* (*Manual of Economic System Reform*) (Beijing: Jingji ribao chubanshe, 1987), 110–14.
86. Denis F. Simon, 'China's Drive to Close the Technological Gap: S&T Reform and the Imperative to Catch Up', *CQ* 119 (Sept. 1989), 620, and SSTC, *Zhongguo kexue jishu zhengce zhinan kexue jishu baipishu dierhao* (*A Guide to China's Science and Technology Policies, Science and Technology White Paper No. 2*) (Beijing: Kexue jishu wenxian chubanshe, 1987), 169.
87. In this manner, the spark plan provided short-term training to over 5 million rural residents between 1986 and 1989 (see SSTC, *Zhongguo kexue jishu zhengce zhinan kexue jishu baipishu disihao* (*A Guide to China's Science and Technology Policies, Science and Technology White Paper No. 4*) (Beijing: Kexue jishu wenxian chubanshe, 1990), 101.
88. e.g. in 1990 the Agricultural Bank and the Industry and Commerce Bank earmarked RMB 600–700 million for loans to spark projects (see SSTC, *Zhongguo kexue jishu zhengce zhinan kexue jishu baipishu disihao*, 105–6). For a discussion of the criteria used to select spark projects, see SSTC, *Zhongguo kexue jishu zhengce zhinan kexue jishu baipishu dierhao*, 162–3. On the whole, the criteria favour projects that (i) have as their central purpose technical development or improvement (e.g. a project that would raise the technical level of a local industry) and (ii) involve the co-operation between a rural enterprise and a technology and science unit.
89. Of the RMB 12.5 billion spent on spark projects during 1986–9, 1 per cent came from the central government, 8 per cent from local governments, 39 per cent from bank loans, and 52 per cent from rural enterprises. See SSTC, *Zhongguo kexue jishu zhengce zhinan kexue jishu baipishu disihao*, 101–2.
90. For discussions of the theory behind the strategy and the internal debate before and following its adoption, see Dali Yang, 'China Adjusts to the World Economy: The Political Economy of China's Coastal Development Strategy', *PA* 64/1 (1991), 42–64, and David Zweig, 'Internationalizing China's Countryside: The Political Economy of Exports from Rural Industry', *CQ* 128 (Dec. 1991), 716–41.
91. See Ch. 3 where an attempt is made to estimate the growth of net non-agricultural output in rural Jiangsu.
92. *ZTN* (1990), 622.
93. Ibid. 623.
94. Rural non-agricultural employment is the labour force at and below the township (commune) level (*xiangcun laodongli*) working outside of agriculture. The growth rate is calculated from data in *ZNTN*, various years.
95. This is not the place to analyse the causes of China's inflation. Suffice it to say that economic reform gave enterprises and local governments greater economic and financial autonomy, reduced the scope of central planning, and gave the market a more prominent role in the economy. But economic and financial autonomy were introduced before the central government had

firmly established the institutions and policy instruments (e.g. monetary policy and fiscal policy) that are needed to manage the economy and before price reform had been introduced. The inevitable result was that the central government lost control of investment and of aggregate demand. With more discretionary funds and assisted by low interest rates and accommodating bankers, enterprises went on a spending spree. Aggregate demand surged, and with it so did prices. For an in-depth examination of China's experience with inflation and economic instability, see World Bank, *China: Macroeconomic Stability and Industrial Growth under Decentralized Socialism* (Washington, DC: World Bank, 1990), particularly 30–61.

96. Between 1978 and 1983, adjustment and consolidation reduced the number of commune-brigade industrial enterprises by 50,000 (see *Zhongguo nongye de guanghui chengjiu, 1949–1984*, 125). One of the most affected industries was machinery, which accounted for 33.5 per cent of the GVIO produced by CBEs in 1978 but only 27.6 per cent in 1983. In total, between 1978 and 1983, some 600,000 CBEs either closed or shifted to other lines of activity. But since some 400,000 new CBEs were also formed, the net decline was only about 210,000 (see *XQZFX* 118).

97. In 1985, GVIO produced by enterprises at the township level increased by 38 per cent and GVIO produced by enterprises at and below the villlage level rose by 67 per cent.

98. The annual rate of growth of the general index of retail prices rose from 7 per cent in 1987 to 19 per cent in 1988 (*ZTN* (1990), 249). Since the prices of many goods were under control, the inflationary pressure was undoubtedly greater than suggested by the growth in the price index. For example, the prices of vegetables and meats, which had been partially or fully decontrolled earlier, increased by 32 per cent and 37 per cent respectively in 1988 (*ZTN* (1990), 251).

99. In 1990, the general index of retail prices increased by only 2.1 per cent (see Zou Jiahua, 'Report on the Implementation of the 1990 Plan for National Economic and Social Development and the Draft 1991 Plan', *BR* (22–8 Apr. 1991), 27).

100. The economic sanctions imposed on China after the government crushed the student demonstration in Tiananmen Square in June 1989 undoubtedly contributed to the depth of the 1989–90 recession. The erosion in consumer confidence caused retail sales to decline by 8 per cent in 1989, the largest annual drop since the downturn following the failure of the Great Leap Forward in the early 1960s (see Niu Renliang *et al.*, '1988 nian yilai jinsuo de zongti xiaoying fenxi (An analysis of the overall effects of the retrenchment since 1988)', *JY* 5 (1990), 14).

101. *ZTN* (1990), 33.

102. Han Baocheng, 'Readjustment Improves Rural Enterprises', *BR* (27 Aug.– 2 Sept. 1990), 25.

103. Many TVEs used short-term bank loans to finance long-term investments, and nearly all TVEs depended heavily on loans for working capital. See Yu Guoyao and Li Yandong, 'Dangqian xiangzhen qiye mianlin de zhuyao wenti (Major problems currently confronting rural enterprises)', *NJW* 10 (1989), 22–7.

104. Scattered reports from various parts of China suggest a shut-down rate of between 5 per cent and 10 per cent. The New York Times Service reported in Nov. 1989 that about 5 per cent of all rural collective enterprises in China were forced to close (*Globe & Mail* (9 Nov. 1989), sect. A, p. 2).
105. *ZTN* (1991), 382.
106. In addition, hundreds of thousands of rural workers who had temporary and contract jobs in urban enterprises were sent back to the countryside.
107. Here, non-agricultural activities include the 'non-productive' services.
108. See *ZNTN* (1988), 18, 20, and 180; and *ZTN* (1988), 153 and 535.
109. *ZTN* (1990), 412.
110. *ZNTN* (1988), 31–4.

3

Rural Non-agricultural Development in Jiangsu
An Overview

THE next three chapters discuss in detail the collective and the private components of Jiangsu's rural non-agricultural sector. Here we present an overview of the growth and structural changes of the sector as a whole and discuss several important aspects of the process of rural non-agricultural development: the uneven spatial distribution of rural non-agricultural activities, the impact of rural non-agricultural development on farm production, the impact of international trade on rural industrialization, and the effects of rural industrialization on the environment.

Growth and Structural Changes

Economic indicators for rural Jiangsu, presented in Table 3.1, show that rural economic growth was extremely rapid after 1978. The growth of total social product[1] (in 1980 prices) in rural Jiangsu averaged slightly less than 18 per cent per year from 1979 to 1990, and growth was particularly rapid in the mid-1980s. The average growth of Jiangsu's rural social product was 13 per cent per year from 1979 to 1983, jumped to nearly 27 per cent per year from 1984 to 1988, and then declined to about 6 per cent from 1989 to 1990 when the national economy went into a severe recession. When rural social production is disaggregated by the major material products sectors, it becomes clear that the primary source of rural growth since 1978 has been the non-agricultural sector, particularly rural industry. From 1984 to 1988, GVAO (including cultivation, animal husbandry, forestry, fishery, and sidelines) increased by a vigorous 7 per cent per year, but in the same period RGVIO increased by over 41 per cent per year, and industrial growth was strong at both the township and the village levels.[2] The growth of rural non-agricultural activities other than industry was less rapid but still averaged well over 25 per cent per year from 1984 to 1988.

However, the pace of rural non-agricultural development may be significantly lower than that indicated by the growth of gross output (RGVNAO). Gross output gives an accurate measure of economic growth

TABLE 3.1. *Growth Performance of Jiangsu's Rural Sector and of its Components, 1979–1990 (% per annum)*

	1979–83	1984–8	1989–90	1979–90
Rural gross output (1980 prices)	13.1	26.7	6.1	17.7
Agriculture[a]	6.8	7.1	1.4	6.1
Non-agricultural sector	22.3	37.7	7.2	26.2
Industry	21.3	41.5	8.6	27.6
Township	21.6	36.9	8.6	25.8
Village and below	20.8	50.2	20.8	30.2
Construction	46.7	21.0	−3.2	27.1
Transport	19.8	44.9	1.1	27.1
Commerce[b]	5.0	28.8	4.6	14.8
Rural value added (1980 prices)	16.0	18.9	2.3	14.9
Agriculture[a]	9.3	4.7	0.0[c]	5.8
Non-agricultural sector	20.7	30.0	3.6	21.7
Industry	21.2	32.6	5.3	23.3
Construction	48.4	19.4	−4.4	27.5
Transportation	27.1	42.6	−3.6	28.5
Commerce[b]	4.7	22.6	5.3	12.2

[a] Includes cultivation, forestry, fishery, animal husbandry, and household sidelines.
[b] Includes trade, food and drink, and services.
[c] Less than 0.05%.

Source: For the sources of data used in the calculation of the growth rates of rural gross output and that of rural value added, see Tables 3.2 and 3.3.

only if the industrial structure and the relationship between value added and gross output are stable. But, in rural Jiangsu, neither was stable during the 1980s. With the introduction of new economic activities and the decline of old industries, economic growth in rural Jiangsu was accompanied by dramatic changes in its industrial structure and in the relationship between its value added and gross output. Specifically, the decline of some heavy industries and the growing importance of processing and simple assembling activities have caused a steady decline in the ratio of industrial value added to industrial gross output. Thus, the growth of RGVNAO, particularly that of RGVIO, suggests a pace of rural non-agricultural development that may be significantly upward biased.[3]

Data are not available to estimate directly the value added produced by Jiangsu's rural non-agricultural sector. However, a crude approximation is possible by assuming that the ratios of intermediate products to gross output for each of the major material products sectors in rural Jiangsu are the same as those for the province as a whole. When estimated in this manner, rural Jiangsu's value added suggests a rate of rural non-agricultural development that is significantly lower than that suggested by the growth of RGVNAO (see Table 3.1). From 1979 to 1990, the real value added produced by the five material products sectors in rural Jiangsu grew at an average annual rate of 14.9 per cent. Growth in rural industrial value added in the five years after 1983 (33 per cent) was higher than in the previous 5 years (21 per cent), but the difference is not nearly as great as that found in the growth of RGVIO (21 per cent in 1979–83 and 42 per cent in 1984–8). Nevertheless, for the period 1979–90, the average growth of rural industrial value added was an extremely impressive 23 per cent a year. To put rural Jiangsu's performance in perspective, during 1964–73, a decade when Taiwan experienced its fastest growth, its manufacturing output index increased on average by 21 per cent a year and its real GDP by 10.4 per cent a year.[4]

Various growth indicators for TVEs, the most important form of rural non-agricultural activity in Jiangsu, are presented in Table 3.2, and they show a pattern consistent with that suggested by the aggregate data in Table 3.1. After a short period of consolidation, TVEs in Jiangsu developed at an incredible pace until the severe recession of 1989–90.

TABLE 3.2.   *Growth Indicators, Township-Village Enterprises in Jiangsu,*
1978–1990 (% per annum)

	1978–80	1981–3	1984–8	1989–90
Number of enterprises	−0.0	−3.2	12.3	−2.8
Township level	n.a.	0.7	5.9	−1.2
Village level	n.a.	−5.4	16.0	−3.5
Number of workers	9.8	5.7	10.1	−4.4
Township level	n.a.	6.7	8.2	−4.1
Village level	n.a.	4.3	13.0	−4.8
Financial indicators				
Gross revenue	31.9	15.7	38.0	7.6
After tax profit	36.6	−1.7	16.8	−19.5
Indirect and direct taxes	21.4	29.6	31.2	1.6
Wage bill	32.9	13.3	27.5	4.0
Fixed capital[a]	n.a.	19.2	32.3	19.9
Net fixed capital	n.a.	17.7	32.6	16.4
Working capital at year-end	n.a.	19.4	42.2	21.5
Outstanding bank loans at year-end	n.a.	23.9	56.2	15.9

[a] At original prices and unadjusted for depreciation.

Sources: *JSN* 136; *JTN* (1991), 114, and data provided by the Ministry of Agriculture, Animal Husbandry, and Fishery.

Employment by TVEs increased at an average annual rate of 5.7 per cent during 1981–3, over 10 per cent during 1984–8, and then declined by 4.4 per cent a year during 1989–90. The financial indicators in Table 3.2 (e.g. gross revenue, wage bill, taxes) also show a similar pattern, i.e. very rapid increase until the late 1980s. The one exception is profits, which increased significantly less rapidly and with greater fluctuation than the other financial indicators, suggesting that during the 1980s the profitability of TVEs had declined. During the 1980s, wage payments grew more rapidly than employment, so the average wage increased; in fact, its rate of growth was in the double-digit range between 1983 and 1988, considerably faster than inflation. The very rapid growth in outstanding bank loans to TVEs during 1984–8 suggests that the rapid expansion of TVEs was largely financed by credits. Outstanding bank loans accounted for about 23 per cent of working capital in the early 1980s but 41 per cent in 1986. Indeed, a major cause for TVEs' slower growth in the late 1980s was the government's tight monetary policy introduced in the autumn of 1988 to bring an accelerating inflation under control.

Because of the vastly differing sectorial growth rates, rural Jiangsu's economic structure changed substantially during the 1980s. In Tables 3.3

TABLE 3.3. *Composition of Total Social Product, Rural Jiangsu, Selected Years* (% of total)

	1978	1980	1982	1984	1986	1988	1990
In current prices							
Total[a]	100	100	100	100	100	100	100
Agriculture[b]	58	51	52	46	35	29	28
Industry	34	40	37	41	52	58	60
Construction	3	5	7	7	8	8	6
Transport	1	1	1	3	3	3	2
Commerce[c]	4	3	3	3	3	3	3
In 1980 prices							
Total[a]	100	100	100	100	100	100	100
Agriculture[b]	62	50	50	41	28	19	18
Industry	29	40	39	46	59	68	71
Construction	3	5	8	7	8	8	6
Transport	1	1	1	3	3	3	2
Commerce[c]	4	3	2	3	3	3	3

[a] Gross output of the material products sectors.
[b] Includes cultivation, forestry, animal husbandry, household sidelines, and fishery.
[c] Includes trade, food and drink, and services.

Sources: Calculated from data in *JSN* 128 and *JTN* (1991), 93.

TABLE 3.4. *Composition of Value Added of the Material Products Sectors in Rural Jiangsu, Selected Years (% of total)*

	1978	1980	1982	1984	1986	1988	1990
Value added in current prices							
Total	100	100	100	100	100	100	100
Agriculture[a]	69	66	71	64	54	50	49
Industry	22	25	21	25	34	37	39
Construction	2	4	4	5	5	6	5
Transport	1	1	1	3	3	3	3
Commerce[b]	6	4	3	3	4	4	4
Value added in 1980 prices							
Total	100	100	100	100	100	100	100
Agriculture[a]	75	66	68	60	47	38	36
Industry	17	25	23	28	40	47	50
Construction	2	4	4	5	6	7	6
Transport	1	1	1	3	3	4	4
Commerce[b]	5	4	3	4	4	5	5

[a] Includes cultivation, forestry, animal husbandry, household sidelines, and fishery.
[b] Includes trade, food and drink, and services.

Sources: The rural value added were estimated on the assumption that the ratios of intermediate product to gross output for the five main sectors were the same for rural Jiangsu as for the province as a whole. The real value added for each sector was obtained by deflating the current price series with the implicit national income deflator for that sector in Jiangsu. The value-added data and the implicit deflators were calculated from statistics in *JTN* (1991), 30, 31, 40, and 93, and *JSN* 128.

and 3.4 rural Jiangsu's industrial structure as measured by gross output and by value added respectively are presented for selected years. By either measure, rural Jiangsu experienced a structural transformation in the 1980s that can be described only as extraordinary. In 1978, reflecting its more developed economy, the share of agriculture (broadly defined) in the rural economy was lower in Jiangsu than in the country as a whole, but it still accounted for 62 per cent of rural Jiangsu's total social product and 75 per cent of its value added (these and the following percentages are all calculated from constant price data). A decade later, in 1988, only 19 per cent of the gross output and 38 per cent of the value added produced by the material products sector in rural Jiangsu came from agriculture. The contribution of farming was, of course, even smaller. In one decade, industry had replaced agriculture as the dominant sector in rural Jiangsu. In 1978, industry and construction accounted for one-third of rural Jiangsu's social product and not quite one-fifth of its value added. By 1988, its share of rural social product and of rural value added had climbed to 76 per cent and 54 per cent respectively. The shift from agriculture to industry was particularly rapid in the mid-1980s. Between 1982 and 1988, the share of agriculture in rural social product and that in rural value added declined by 31 and 30 percentage points respectively. Because agricultural prices increased more rapidly than industrial prices in this period, the changes in economic structure when output is measured in current prices are somewhat less but still impressive.

The extraordinary rate of non-agricultural output growth in rural Jiangsu was largely brought about by a rapid growth in factor inputs made possible by the policy and regulation changes at the national level discussed in Chapter 2. Possibly the single most important consequence of these changes was that rural residents were given more economic choices and greater freedom in decision-making. In the more permissive and flexible environment of the 1980s, they responded aggressively to the new opportunities by shifting resources from agriculture in general and cultivation in particular to more profitable non-agricultural activities.

Between 1980 and 1988, Jiangsu's rural labour force expanded from about 22 million to over 27 million, and none of the increase was absorbed by agriculture.[5] In fact, agricultural employment (defined broadly to include workers in farming, forestry, animal husbandry, sidelines, and fishery) declined from about 19 million in 1980 to 16.2 million in 1988. The implication is that between 1980 and 1988 Jiangsu's rural non-agricultural sector absorbed from 7 to 8 million workers. Many of these workers went to work for rural enterprises. The employment in industrial TVEs, which represents most but not all rural industrial employment, increased rapidly, with rates in or near the double-digit range from 1984 to 1987.[6] Growth was particularly rapid in 1984 and 1985 when it exceeded 17 per cent per year. Furthermore, the evidence shows that while agriculture gave up workers in good years, when rural industry expanded rapidly (e.g. 1984–8), it also took them back in bad years when rural industry declined (e.g. 1989–90).[7] In other words, agriculture was the residual employer in rural areas, and having it as the residual employer gave rural enterprises a degree of flexibility not available to state enterprises.

Unfortunately there is not a reliable capital stock series for Jiangsu's rural non-agricultural sector. The closest measure available is the fixed capital stock series for TVEs. Fixed capital is measured in original prices and is available in both gross and net (adjusted for depreciation) terms. Both series suggest very rapid growth. Between 1980 and 1988, fixed capital owned by TVEs, whether in gross or in net terms, increased more than sixfold.[8] Since neither series is in constant prices and TVEs include a small but not insignificant number of agricultural enterprises,[9] the data need to be treated with caution. Nevertheless, even after mentally adjusting the data for price increases and for the inclusion of some agricultural enterprises, these growth rates still suggest an impressive increase in rural non-agricultural capital stock. This massive infusion of labour and capital into Jiangsu's rural non-agricultural sector goes a long way towards explaining the rapid growth of rural non-agricultural output.

The shift from agriculture to other activities is clearly reflected in the deployment of Jiangsu's rural labour. Table 3.5 brings together the available data on the distribution of rural labour in Jiangsu by industry for selected years between 1978 and 1990. Although the data for the earlier years are less complete, Table 3.5 nevertheless provides a useful comparison of how labour deployment changed in rural Jiangsu after 1978. The most significant change is the decline of 27.6 percentage points in the share of the rural labour force employed in agriculture, falling from

TABLE 3.5. *Structure of Labour Force in Rural Jiangsu, Selected Years (% of total)*

	1978	1980	1985	1988	1990
Total rural labour force[a] (million)	n.a.	22.64	25.98	27.37	27.89
Distribution (% of total)					
Total	100	100	100	100	100
Agriculture	86.8	84.1	65.6	59.2	61.5
Cultivation	79.7	75.2	58.6	52.2	54.7
Other agriculture[b]	7.1	8.9	7.0	7.0	6.8
Industry	10.6	11.8	18.5	20.5	18.7
Township industry			9.9	10.9	9.9
Village industry			7.2	7.4	6.6
Industry below village[c]			1.5	2.2	2.1
Construction			5.8	7.2	7.0
Transportation			1.8	2.7	2.5
Commerce[d]	2.6	4.1	1.3	1.7	1.8
Education, health, and other services			2.0	1.9	2.0
Other			5.1	6.8	6.4
(working on contract in urban units above the county level)			1.8	2.8	2.9

[a] *Xiangcun laodongli*, i.e. labour force with residence at and below the township (commune) level excluding those on state farms.
[b] Includes forestry, fishery, animal husbandry, and sidelines.
[c] Includes industrial enterprises operated by teams, partnerships, and individuals.
[d] Includes food and drink.

Sources: The figures for 1985, 1988, and 1990 were calculated from data in *JTN* (1991), 91, and those for 1978 and 1980 were calculated by the JPASS using data provided by the JPSB.

86.8 per cent in 1978 to 59.2 per cent in 1988.[10] Nearly the entire decline was in cultivation, which as a percentage of total employment fell from 79.7 per cent to 52.2 per cent.[11] Workers employed by rural industry accounted for about 10 per cent of rural employment in 1978 and over 20 per cent in 1988. Although rural labour became much more mobile after 1978 and some moved to work in collective and state units in urban areas, the number of such migrant workers is relatively small. In 1988, rural workers employed in large collective and state units numbered perhaps 1.23 million, and those working as contract workers in collective and state units above the county level numbered only 0.76 million (less than 3 per cent of all workers with rural residence).[12] Thus the redeployment of rural labour in Jiangsu was primarily the result of changes in labour allocation within the rural sector, particularly at the village and household levels.

How was family labour deployed in rural Jiangsu in the late 1980s? A one-year household record-keeping survey conducted in six Jiangsu villages (two in each of three townships—Sigang, Hufu, and Haitou) in 1987 sheds some light on this question. Before discussing the findings, presented in Table 3.6, it should be noted that the surveyed villages are in townships that had higher than average pressure on land and were undergoing rapid economic changes in the mid- and late 1980s.[13] For example, the shift of labour from agriculture to non-agricultural activities was substantially more dramatic in two of the surveyed townships than in Jiangsu as a whole. The survey results, therefore, must be interpreted with caution and not be generalized too readily. Certainly, the deployment pattern found in the surveyed villages cannot be taken to represent Jiangsu as a whole. Despite this limitation, the data are nevertheless important for what they reveal about labour deployment in those regions where rural non-agricultural development has been rapid.

The most remarkable survey finding is how quickly and aggressively rural households, in the economically more permissive environment of the 1980s, have turned away from agriculture in some parts of Jiangsu.[14] Take for example the two villages surveyed in Sigang, a township in one of Jiangsu's more industrialized counties (see Table 3.6). In Jingxiang, the less developed of the two villages, only 18.5 per cent of family labour time (measured in standard workdays) was allocated to agriculture (defined broadly) in 1987, while in the economically more developed village of Zhashang, the percentage allocated to agriculture was an insignificant 4.9 per cent. The share allocated to own-farm cultivation (i.e. the cultivation of one's 'subsistence' and 'responsibility' land) was even smaller—11.5 per cent in Jingxiang and 4.4 per cent in Zhashang. The two villages in Hufu, a township in a less developed but still fairly prosperous county, reported a slightly higher level of participation in

TABLE 3.6. *Allocation of Family Labour, Six Villages in Jiangsu, 1 January–31 December 1987*

	Haitou		Hufu		Sigang	
	Haiqi	Haiqian	Yangquan	Xiaojian	Jingxiang	Zhashang
Number of households in village	637	947	341	281	373	714
Cultivited area per person (*mu*)	0.36	0.37	0.55	0.30	1.01	0.79
Survey findings						
Number of households in sample	20	30	32	26	19[a]	33
Standard days[b] of family labour used per household per year	506	700	678	891	632	709
Standard days worked per family worker per year	160	223	319	282	273	285
Allocation of family labour (% of total)						
Agriculture[c]	56.4	22.7	17.8	22.0	18.5	4.9
Own-land cultivation[d]	12.7	8.3	8.7	12.4	11.5	4.4
Agricultural processing	2.2	1.6	2.6	0.0	0.4	0.1
Handicraft production	13.6	27.4	1.4	2.7	0.0	0.0
Commerce[e]	5.2	20.7	2.4	1.0	0.0	2.5
Rural enterprises	7.4	14.4	69.8	71.9	62.8	88.0
Transportation	0.1	1.1	4.3	0.0	5.3	0.6
Construction	5.4	5.2	0.8	0.7	5.7	0.2
Own-house construction	2.8	0.7	0.7	0.6	0.0	0.0
Other activities[f]	9.7	6.8	0.9	1.6	7.3	3.1

[a] There were 19 usable observations out of 20 households surveyed.

[b] A standard day is 8 hours.

[c] Includes cultivation, animal husbandry, aqua-culture, forestry (e.g. gathering), exchanged-out, hired-out, and other own labour used in agriculture.

[d] Own-land cultivation is the cultivation of one's subsistence and responsibility land.

[e] Trade, restaurant and catering, repair services, personal services.

[f] Social labour and labour allocated to unclassified non-agricultural activity.

Source: Household Record-Keeping Survey.

agriculture—17.8 per cent in Yanquan and 22 per cent in Xiaojian. However, as in the two villages in Sigang, a significant portion of this was in activities other than own-farm cultivation (e.g. animal husbandry, forestry). It would appear that by the late 1980s most peasants in Sigang and Hufu (and probably also in many other regions in south Jiangsu) had become part-time farmers.

Even in a relatively underdeveloped region such as Haitou township, one village (Haiqian) reported only 22.3 per cent of its family labour time allocated to agriculture. Haiqi, the other village surveyed, reported 56.4 per cent of its family labour time allocated to agriculture. However, because cultivated land was very scarce in Haiqi and Haiqian, the amount of work time allocated to cultivation was still extremely limited, 8.3 per cent in Haiqian and 12.7 per cent in Haiqi. The bulk of the family labour time allocated to agriculture was spent on aqua-culture and fishing, a reflection of the coastal location of the two villages.

Rural enterprises have emerged as the main rural employer in many regions in Jiangsu, particularly in the south. In Zhashang, the most developed of the six villages surveyed, 88 per cent of its family labour time was absorbed by rural enterprises. To a lesser extent, rural enterprises were also the dominant employer in Jingxiang, and in the two villages surveyed in Hufu township. Although Haitou was, in 1987, the most prosperous and probably also the most industrialized township in Ganyu, it was poorer and significantly less industrialized than the townships in south Jiangsu.[15] The lack of industry explains why rural enterprises absorbed only 14 per cent of rural household labour time in Haiqian and 7.4 per cent in Haiqi. It also explains why individual non-farm activities (e.g. handicraft production and commerce) were so much more important in these two villages.

The available statistics show that Jiangsu's rural non-agricultural sector has developed at an extraordinary pace since 1978 and has caused dramatic changes in rural product structure and in how rural resources (particularly labour) are used. In one decade industry has replaced agriculture as rural Jiangsu's most important component in terms of output value, and many Jiangsu peasants, while remaining in their villages, have become part-time farmers and full-time workers or entrepreneurs (or what the Chinese call peasant-worker and peasant-entrepreneur).

The Spatial Distribution of Rural Non-agricultural Activities

As in other developing countries, rural non-agricultural activities are not distributed evenly in China. In general the rural non-agricultural sector is more developed in the economically more advanced than in the less advanced regions. For example, in 1982, in provinces where the per capita GVAIO was above RMB 2,500, TVEs generated revenue per rural person that was nearly sixteen times the level in the least developed areas (provinces where the per capita GVAIO was below RMB 600).[16] The uneven distribution of rural non-agricultural activities found at the national level also exists at the provincial level.

In 1986, Jiangsu was composed of 11 *diji shi* (regional-level administrative cities), which were in turn divided into 41 *shixia qu* (urban or suburban districts under the direct jurisdiction of the city proper or municipality), 60 *xian* (counties), and 4 *xianji shi* (county-level administrative cities).[17] Because municipalities include agricultural areas, agricultural and rural economic data are available for 75 separate localities (the 64 *xian* and *xianji shi* plus the suburbs of the 11 municipalities). When these 75 localities are grouped by the intensity of rural non-agricultural activity, defined as the ratio of gross output produced by the four rural non-agricultural material goods sectors to rural population,[18] they fall roughly into five categories: 'extremely active' (RGVNAO/RP > RMB 6,000), 'very active' (RMB 3,000 < RGVNAO/RP < RMB 6,000), 'active' (RMB 1,500 < RGVNAO/RP < RMB 3,000), 'moderately active' (RMB 1,000 < RGVNAO/RP < RMB 1,500), and 'inactive' (RGVNAO/RP < RMB 1,000). Since industry dominates the rural non-agricultural sector in most parts of Jiangsu, it is not surprising that, when the 75 localities are classified by the intensity of rural industrial activity

TABLE 3.7. *Spatial Distribution of Rural Non-agricultural and Rural Industrial Activities, Jiangsu, 1986*

	Jiangsu	Regional RGVNAO/rural population in RMB				
		0–1,000	1,000–1,500	1,500–3,000	3,000–6,000	6,000 and over
Number of region	75[a]	39	11	13[b]	9[c]	3[d]
RGVIO/rural population (*yuan*)	967	278	834	1,798	3,805	8,436
RGVNAO/rural population (*yuan*)[e]	1,397	597	1,336	2,360	4,629	8,787
Rural industrial labour/rural labour (%)	19	9	20	32	47	68
Rural non-agricultural labour/rural labour (%)	38	25	43	55	67	80
% of provincial						
Rural population	100	60	15	15	9	1
RGVIO	100	17	13	29	36	5
RGVNAO	100	26	14	26	31	4
Rural industrial labour	100	26	17	28	27	2
Rural non-agricultural labour	100	36	19	25	19	1

TABLE 3.7. *Continued*

	Jiangsu	Regional RGVIO/rural population in RMB				
		0–800	800–1,250	1,250–2,500	2,500–5,000	5,000 and over
Number of region	75	43	9	12[f]	8[g]	3[d]
RGVIO/rural population (*yuan*)	967	297	938	1,958	3,958	8,436
RGVNAO/rural population (*yuan*)[e]	1,397	636	1,384	2,550	4,795	8,787
Rural industrial labour/rural labour (%)	19	10	22	33	48	68
Rural non-agricultural labour/rural labour (%)	38	25	48	54	68	80
% of provincial						
Rural population	100	63	14	14	8	1
RGVIO	100	19	13	28	34	5
RGVNAO	100	29	13	25	29	4
Rural industrial labour	100	29	18	26	25	2
Rural non-agricultural labour	100	39	20	22	18	1

[a] The 75 administrative localities include the suburban districts of the 11 municipalities and the 64 counties (*xian* and *xianji shi*).

[b] The 13 include 10 counties (Wuxian, Wujiang, Hanjiang, Wujin, Yixing, Danyang, Taizhou, Jingjiang, Jiangning, Jiangdu) and the suburban districts of Nanjing City, Xuzhou City, and Yangzhou City.

[c] The 9 include 7 counties (Zhangjiagang, Jiangyin, Wuxi, Changshu, Taicang, Yangzhong, Kunshan) and the suburban districts of Nantong City and Zhenjiang City.

[d] The three are the suburban districts of Changzhou City, Wuxi City, and Suzhou City.

[e] RGVNAO is the gross output of the four rural non-agricultural material products sectors (industry, construction, transport, and commerce).

[f] The 12 include 9 counties (Wuxian, Kunshan, Wujiang, Wujin, Yixing, Danyang, Taizhou, Jingjiang, Jiangdu) and the suburban districts of Nanjing City, Xuzhou City, and Yangzhou City.

[g] The 8 include 6 counties (Wuxi, Changshu, Jiangyin, Yangzhong, Zhangjiagang, Taicang) and the suburban districts of Nantong City and Zhenjiang City.

Sources: Calculated from data in *JJN* (1987) and data provided by JPASS.

(RGVIO/rural population), they also tend to group into five categories: 'extremely active' (RGVIO/RP > RMB 5,000), 'very active' (RMB 2,500 < RGVIO/RP < RMB 5,000), 'active' (RMB 1,250 < RGVIO/RP < RMB 2,500), 'moderately active' (RMB 800 < RGVIO/RP < RMB 1,250), and 'inactive' (RGVIO/RP < RMB 800). The classification of Jiangsu's localities by intensity of rural non-agricultural activity and by intensity of rural industrial activity is presented in Table 3.7.

In 1986, in 12 of the 75 localities, rural non-agricultural activity was either 'extremely active' or 'very active', and in 39 localities it was 'inactive'. (The discussion here focuses only on rural non-agricultural activity since roughly similar results were obtained when localities were

grouped by intensity of rural industrial activity—see Table 3.7.) The
localities in the 'very active' category had an intensity of rural non-
agricultural activity that was on the average 7.7 times that of the 'inactive'
localities. Compared to the 'inactive' localities, the 'very active' had a
distinctly different economic structure. Whereas rural industries absorbed
only 9 per cent of the rural labour force in the 'inactive' localities, they
absorbed 47 per cent in the 'very active' localities. The 12 'very active'
and 'extremely active' localities, with 10 per cent of Jiangsu's rural
population, 29 per cent of its rural industrial labour force, and 20 per cent
of its rural non-agricultural labour force, produced 41 per cent of its
RGVIO and 35 per cent of its RGVNAO. On the other hand, the 39
'inactive' localities, with 60 per cent of Jiangsu's rural population, 26 per
cent of its rural industrial labour force, and 36 per cent of its rural non-
agricultural labour force, produced only 17 per cent of its RGVIO and 26
per cent of its RGVNAO. In other words, not only were Jiangsu's rural
non-agricultural activities concentrated in a small number of 'extremely
active' and 'very active' localities, but the rural establishments in these
localities were also significantly more productive, in the sense of having a
much higher labour productivity, than those elsewhere.

Where are the 'extremely active', the 'very active', and the 'inactive'
localities? Geographically and economically, Jiangsu divides naturally
into five regions: the north coastal region (NCR), the north mountainous
region (NMR), the region north of the Yangzi River (Changjiang) or the
north delta region (NDR), the region south of the Yangzi River or the
south delta region (SDR), and the south mountainous region (SMR).[19]
For convenience, hereafter, we shall call NDR, NMR, and NCR north
Jiangsu, and SDR and SMR south Jiangsu. Compared to south Jiangsu,
the north is less populous and less developed. NCR, NDR, and SDR
consist largely of lowland areas, whereas NMR and SMR are more hilly.
In fact, 1,000 years ago, much of the land in NCR and NDR was covered
by the sea. With good weather and a reliable irrigation and drainage
system, SDR, which encompasses the Yangzi Delta, has the most favourable agricultural environment. Served by a network of canals and rivers
and with easy access to the Shanghai–Nanjing rail line, the region also
has an effective and low-cost transportation system. The transportation
system in the rest of Jiangsu is not as good as in SDR but still reasonably
good by Chinese standards.

Fig. 3.1 shows that SDR is where rural non-agricultural activity is most
active. Rural non-agricultural activity is most intense in the suburbs of the
three historically commercial and industrial cities of Changzhou, Wuxi,
and Suzhou. The areas with the next highest intensity are those adjacent
to these cities. As we move north and west, away from SDR, the intensity
of activity declines. Behind SDR by a considerable distance but still

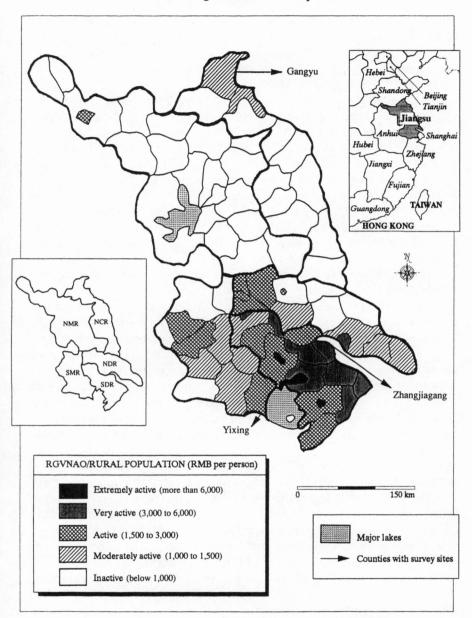

Fig. 3.1. Spatial Distribution of Rural Non-agricultural Activities, Jiangsu, 1986

relatively active are SMR and NDR. To the north of NDR, the intensity of activity drops sharply so that nearly all of north Jiangsu is 'inactive'. Xuzhou City and the two coastal regions of Ganyu and Lianyungang are the exceptions. However, if only rural industrial activity is considered, Ganyu and Lianyungang would be classified also as 'inactive'.

Comparative statistics from the five regions are presented in Table 3.8 and they show the regional differences more precisely. South Jiangsu, with 49 per cent of the rural population in the province and 42 per cent of the cultivated land, accounted for 86 per cent of the RGVIO and 81 per cent of the rural industrial labour. In fact, SDR alone, with only 17 per cent of Jiangsu's rural population, accounted for 56 per cent of its RGVIO and 44 per cent of its rural industrial labour. North Jiangsu, on the other hand, with more rural population than south Jiangsu, had only 14 per cent of the RGVIO and 19 per cent of the rural industrial labour. Thus the intensity of rural non-agricultural activity in south Jiangsu, particularly in SDR, is significantly higher. For example, RGVNAO per capita in SDR is more than six times greater than in north Jiangsu and two to three times that in NDR and SMR.

Jiangsu's rural non-agricultural sector is dominated by collective enterprises, and this is reflected in the similarity between the spatial distribution of rural non-agricultural activities and that of rural collective non-agricultural activities (compare Table 3.8 and Table 3.9).[20] In terms of either employment or gross income, the impact of collective enterprises is substantially greater in south Jiangsu than in north Jiangsu.[21] The region with the most developed collective enterprises is the south delta region (SDR), which in 1986 accounted for 40 per cent of the workers employed and 53 per cent of the gross income generated by Jiangsu's collective enterprises.

Another manifestation of uneven development is that collective enterprises in south Jiangsu are substantially larger than those in north Jiangsu. Indeed village enterprises in the south are as large and often larger than township enterprises in the north. For example, on average, the size of village industrial enterprises in the SDR (59 workers per enterprise) is more than twice that of village industrial enterprises (26 workers) and nearly equal to that of township enterprises (70 workers) in north Jiangsu.[22] Equally striking is the tremendous regional difference in labour productivity. Rural industry in SDR has a labour productivity in 1986 that was 80 per cent higher than that in north Jiangsu and 50–60 per cent higher than in SMR and NDR (see Table 3.8). The same difference in labour productivity exists for the rural non-agricultural sector as a whole. In other words, not only are rural non-agricultural activities concentrated in SDR but they appear to be significantly more productive in SDR as well.

TABLE 3.8. *Rural Non-agricultural Activity and Rural Non-agricultural Labour Productivity by Region, 1986*

	Jiangsu	South Jiangsu[a]				North Jiangsu[a]		
		Total	SDR	NDR	SMR	Total	NCR	NMR
Distributed by region (% of Jiangsu)								
RGVIO	100.0	86.4	56.3	18.9	11.2	13.6	6.2	7.4
RGVNAO	100.0	78.6	47.9	19.1	11.6	21.4	8.7	12.7
Rural industrial labour	100.0	80.9	43.8	24.5	12.6	19.1	8.5	10.5
Rural non-agricultural labour	100.0	73.8	32.9	29.0	12.0	26.2	11.0	15.1
Rural labour	100.0	55.1	20.3	24.1	10.7	44.9	15.6	29.3
Rural population	100.0	48.7	17.4	20.4	10.8	51.3	17.6	33.7
Cultivated area	100.0	42.3	15.1	16.5	10.7	57.7	20.0	37.7
RMB								
RGVNAO/Rural population	1,397	2,255	3,842	1,306	1,490	584	691	527
RGVIO/Rural population	967	1,716	3,123	893	1,005	257	339	213
RGVNAO/Rural non-agricultural labour	7,126	7,582	10,388	4,696	6,866	5,839	5,642	5,983
RGVIO/Rural industrial labour	9,636	10,285	12,368	7,432	8,578	6,881	6,974	6,805

[a] See text for definitions of the different regions.

Sources: Calculated from data in *JJN* (1987) and data provided by JPASS.

TABLE 3.9. *Distribution of Employment and of Gross Income, Collective (Township-Village-Team) Enterprises, 1986, by Region (% of provincial total)*

	South Jiangsu[a]				North Jiangsu[a]		
	Total	SDR	NDR	SMR	Total	NCR	NMR
Employment	76.2	39.6	23.2	13.4	23.8	9.6	14.2
Agricultural (non-farm)	70.8	36.7	14.2	19.9	29.2	19.9	9.3
Industry	79.9	43.8	23.0	13.2	20.1	8.5	11.6
Other non-agricultural	56.9	17.7	24.8	14.4	43.2	15.2	28.0
Gross income	83.8	52.5	19.6	11.7	16.2	7.0	9.1
Agricultural (non-farm)	59.8	32.0	10.0	17.9	40.2	30.1	10.1
Industry	86.6	55.6	19.3	11.6	13.4	5.9	7.5
Other non-agricultural	64.3	29.4	22.3	12.6	35.7	14.4	21.4

[a] See text for definitions of the different regions.

Source: Calculated from data in *JXQTZ* 24–63 and 126–65.

Without data on profit and capital, it is impossible to draw firm conclusions. However, the above evidence strongly suggests that, compared to the other regions, rural non-agricultural establishments in SDR were either more efficient or were engaged to a greater extent in those non-agricultural activities that were more productive, or, most likely, a combination of the two possibilities. What is indisputable, however, is that even in Jiangsu, which has the most developed rural non-agricultural sector in China, the vast majority of its rural population still lives in areas where there is little rural non-agricultural activity.

Although rural non-agricultural development has occurred in all parts of Jiangsu since 1978, its pace has been much more rapid in south Jiangsu, particularly in SDR, and thus regional differences have widened. Table 3.10 shows that in 1978 rural industrial intensity in south Jiangsu was 3.75 times greater than in north Jiangsu, but by 1988 the difference had widened to 6.33 times. If the comparison is made between SDR and north Jiangsu, then rural industrial intensity in SDR was 5.83 times greater than that in north Jiangsu in 1978, and 12.07 times greater in 1988. The south delta region accounted for about 58 per cent of the increase in Jiangsu's RGVIO (in 1980 prices) between 1978 and 1988, and its share of Jiangsu's RGVIO increased from 44 per cent in 1978 to 56 per cent in 1988.

There is no one simple explanation why rural non-agricultural development occurs in one locality and not in another, since the incidence of rural non-agricultural activity is influenced by many factors—historic,

TABLE 3.10. *1978 and 1988 Per Capita RGVIO, Distribution of 1978 and 1988 RGVIO and of Increases in RGVIO between 1978 and 1988 by Region*

	Jiangsu	South Jiangsu[a]				North Jiangsu[a]		
		Total	SDR	NDR	SMR	Total	NCR	NMR
RMB in current prices								
1978 RGVIO/rural population	124	194	302	139	126	52	64	45
1988 RGVIO/rural population	1,904	3,335	6,255	1,613	2,039	566	638	527
RMB in 1980 prices								
1978 RGVIO/rural population	116	180	280	129	117	48	60	42
1988 RGVIO/rural population	1,709	3,027	5,768	1,399	1,830	478	547	442
Distributed by regions (% of Jiangsu)								
1978 RGVIO in current prices	100.0	79.6	43.9	24.7	11.0	20.5	8.8	11.7
1988 RGVIO in current prices	100.0	84.6	55.6	17.5	11.5	15.4	6.0	9.4
Distribution of increases in RGVIO in 1980 prices between 1978 and 1988	100.0	85.9	58.1	16.3	11.5	14.1	5.5	8.6

[a] See text for definitions of the different regions.

Sources: Calculated from data in *JJN* (1989), 3. 109–20, and *1978 nian Jiangsu sheng nongye tongji ziliao (Statistical Material on Agriculture, Jiangsu Province, 1978)* (1979), 2–11 and 34–8.

geographic, and economic. One can think of numerous historic and geographic reasons for the high incidence of rural non-agricultural activity in the Yangzi Delta (SDR)—e.g. its importance in history as a centre of commerce and artisan industries, its proximity to the great port city of Shanghai, its effective and low-cost transportation system, and its well-developed network of small towns and cities. However, basic economic forces do exert strong influences on rural non-agricultural development. Below we explore, in the Jiangsu context, some of the more important influences.

Economic forces both push and pull rural resources from farming to rural non-agricultural activity. Perhaps the most important push factor is the limited capacity of agriculture to absorb labour productively on a given amount of arable land. Thus, when population density rises and farm size falls, farm households come under increased pressure to find ways of supplementing farm income.[23] A widely observed phenomenon in Asian agriculture is the strong inverse relationship between farm size and off-farm activities; that is, as the farm size declines, the farm household becomes more involved in off-farm activities.[24]

The pull factors are related to the availability of attractive non-agricultural employment and the profits that can be made from off-farm activities relative to that from farming. Because rural demand for non-food goods and services increases with the level of rural income and the level of agricultural development, a prosperous and progressive agriculture is more conducive to the growth of rural non-agricultural activities.[25] Agriculture is related to rural non-agricultural activities directly through its forward and backward production linkages, and indirectly through the consumption demand of farm households. Many rural non-agricultural activities (e.g. the fabrication and repair of farm tools) exist primarily to serve the needs of agriculture. In less developed countries, farm households typically spend 30–40 per cent of their incomes on non-food goods and services and another 10–15 per cent on processed agricultural goods, and these expenditures also contribute significantly to rural non-agricultural activities. Service industries are location specific, and many goods with relatively high income elasticities (e.g. furniture, household furnishings, and clothing) can be efficiently produced in small-to-medium-sized establishments that are suitable for rural areas. In addition, agricultural development stimulates the growth of agricultural processing industries, which, to be close to their sources of raw materials, are usually located in rural areas. Because of these production and consumption linkages, the development of rural non-agricultural activities, particularly in its early stages, is likely to be closely linked to the level of agricultural development.

However, in the long run, the evidence from Asia suggests that the fortunes of the rural non-agricultural sector are more closely tied to its ability to become less dependent on rural-type and agriculture-related employment and more dependent on urban-type and urban-related jobs. Urban-type jobs are usually high-income jobs, so employment in urban-type jobs is likely to increase the absolute level of rural non-agricultural income. There are at least two reasons for believing that distance from urban areas is perhaps the single most important determinant of access to urban-type jobs. First, proximity to urban areas makes it possible for rural residents to commute to urban jobs while still participating in farm work on a part-time basis. Second, proximity to urban areas increases the opportunity for lateral linkages between enterprises in rural areas and those in cities, and it is such linkages that bring urban-type jobs to the countryside. In China, however, the first reason is probably less important since rural workers are restricted from working in urban enterprises except under fairly stringent conditions. However, with the liberalization of the economy since 1978, it has become easier for rural residents to work in towns and cities.[26]

Are the same economic forces at work in Jiangsu? For example, do we observe in Jiangsu an inverse relationship between rural non-agricultural (or industrial) development and farm size and a positive relationship between it and proximity to urban areas? Since farm land is collectivized, farm size statistics are not available in Jiangsu. However, cultivated area per rural person can be readily calculated, and as a measure of population pressure on land it is as good as, if not better than, farm size. It is more difficult to find a good proxy for distance to urban areas (or access to urban-type employment). For lack of a better measure, the ratio of the GVIO in urban areas to total population is used here. That rural residents have better access to urban areas (or urban-type employment) in regions where the ratio of GVIO in urban areas to total population is high than where it is low is, I believe, a defensible assumption. Summary measures of these variables are presented in Table 3.11, and they show that in Jiangsu there does appear to be a strong inverse relationship between the intensity of rural non-agricultural (or industrial) activity and cultivated land per rural person. As well, there is a strong positive relationship between the intensity of rural non-agricultural (or industrial) activity and per capita urban GVIO (here used as a proxy for proximity to urban areas). It seems that some of the same push and pull forces that are moving resources from farming to rural non-agricultural activity in other less-developed countries in Asia are also at work in Jiangsu.

These relationships can be explored in a preliminary way by regressing the following equation using 1986 data:

$$Y_i = a + b_1X_{1i} + b_2X_{2i} + b_3X_{3i}$$

where Y_i = RGVNAO per rural person (RMB/person) in the ith locality in 1986,

TABLE 3.11. *Land/Rural Population and Urban GVIO/Total Population for 75 Localities by Intensity of Rural Non-agricultural Activity and by Intensity of Rural Industrial Activity, Jiangsu, 1986*

Range of RGVNAO/RP (RMB)	Number of locality	Average in range		
		RGVNAO/RP (RMB)	Land/RP (*mu*)	Urban GVIO/P (RMB)
5,000+	5	6,064	0.89	5,159
3,000–5,000	7	4,113	1.19	2,354
2,000–3,000	9	2,631	1.16	2,278
1,000–2,000	15	1,413	1.16	606
750–1,000	13	850	1.35	681
500–750	11	635	1.61	274
0–500	15	362	1.49	232
Jiangsu	75	1,397	1.34	1,140

Range of RGVIO/RP (RMB)	Number of locality	Average in range		
		RGVIO/RP (RMB)	Land/RP (*mu*)	Urban GVIO/P (RMB)
4,500+	5	5,254	0.89	5,159
2,500–4,500	6	3,493	1.08	2,518
1,500–2,500	10	2,062	1.22	2,222
800–1,500	11	988	1.15	543
400–800	15	511	1.30	648
200–400	12	303	1.56	457
0–200	16	127	1.52	222
Jiangsu	75	967	1.34	1,140

Sources: Calculated from data in *JJN* (1987) and data provided by JPASS.

X_{1i} = cultivated land per rural person (*mu*/person) in the ith locality in 1986,

X_{2i} = per capita urban GVIO (RMB/person) in the ith locality in 1986,

X_{3i} = GVAO per rural person (RMB/person) in the ith locality in 1986, and

a, b_1, b_2, b_3 are parameters to be estimated.

Our expectation is that b_1 would be negative and b_2 and b_3 would be positive. The estimated ordinary least squares equation is:

(1) $\qquad Y_i = 1311 - 2543\, X_{1i} + 0.43\, X_{2i} + 6.54\, X_{3i}$
$\qquad\qquad t = (1.71)\quad (-4.74)^*\quad (5.37)^*\qquad (4.57)^*$

*statistically significant at the 5 per cent level.

$$R^2 \text{ (adjusted)} = 0.65, \quad F_{3,71} = 46, \quad df = 71.$$

The regression results are consistent with our earlier observation that the basic economic forces that have pushed and pulled rural resources from agriculture to rural non-agricultural activities in other Asian developing countries have also been at work in Jiangsu. The estimated slope coefficients have the right signs and all are statistically different from zero at the 5 per cent level of significance. The adjusted R^2 indicates that, after taking account of the degrees of freedom, about two-thirds of the variation in regional RGVNAO per rural person can be explained by differences in the strength of the push (cultivated area per rural person) and the pull (opportunities for rural non-agricultural employment) forces in the various localities. Given the large F value, the estimated regression is obviously significant, and we may reject the null hypothesis that rural non-agricultural development is not linearly related to the specified push and pull forces.

Too much should not be made of these regressions findings. For one thing, the analysis is flawed by measurement problems. For example, instead of per capita urban GVIO, one would like to have a more direct measure of distance to urban areas (and the opportunities for lateral linkages with urban enterprises), but unfortunately such data are not available. It would also be better if, instead of gross output, we used net output. However, despite the data problems, the basic findings that the level of rural non-agricultural development is inversely related to population pressure on cultivated area and positively related to the level of agricultural development and to distance to urban areas are likely to stand up to further analysis.[27] In other words, the extremely uneven spatial distribution of rural non-agricultural activities in Jiangsu can be explained to a large extent by regional differences in population density, in agricultural development, and in the development of urban industries.

Rural Non-agricultural Development and Farm Production

There are several reasons to believe that increased involvement in non-agricultural activities by farm households may affect farm production adversely: (1) Non-agricultural activities may attract the more able and productive family members and leave only the old and the very young to do the farming, (2) they may divert scarce supervisory skills from farming, (3) by attracting labour from farming, they may also reduce the application of agricultural inputs that are complementary to labour, and (4) land and capital investment may also be diverted from agriculture to non-agricultural uses. In other words, the concern is that rural non-agricultural development and the increased participation of farm households in non-agricultural activities may reduce both the quantity and the quality of inputs allocated to agriculture.

In Jiangsu, the withdrawal of land and labour from agriculture began in the late 1970s and the pace quickened after 1983. Between 1978 and 1990, cultivated area declined by 1.54 million *mu* (over 2 per cent of the 1978 total), and perhaps as many as 3 million workers left agriculture, and most of them were from farming. One would expect a reduction of resources of this magnitude, other things remaining equal, would cause agricultural production to decline. Of course, the decline may not occur if there are sufficient off-setting changes. For example, farm households may invest a part of their earnings from non-agricultural employment in labour-saving and yield-raising agricultural inputs. Did rapid rural non-agricultural development adversely affect Jiangsu's agriculture?

Table 3.12 shows that GVAO in 1980 prices increased from about RMB 14 billion in 1978 to nearly RMB 28 billion in 1990. However, much of this increase came not from the mainstay of agriculture, cultivation, but from activities (in particular fishery, animal husbandry, and household sidelines) that had been suppressed before 1978 but were now permitted and encouraged under the new policy of all-round development. Still, between 1978 and 1990, crop production increased at an average annual rate of 3.36 per cent, although most of the growth occurred in the early 1980s. Crop output grew at an average annual rate of 6.51 per cent between 1978 and 1984, substantially faster than in the previous decade, but only 0.21 per cent from 1984 to 1990. The contrast is even more striking for grain production, where the average annual growth was 5.91 per cent from 1978 to 1984 and −0.37 per cent from 1984 to 1990.

What happened to agriculture at the provincial level is of course also reflected in the surveyed villages.[28] For example, all six surveyed villages reported a higher multiple cropping index in 1987 than in 1978, and in all but one of the six villages, grain production remained either stable or increased (in some instances, substantially) between 1978 and 1987. This occurred despite a dramatic decline in labour allocated to farming and a marginal decline in cultivated area (except in the two villages in Hufu, where, for special reasons, the decline was substantial).[29]

In the face of the substantial reduction in labour and land, why did

TABLE 3.12. *Cultivated Area, Fixed Capital Investment in Agriculture, and Agricultural Production, Jiangsu, 1978–1990*

Year	Cultivated area (million *mu*)[b]	Fixed capital investment in agriculture (RMB million)	GVAO[a] (RMB 1980 billion)	Crop production (RMB 1980 billion)	Grain output (million tons)
1978	69.91	189	14.09	11.27	24.01
1979	69.76	286	15.62	12.26	25.75
1980	69.62	196	14.77	11.26	24.18
1981	65.56	67	15.93	12.30	25.12
1982	69.47	83	18.31	13.86	28.55
1983	69.45	145	19.39	14.74	30.53
1984	69.32	132	22.58	16.24	33.54
1985	69.06	171	23.28	15.65	31.27
1986	68.86	172	24.74	16.28	33.40
1987	68.70	231	25.51	16.25	32.58
1988	68.53	268	27.20	16.60	32.43
1989	68.43	252	27.29	16.49	32.83
1990	68.37	324	27.98	16.42	32.64

			Average annual growth rate (%)		
1978–90			6.04	3.36	2.77
1978–84			8.43	6.51	5.91
1984–90			3.66	0.21	−0.37

[a] Gross value of output from cultivation, forestry, animal husbandry, sidelines, and fishery.
[b] 1 *mu* = 0.0667 ha.

Sources: *JSN* 127 and 131, and *JTN* (1991), 92, 97, 102, and 169.

farm output in Jiangsu, far from declining, in fact increase sharply in the early 1980s, reaching a peak in 1984, and then remain roughly at this higher level into the late 1980s? In brief, three factors have worked to offset the decline in agricultural labour and land: changes in the incentive environment, the presence of under-utilized resources, particularly labour, in rural Jiangsu, and increased application of labour-saving and yield-raising inputs.[30]

When China moved into the 1980s, it adopted two policy changes of enormous importance to agriculture. First, in 1979, it raised, by substantial margins, the prices paid for agricultural and sideline products, in particular grain, cotton, and oilseeds.[31] Second, in the early 1980s, it promoted the widespread adoption of the household responsibility system. In Jiangsu, the 1979 price adjustments increased the ratio of prices received by farmers to the prices paid by farmers for industrial goods by nearly 20 per cent,[32] thus making it profitable for farmers to produce more agricultural goods. The household responsibility system, under which farmers were permitted to keep all their output after meeting their fixed obligations to the team, also increased their incentive to produce more. Thus decollectivization and higher agricultural prices gave Jiangsu's farmers a tremendous incentive boost to increase production.

The second reason why increased activity in the rural non-agricultural sector in the 1980s did not affect agriculture in a more noticeably adverse manner is that, as in other parts of China, extensive underemployment existed in rural Jiangsu.[33] In the 1970s, work-sharing in Jiangsu's agriculture was widespread and, probably, so was work-shirking. Once the government liberalized the rural economy, raised agricultural prices, and strengthened the link between reward and performance, rural households, given the opportunities and the incentives to earn more income, participated in both farming and non-agricultural activities simply by working longer hours and more efficiently. In other words, because agricultural workers were underemployed in the late 1970s, the number of workers in agriculture could be reduced without reducing the actual labour input (measured in workdays or work effort) to agriculture. The simultaneous and rapid expansion of both agricultural and non-agricultural output in rural Jiangsu from the late 1970s to the mid-1980s is an indication of the extent to which parts of rural Jiangsu were prevented by previous economic policies from utilizing their labour fully and efficiently.

The achievement of increases in farm output with less land and labour suggests a substantial increase in yields. Indeed, between 1978 and 1988, grain yield increased by 41 per cent.[34] In other words, the reduction in land and labour was in part offset by increases in labour-saving and yield-enhancing modern inputs. In the 1970s, Jiangsu started to utilize increasing amounts of modern industrial inputs in its agriculture. Between 1970 and 1978, agricultural machinery, as measured by motive power, increased from 1.46 million kW to 8.55 million kW; tractor-ploughed land increased from 9.81 million *mu* to 38.88 million *mu*; and the application of chemical fertilizers increased from 16.16 kg. per *mu* of cultivated area to 53.17 kg.[35] After 1978, the move towards greater mechanization and more intensive fertilization continued, though the growth was at a somewhat slower rate. By 1988, rural Jiangsu had access to agricultural machineries with a total motive power of 21 million kW, had increased its tractor-ploughed area to 52.66 million *mu* (out of a total 68.53 million *mu*), and was applying chemical fertilizers at the rate of 126.46 kg. per *mu* of cultivated land. The addition of more labour-saving and yield-raising inputs undoubtedly freed some resources from farming for non-agricultural uses.

By the mid-1980s, however, rural conditions had changed considerably. Jiangsu was closer to full employment and the benefits of the new incentive environment had also been exhausted. We have no reliable measure of underemployment in rural Jiangsu but the evidence suggests that while some rural underemployment continued to exist into the late 1980s,[36] underemployment was much reduced. In fact, after a decade of rapid employment growth, labour was in short supply in some parts of rural Jiangsu. Decollectivization gave a powerful one-time stimulus to production, but by the mid-1980s it was no longer the important source of growth that it once was.

However, lower farm profits caused by production costs rising more rapidly than output prices also contributed to the slower agricultural growth.[37] Because it feared that higher food prices would cause urban unrest, the government did not pass the higher procurement prices for agricultural products on to urban consumers. Consequently, in the years following the 1979 price reform, the government had to absorb huge food subsidies which it could ill afford. Unable to shoulder additional large food subsidies, the government was limited in what it could do to stimulate agricultural production by raising agricultural prices at the farm level. In 1985, when the government replaced unified procurement (*tonggou*) with contractual procurement (*hetong dinggou*), it also introduced a new price arrangement that in effect reduced the farm-gate price for grain.[38] However, the prices of chemical fertilizers and other industrial inputs to agriculture continued to rise rapidly. The inevitable consequence was that the terms of trade turned against agriculture. Not surprisingly, in this altered economic environment, farmers were reluctant to increase agricultural output. In fact, compulsory grain sales to the state at prices barely sufficient to cover costs was a major source of tension between peasants and rural cadres in the late 1980s.[39]

With non-agricultural activities more profitable than farming, an increasing number of peasants left the land. To maintain agricultural production and to slow down the outflow of farmers from agriculture, local governments were instructed to use industry to subsidize agriculture (*yi gong bu nong*) and to use industry to build agriculture (*yi gong jian nong*). For example, in the mid-1980s, all six of the villages we surveyed subsidized agricultural production and invested a part of the profits from industry in agriculture.[40]

It also appears that, without additional agricultural investment, rapid agricultural growth is not likely to resume (particularly if the shift of land and labour from farming to non-farm activities continues). Although decollectivization improved incentives for production, its effect on agricultural investment was less laudable. After land was contracted to households, collective investment in agriculture dropped sharply, and the decline was not offset by increases in private investment. In consequence, as the data in Table 3.12 show, total investment in agricultural fixed capital in Jiangsu declined from a high of RMB 286 million in 1979 to RMB 67 million in 1981 and RMB 87 million in 1982 (all in current prices). Land with effective irrigation, which had increased steadily since the 1950s, began a gradual decline in the mid-1980s, falling from 53.94 million *mu* in 1984 to 52.46 million *mu* in 1988.[41] Apparently, in some parts of Jiangsu, existing irrigation facilities were not properly maintained. That rural residents were reluctant to invest in land and agricultural machinery is hardly surprising in view of the fact that privately owned productive assets had been collectivized without compensation in the past. In the early and the mid-1980s, most of the rural private savings were used to finance investment in private housing, the one asset for which the ownership rights have not been altered repeatedly since 1949. Between 1980 and 1988, living space per capita in rural Jiangsu increased from 9.9 square metres to 22.2 as thousands and thousands of thatched roof houses were replaced by multi-level brick dwellings.[42]

Since the early 1980s, the government has urged individuals and collectives to fund agricultural investment more generously. The incentives for private investment in agriculture were strengthened in 1984 when the length of land contracts under the household responsibility system was formally extended to a minimum of fifteen years.[43] Subsequently, the government also made it possible to inherit land contracts. In the early 1980s, TVGs were also told to use industry to support agriculture, e.g. by reinvesting TVE profits in agriculture or using TVE profits to subsidize agricultural production. This was done in many rural areas, particularly in south Jiangsu. However, the available evidence suggests that less and less of the profits from TVEs were used to support agriculture after 1983 (see Table 7.6). Furthermore, in those regions (e.g. north Jiangsu) where rural industry was undeveloped, using industry to support agriculture was not a viable option. Nevertheless better incentives for private investments and the contribution of funds from TVEs did bring about an increase in agricultural investment (see Table 3.12), but its annual value in the late 1980s, at around RMB 250–270 million, was still below that in 1979, and in real terms it was even lower. As a share of total provincial fixed capital investment, agricultural investment in Jiangsu averaged 8.1 per cent in 1978–80 but only 1.4 per cent in 1986–8.

During the 1980s, despite the transfer of land and labour from farming to non-farm uses, agricultural production did not decline in Jiangsu partly because surplus labour existed in the countryside, partly because peasants, responding to the improved incentive structure, worked longer and harder, and partly because of increased application of labour-saving and yield-raising inputs. However, in the long run, significant increases in agricultural output will come only from investment and technological change (i.e. more and better modern inputs). Investment in agriculture is not likely to rise significantly unless the central government allocates a larger share of its investment to agriculture or increases its pressure on local governments to invest more of their resources (e.g. profits from TVEs) in agriculture. If more investment is not forthcoming, agricultural productivity growth will slow down and it will be more difficult to transfer additional labour and land from farming to non-farm uses without causing disruption to farm production.[44]

The Open Policy and Rural Industrialization

When China adopted its rural reforms in the late 1970s, it also decided to turn outward for trade, technology, and capital, the so-called *kaifang zhengce* (variously translated as 'open policy' or 'a policy of opening to the outside world'). What role, if any, has the open policy played in China's rural industrialization?

Initially, the open policy had little impact on rural industry in most parts of China, including Jiangsu. This was in sharp contrast to what happened in south China, particularly the province of Guangdong. Between 1978 and 1985, Guangdong's export more than doubled, rising from US$1.39 billion to US$2.95 billion, a performance far superior to the growth achieved by other provinces,[45] and exports from its rural enterprises, particularly those in the Pearl River Delta just north of Hong Kong, accounted for a significant share of this increase.[46] Taking advantage of their location and the special incentives given to Guangdong, rural enterprises in the Pearl River Delta focused on the development of two types of economic arrangements with Hong Kong firms to produce simple labour-intensive products (e.g. clothing, textiles, knitted goods, toys, embroidered goods, electric components) for export: (i) 'special export arrangements'[47] and (ii) the so-called *sanzi* (three (foreign) capitals) enterprises, i.e. enterprises that make use of foreign capital.[48] In Guangdong, exports generated by special export arrangements and by *sanzi* enterprises as a share of total exports increased from below half a per cent in 1978 to 17 per cent in 1985, and much of the increase involved rural enterprises.[49] By 1991, *sanzi* enterprises accounted for two-thirds of Guangdong's exports.[50]

The open policy stimulated rural industrialization in Guangdong and Fujian because the government gave rural enterprises there strong incentives to export.[51] But the open policy did not significantly improve export incentive for rural enterprises in other parts of China. Despite a decentralization of the foreign trade system that saw the number of foreign trade companies (FTCs) increase from about a dozen in the late 1970s to several hundreds in the mid-1980s, rural enterprises in most parts of China remained isolated from international markets. In Jiangsu, for example, rural enterprises that exported did so on contracts from local branches of FTCs since they alone had access to market information and the authority to grant export licences. Consequently whether or not an enterprise exported depended largely on whether it had friendly relations (*guanxi*) with the foreign trade system.[52]

Since in most rural regions the local branch of the national (or provincial) foreign trade company was the only source of market information and the primary buyer of goods for export, it used its monopoly power to exploit rural enterprises by paying them low prices that did not reflect market conditions abroad. Nor did the foreign trade system pass on to rural enterprises benefits of government measures adopted to promote export. For example, at the beginning of 1981, the government introduced an internal settlement rate for trade transactions that was nearly twice the official exchange rate, but most rural producers did not receive the benefits of the *de facto* devaluation of the *yuan* since foreign trade corporations continued to purchase goods for exports at the same low prices.[53] In 1984, further to encourage export, the government decided to rebate some of the indirect taxes paid by producers on the goods they exported. However, in Jiangsu, the tax rebates on goods exported by rural enterprises were returned not to the producing enterprises but to the foreign trade companies.[54] Finally, neither the exporting enterprises nor the township-village governments that controlled them were entitled to retain much, if any, of the foreign exchange earnings from export for discretionary use. In other words, outside of Guangdong and Fujian, the incentive for rural enterprises to increase their export remained discouragingly low.

The export environment improved somewhat in the mid-1980s. In 1984 the government opened fourteen coastal cities to the outside world and granted them some of the same incentives to promote export and foreign investment given earlier in 1979 to the special economic zones (SEZs) in Guangdong and Fujian.[55] Then, in 1985, similar incentives were given to five newly designated coastal economic development zones. Two of the fourteen coastal cities (Nantong and Lianyungang) and one of the five coastal economic development zones (the Yangzi Delta) were in Jiangsu.[56]

Following the spirit of the open policy, Jiangsu gave its open areas increased authority to approve joint ventures, the authority to administer export goods other than those under national unified management, and permission to establish local foreign trade companies.[57] The open areas were told to make use of foreign investment to upgrade existing enterprises so they could export and that hereafter they would be judged primarily by their export performance. However, unlike in Guangdong, rural enterprises in the open areas in Jiangsu did not at first receive the same incentives to export and to attract foreign capital and technology as urban enterprises. Even though the Yangzi Delta included the entire *sunan* region (i.e. Suzhou *shi*, Wuxi *shi*, and Changzhou *shi*), only the city proper of Suzhou, Wuxi, Changzhou, and a few smaller county cities were designated 'areas open to foreign investment and where joint ventures have special incentives (*duiwai kaifang qu*)'.[58] Nevertheless, because export was now a priority of the regional governments in the open areas, they applied pressure on rural enterprises in their jurisdiction to increase export. Between 1985 and 1986, Jiangsu's rural enterprises doubled their 'direct exports',[59] and in 1986 they accounted for nearly one-quarter of all the goods purchased for export in the province.[60] Of the goods sold to the foreign trade system by rural enterprises, 71 per cent were from the open areas. In other words, outside the relatively confined open areas, few rural enterprises in the mid-1980s, even in an industrialized coastal province like Jiangsu, were involved in foreign trade.

Only in 1987 when Zhao Ziyang, then General Secretary of the Chinese Communist Party, publicly advocated what was referred to in China as the coastal development strategy did the idea of linking large numbers of rural enterprises to the export sector gain national attention.[61] In brief, the strategy called for the development of the coastal areas as an export base, and for rural enterprises to become part of that base. Specifically, the strategy advocated the use of rural labour in the coastal areas to produce labour-intensive light industrial products for export. The export earnings would then be used to import capital goods and technology needed to modernize the medium and large-scale state enterprises in the heavy industries so they could in turn supply the labour-intensive industries with more capital goods, thus enabling them to absorb more rural workers from agriculture. In this manner, rural workers would be transferred out of agriculture while at the same time state enterprises in the heavy industries would be modernized. The strategy was guided by the principle of 'attract from outside and connect within (*wai yin nei lian*)', i.e. to use the coastal areas to attract and use foreign technology and capital to produce goods mainly for export, to disseminate the imported technology and the successful experiences to interior regions (the less developed parts of the coastal provinces and the interior provinces), and to develop lateral links (e.g. through supply and subcontracting) with interior regions. With the support of Deng Xiaoping and the coastal provinces, Zhao's proposal was approved by the Politburo in February 1988 and officially launched at a national conference in March. After the Tiananmen tragedy and the downfall of Zhao Ziyang in June 1989, the coastal development strategy came under attack, but the most important features of the strategy have survived.[62]

To increase exports, the coastal development strategy proposed that enterprises in the coastal areas emulate their counterparts in Guangdong by putting 'both ends on the outside (*liangtou zaiwai*)', i.e. importing raw materials and using foreign capital to produce goods primarily for foreign markets.[63] 'Using imports to feed exports (*yi jin yang chu*)' was thought necessary because the coastal areas were seriously short of raw materials. In the early and mid-1980s, stimulated by price distortions that allocated most of the profits to the final stage of production, many processing enterprises were established by townships and villages in the coastal areas and they competed with state enterprises for agricultural raw materials. When interior regions that had traditionally transferred agricultural raw materials to the coast for processing also developed their own processing industries in order to capture more profits and tax revenues locally, the competition for raw materials and markets between rural and state enterprises and between coastal and interior regions became intense, and many enterprises operated below capacity for lack of raw materials.[64] For example, the Agricultural Bank of China estimated that in the mid-1980s about 15 per cent of China's 400,000 TVEs were closed because of raw material shortages.[65] The hope was that by putting both ends on the outside enterprises in the coastal areas would solve their raw material problem and at the same time lessen the tension between rural and state enterprises and that between raw material producing and processing regions.

When the central government implemented the coastal development strategy, it also adopted new foreign trade reforms. The control of local FTCs and foreign trade bureaux were transferred from the Ministry of Foreign Economic Relations and Trade to local governments and given responsibility for profits and losses in domestic currency. Furthermore, provincial governments entered into foreign trade contracts with the state under which they agree to remit annually specific amounts of foreign exchange earnings (fixed for three years) to the central government and were given entitlement to a large share of any foreign exchange earned above the contractual amount.[66] In turn, the provincial governments, at least in Jiangsu, introduced an analogous contract system with governments below the provincial level. As it did earlier in Guangdong and Fujian, the foreign trade contract system gave Jiangsu and its local governments increased incentives to export.[67]

With the adoption of the coastal development strategy, exports became one of the criteria by which the performance of rural enterprises and that of rural cadres were judged. In 1988, Jiangsu adopted the double line contract (*shuang xian cheng bao*) system under which export quotas were issued through the industrial system (*gongye xitong*) down to the production unit, and at the same time export purchase quotas were issued through the foreign trade system (*waimao xitong*) down to the branch office (in some cases down to individual members) of foreign trade companies.[68] Under this arrangement, townships were allotted export quotas which were in turn distributed among the TVEs, and leading cadres in TVGs and TVEs were evaluated in part by the export performance of the region or that of the enterprise.

If the export quota was to pressure local cadres in TVGs and TVEs to push exports, there were also incentives to encourage exports. In 1988, rural enterprises in Jiangsu that exported were given the following privileges:[69] (1) a lower electric power rate, (2) access to bank loans at preferential rates, (3) preferential access to raw materials at lower state prices, (4) entitlement to 12.5 per cent of the foreign exchange earnings generated,[70] (5) subsidies to cover any losses incurred by selling to FTCs instead of on the domestic market, (6) tax rebates on exported goods (previously given to FTCs), and (7) preferential access to locally controlled foreign exchange.[71] Furthermore, TVEs involved in special export arrangements were entitled to use an additional 47.5 per cent of their exchange earnings to import equipment, technology, or raw material. Rural enterprises involved in joint ventures (*sanzi* enterprises) were granted tax holidays, direct export rights, the right to import two duty free automobiles, and preferential access to raw materials, credits, water, and electric power.

It is still too soon to gauge the full impact of the decision to link rural enterprises to the export sector. However, the early results are encouraging. Since 1988, Jiangsu has significantly increased the involvement of its rural areas in foreign trade and investment. An increasing number of rural towns have been designated focal point industrial satellite towns (*zhongdian gongye weixing zhen*) that are opened to the outside world (*duiwai kaifang qu*) where foreign investment is given special incentives. In 1989, Jiangsu's open areas included 962 such industrial satellite towns.[72] To keep its export enterprises supplied with raw materials, the province has imported large amounts of steel, cotton, chemical fibre, wool, etc.[73] Many counties have established their own FTCs to facilitate local export. Between 1986 and 1990, the number of rural enterprises in Jiangsu that exported increased from 1,025 to 4,208, and the value of goods sold by rural enterprises to the foreign trade system increased from RMB 1.77 billion to RMB 10.66 billion, raising their share in the total value of goods purchased by the foreign trade system from 24 to 47 per cent.[74] Jiangsu is just beginning to tap the export potential of its rural industry. In 1990, rural enterprises in Jiangsu exported less than 9 per cent of their gross output, so the potential for growth is considerable.[75]

An increasing number of foreign businesses, attracted by the low cost and productive labour and improved investment incentives, have also invested in Jiangsu's rural enterprises.[76] At the end of 1990, of the 1,112 co-operative ventures involving foreign capital (*sanzi* enterprises) approved by Jiangsu, 423 had TVEs as partners,[77] and of the 1,138 approved in 1991, 740 were co-operative ventures with rural enterprises.[78] To rural enterprises, the importance of having a foreign partner is not just the foreign investment (which is usually quite small) but also the better access to market information and foreign distribution networks, the higher profits from exporting directly, and the greater opportunity to acquire new technology that come with the partnership.[79]

The success of Jiangsu's rural industry in expanding export and attracting foreign investment has been duplicated in other coastal provinces outside of south China, e.g. Zhejiang, Shandong, and Liaoning. Largely because of the performance of these provinces and those in south China, exports by rural enterprises increased from US$5.1 billion in 1987 to US$12.5 billion in 1990, and exports by rural enterprises as a share of total exports rose from below 13 per cent in 1987 to 21 per cent in 1990 and 30 per cent in 1991.[80] Rural industry in the coastal areas has emerged not only as the fastest growing export sector in China, accounting for an impressive share of the increase in total exports since the mid-1980s, but also as the destination for an impressive amount of foreign investment.

Because the incentives to export were extended only gradually to rural enterprises outside of south China, the impact of the open policy on rural industrialization outside of Guangdong was minimal before the mid-1980s. With the adoption of the coastal development strategy in the late 1980s, international trade has become an important stimulus to rural industrialization in Jiangsu and other coastal regions. Indeed, rural industry has emerged as China's leading exporter. This is not surprising since China's comparative advantage is in labour-intensive light industrial products, which is also the strength of its rural industry. In the late 1980s, textiles, garments, and handicraft products accounted for about two-thirds of the exports by Jiangsu's rural enterprises.[81] Since the late 1980s, labour-intensive light industrial products such as textile products and articles in daily use have been among China's fastest growing exports, and, in 1991, rural enterprises accounted for 72 per cent of China's garment exports, 45 per cent of its handicraft exports, 29 per cent of its export of articles in daily use, and 24 per cent of its export of silk fabrics.[82] Furthermore, because rural enterprises are market-orientated and do not depend on the state for resources, they have a stronger sense of competition and are less encumbered by government bureaucracy than state enterprises. Thus, compared to state enterprises, rural enterprises are more flexible and more dynamic and therefore better able to adapt to the changing needs and the demanding requirements of international markets. An abundant supply of low-cost labour, cheap land, less bureaucracy, greater flexibility, and adaptability are, of course, also some of the reasons why foreign businesses find investing in rural industry attractive. If the current open policy is maintained and the government continues to encourage rural enterprises to export and interact with foreign businesses, China's rural industry (particularly in Jiangsu and other coastal areas), in the decades ahead, will change in dramatic ways and become increasingly important in the national economy.

Rural Industrialization and the Environment

While rural industry has increased employment and income in the country-side, it has also brought environmental problems, in particular the loss of prime farm land and a significant rise in industrial pollution. The evidence further suggests that these problems did not begin in 1978 but in the 1950s. Although it is not possible to quantify precisely the extent of environmental damage that has been caused by rural industrialization, the following facts are suggestive.

Land Loss

Cultivated land in China declined dramatically from 111.8 million ha. in 1957 to 103.6 million ha. in 1965, and 99.7 million ha. in 1975, or a total decline of 12.1 million ha. in less than two decades (10.8 per cent of the 1957 total).[83] The decline in Jiangsu (and other more industrialized regions) was even more precipitous, falling from 5.83 million ha. in 1957 to 4.67 million ha. in 1977, or a decline of nearly 20 per cent in two decades.[84] These figures represent net losses, i.e. the differences between land taken from cultivation and that replaced by land reclamation. Because the land taken from cultivation was generally of higher quality than reclaimed land, the more revealing statistics are the gross loss figures. While a gross loss figure is not available for Jiangsu, several have been reported for China. Between 1957 and 1977, China lost a total of 33.3 million ha., or 29.8 per cent of the 1957 total, a staggering figure in view of how little arable land there is in China.[85]

Of course, not all the land loss between 1957 and 1977 was caused by rural industrialization. A great deal was lost to urban construction—indeed land losses were particularly serious in the suburbs of large cities—and some were lost to natural disasters. But the rapid growth of rural industries was undoubtedly one, and perhaps the most important, cause of the widespread loss of farm land before 1978. A huge number of rural enterprises were launched during the GLF. Then, in the 1970s, the campaign to develop rural industries to support agriculture and to provide local self-sufficiency created hundreds of thousands more rural enterprises. Established with little planning or economic calculations, many of these enterprises were built on flat and easily accessible farm land. A survey of 222 administrative villages (*xingzhengcun*) in eleven provinces found that rural enterprises occupied 0.09 *mu* of cultivated land for every worker they employed in 1978 (0.19 *mu* in those villages surveyed in Jiangsu).[86]

Prior to 1978 slow income growth and tight political control kept residential construction in rural areas to a minimum, but when reform and rural industrialization brought prosperity and a more relaxed political environment to the countryside, the pent-up demand for better housing set off a major housing boom. In the frenzy to construct new and larger houses, rural cadres had difficulties controlling the size of new houses and preventing them from encroaching on farm land because they were among the first to build larger houses, because building sites were often near farm land, and because individual builders did not bear the full cost of the collectively owned land. The problem was most severe in the more prosperous parts of rural China. 'Between 1978 and 1980 more than two-fifths of the cropland taken out of cultivation in the southern provinces was used for new rural housing', and in Hangzhou, Zhejiang, housing took 90 per cent of the farm land lost in 1979.[87] Reports from Yangzhou, Jiangsu, indicated that, before 1983, housing construction took on average more than 25,000 *mu* (1,667 ha.) of farm land a year.[88] Since the early 1980s, rural enterprises and residential construction have continued to occupy farm land despite repeated injunctions against it.[89] A 1986 government circular complained that in some provinces the annual decline in farm land was equivalent to the cultivated area in a middle-size county.[90] From 1979 to 1989, the cultivated area shrank by 3.8 million ha. in China, or nearly 4 per cent of the 1979 total, and by about 88,000 ha. in Jiangsu, or slightly less than 2 per cent.

Industrial Pollution

Industrial pollution has emerged as a serious problem in some parts of rural China, the consequence of careless development, ignorance, poverty, and lax enforcement of environmental and health standards. Many of the rural enterprises established in the 1960s and the 1970s to promote local self-sufficiency and in support of agriculture were in industries with well-known pollution problems, e.g. metals, cement, and agricultural chemicals. Some of the light industries established in rural areas, e.g. tanning, bleaching, and dyeing of fabrics, were also serious polluters. Industrial pollution was most serious where rural industry was most developed, i.e. in the suburbs of large cities and the more urbanized and developed townships. For example, in the late 1970s, 40 per cent of the water surfaces in Wujin, one of the most developed counties in Jiangsu, were seriously polluted.[91] Of course, rural enterprises were not the only source of industrial toxins, but since Wujin was one of the most developed rural industry zones in Jiangsu, they undoubtedly were major contributors to the problem.

In the early 1980s, in part as a consequence of attempts to clean up some of China's larger cities, more polluting enterprises were added to the countryside. In the late 1970s, China admitted that its environment had deteriorated seriously since 1949, and that the country faced enormous environmental problems including widespread industrial pollution in cities.[92] In response to this crisis, the government promulgated new environmental standards and regulations and initiated efforts to clean up its urban environment.[93] Under pressure to meet new environmental and health standards, urban enterprises discarded the more dangerous and dirty jobs, and many of them were taken over by nearby villages where people, anxious for employment, were willing and eager to accept any factory job.

Despite a 1979 regulation that prohibited the transfer of products, the production of which involve poisonous and dangerous substances, to rural areas unless effective pollution control equipment was also transferred, many harmful activities were moved to the countryside without such equipment.[94] For example, in the late 1970s and early 1980s, Suzhou moved some of its most polluting enterprises from their urban sites, presumably to suburbs and nearby townships.[95] Of those TVEs established in Shanghai's suburbs with equipment from urban enterprises, 84 per cent produced products that involved poisonous and dangerous substances.[96] The most common example of this is electroplating. The number of electroplating establishments in Shanghai's suburbs rose from 33 in 1978 to 425 in 1982, and 'only one in six could meet [environmental] standards'.[97] In 1982, there were 293 electroplating enterprises in Tianjin's suburbs discharging more than 1,790 tons of untreated waste water containing heavy metals per day.[98] Not all the polluting enterprises transferred to rural areas were from Chinese cities; in Guangdong's Pearl River Delta they also came from Hong Kong to escape from that city's tougher anti-pollution regulations.[99]

By the mid-1980s, many polluting enterprises could be found in rural areas, and in some areas emissions had exceeded the point at which they could be easily assimilated by the environment and were endangering the health, and in some cases taking the lives, of workers and residents. A province-wide survey in Jiangsu in 1989 found 37,342 of the 115,986 rural industrial enterprises surveyed (or 32 per cent) to be polluters, of which 33,013 (or 29 per cent of the total) were moderate (*yiban*) polluters and 4,329 (4 per cent) were serious (*zhong*) polluters.[100] Industrial pollution continued to be most serious in the suburbs of large cities and in the more developed townships. For example, in the mid-1980s, one-third of the water surfaces in *sunan* (i.e. Suzhou *shi*, Wuxi *shi*, and Changzhou *shi*) was polluted, with water quality at or below grade five, and there was hardly any water in Wuxi County (one of three counties in Wuxi *shi*) above grade four.[101] In Shanghai, industrial toxins had so severely contaminated the soil that suburban land had to be taken out of cultivation because crops produced on it were unsafe to consume.[102] Furthermore, when enterprises in the same industry were compared, those in rural areas were found to emit 10 to 20 times more pollutants than those in urban areas, reflecting the absence of effective pollution control in rural enterprises.[103] This is not surprising since many TVEs used old equipment inherited from urban enterprises.

Controlling Land Losses and Industrial Pollution

The loss of farm land and increased industrial pollution in rural areas, like other types of environmental degradation in China, occurred because those who made decisions about resource use either ignored or underestimated the social costs of environmental damage. The usual cause for this is either policy failures, market failures, or a combination of the two. In China, since resources were allocated either by the government or under its direction, the primary culprit must be policy failures. It was the promotion of questionable economic policies to extremes during the GLF and in the 1970s that accelerated the loss of cultivated land, did severe damage to China's forests, and permitted excessive industrial pollution in rural areas.[104] The government's ineffective management of lands and the nonaccountability of polluters directly under its management and control also contributed to the problem. The damages caused by policy failures were exacerbated by poverty and ignorance. The rural poor, with little understanding of science or experience in handling industrial wastes, cared more about the income they earned from hazardous industrial jobs than about long-term damage to health or environment.

Since the late 1970s, the government has tried to correct its past mistakes. In 1982, recognizing that the country could ill afford to lose more farm land to rural construction, the government announced new application, inspection, and approval procedures for rural construction, established standards and limits on residential building sites, and urged townships and villages to develop plans for the rational use of land.[105] However, since many rural communities lacked the skills and resources to develop plans, progress has been uneven. In 1984, only 32 per cent of China's villages and 22 per cent of its market towns had formulated land-use plans.[106] Jiangsu's performance was somewhat better—50 per cent of its nearly 2,000 market towns had developed plans in 1984, and, by 1989, all market towns and 84 per cent of the townships in Jiangsu had completed land-use plans.[107] Still, in that year, one-fourth of the villages in Jiangsu had not completed their plans.

The government has also tightened its control of land management. For example, before 1986, to use land in Jiangsu for non-agricultural purposes, only the approval of the township (the government immediately below the county) was needed. In 1986, such decisions were removed from township governments and transferred to higher levels of government.[108] In the late 1980s, cities, counties, and townships were also told to set control figures for the amount of cultivated land that could be transferred from agriculture to non-agricultural uses at around 0.15 per cent of cultivated land.[109] Then, in 1989, to encourage users to economize on village land, selected areas in rural Jiangsu began to charge a land-use fee on new industrial and residential constructions.[110]

Since the early 1980s, numerous measures to control industrial pollution in rural areas have been adopted. For example, in 1984, the State Council announced new stipulations that prohibited new investment in activities with serious pollution problems (e.g. electroplating, tanning, and manufacturing of asbestos products) and required those plants already in operation to renovate so as to meet national or local emissions standards or, if they could not do so in the stipulated time, to transform, merge, or close.[111] In addition, the government also required (i) polluting TVEs to pay effluent charges in accordance to the 1979 Environmental Protection Law, (ii) the filing of environment assessment reports before TVEs could expand or invest in new projects, and (iii) rural industrial projects to proceed strictly according to the rule of 'three simultaneously (*san tongshi*)'—to design project and pollution control measures simultaneously, to construct them simultaneously, and to put them in operation simultaneously (*tongshi sheji, tongshi shigong, tongshi touchan*).

Of the two environmental problems discussed above, the loss of farm land is probably the more serious, judging by the more frequent government warnings against the misuse of farm land.[112] With little unused arable land and a population that is still growing, it is understandable why the government's priority is food first, construction second (*yi yao chifan er yao jianshe*). But despite better planning and criticism of corrupt land management practices, the loss of farm land to rural construction has not stopped. The main reason is the uneven enforcement of land-use regulations at the local level in the face of continued strong pressure to use farm land for non-agricultural purposes.[113]

The housing boom has cooled down considerably in the most prosperous rural regions where most households have already constructed new homes, but in many areas where development and prosperity have come more slowly it is still going strong or is only in its early stages.[114] If spreading prosperity and the accompanying housing boom are not to remove more farm land from cultivation, local governments in the less developed areas must learn from the mistakes of other regions and take control of land management quickly by strictly enforcing land-use regulations from the start. In the more developed regions, even though the housing boom is nearing an end, the growth of satellite cities, rural towns, and industrial estates has continued to put pressure on farm land.[115] In recent years, to ensure that rural land is used wisely and economically, the central government has called on cities, townships, and rural towns to formulate development plans. But regions do not always enforce their own plans. While laws and regulations that can slow down and perhaps even end the decline in cultivated areas are now in place, enforcement remains uneven. To stop the decline in cultivated land, the government will need to apply even more pressure on local cadres to enforce existing laws and regulations concerning land use.

Since the early 1980s, limited progress has been made in the control of industrial pollution in rural areas. Some serious polluters have been closed or merged with larger enterprises in which pollution emission is more strictly controlled.[116] Some TVEs have installed pollution control equipment, and a few areas (e.g. Suzhou *shi*, Wuxi *shi*, and Changzhou *shi* in Jiangsu) have also started to collect effluent charges from TVEs. But, on the whole, the enforcement of environmental regulations has been sporadic and uneven. Even in south Jiangsu, where environmental regulations have been enforced more conscientiously, we are told that the control of industrial pollution in rural areas has not kept pace with the increase in pollution.[117]

The government admits that, because of the large number of rural enterprises, because they are so widely scattered, and because they have limited resources to invest in cleaner technology and pollution control, the control of industrial pollution in rural areas will take a long time.[118] Currently, to reduce industrial pollution, China is relying heavily on effluent charges to correct market failures and alter the behaviour of resource users. With economic activities in rural China increasingly market orientated and decision-making more decentralized, it was sensible for the government to extend the existing market-based environmental policies to rural enterprises.

However, for either market-based policies or regulations to work, they must be enforced. Unfortunately, enforcement is where the system is at its weakest. Rural enterprises have little incentive to comply with environmental regulations. They frequently change their product lines, so are understandably reluctant to invest in pollution abatement equipment that may soon be useless. TVEs are often operated by managers on short-term (1–3 year) contracts with little interest in pollution control since their earnings are linked directly to profits. Township and village governments also have little interest to make sure that TVEs comply with emission standards and other environmental regulations since TVE profits are their main source of income. Besides, they are under considerable pressure from senior governments to achieve rapid production growth, and therefore have little interest to control industrial pollution. This leaves the Environment Protection Agency as the primary enforcer of China's environmental policies. But the Environment Protection Agency is short of both personnel and funds, and has barely penetrated the rural sector. When it has tried to enforce environmental regulations in rural areas, it has received little or no co-operation from local governments since the latter consider it a nuisance and see its activities as just another attempt by a senior government agency to extract monies that are much needed at the local level.[119]

Whether the government should increase its efforts to control industrial pollution in rural areas requires a comparison of the costs of damage with the costs of preventing it. Because monitoring emissions is expensive, one suspects that it is economical to collect effluent charges only from those rural enterprises that can be monitored at reasonable cost, i.e. the small number of large TVEs that are mostly located in the suburbs of cities and the more developed townships. To influence the behaviour of small rural enterprises, the government will need to resort to blunter administrative methods, i.e. putting pressure on those who control rural enterprises, the township and village cadres, to reduce the discharge of industrial wastes. But the government is not likely to increase the pressure significantly since here too it must compare the cost of environmental damage and the cost of strict enforcement, i.e. slower industrial growth and the loss of industrial jobs in rural areas. One suspects that governments at all levels in China are likely to give greater weight to jobs and growth than to the environment.

It should be noted that, besides government regulations and market-based incentives, the control of industrial pollution may be helped by the economic transformation already under way in rural China. Because a large share of the industrial wastes come from a small number of industries—e.g. metallurgy and chemicals—changes in the industrial composition of rural industry will influence the level of industrial pollution. Since the early 1980s, responding to competition and changing demand, rural industrial activities have shifted away from some heavy industries known for their pollution to more labour-intensive light industries. Of course, not all labour-intensive light industries are environmentally friendly, but one suspects on balance the ongoing structural change will help reduce industrial pollution in rural areas. To be more competitive, rural enterprises have had to invest heavily in new technology and equipment, and over time this should also bring about a reduction of pollution per unit of output since much of the industrial pollution in rural areas is currently caused by outdated technology and equipment. To economize on infrastructure, some townships are gradually moving their TVEs to more centralized locations, e.g. in Kunshan County, Jiangsu, fourteen of its twenty townships in 1989 had established industrial estates (*gongye xiaoqu*) for TVEs.[120] The concentration of TVEs will make it easier and cheaper to monitor and control industrial emissions. Finally, economic reforms have reduced some of the most serious price distortions that in the past have encouraged enterprises in China to use resources wastefully, and this too will encourage rural enterprises to adopt production methods that are more environmentally friendly.

This chapter has dealt with five aspects of rural non-agricultural development in Jiangsu: its pace, its spatial distribution, its impact on agricultural production, its relations to international trade, and its impact on the environment. Since the late 1970s, Jiangsu's rural non-agricultural sector has grown at an extraordinary pace, bringing prosperity and rapid structural changes to the rural economy. In one decade, rural industry has replaced agriculture as the dominant sector. However, growth has not been even, so regional differences in the level of rural non-agricultural development have widened. In general rural non-agricultural activities have been concentrated in the more densely populated areas where agriculture is more developed and where villages are closer to urban centres. Despite the fact that large numbers of workers moved from farming to non-agricultural activities, agricultural growth did not at first decline partly because of the presence of surplus labour in farming, partly because those who remained in farming responded to the incentives introduced by rural reform and worked harder, and partly because of the increased application of labour-saving and yield-raising inputs. In contrast to TVEs in Guangdong, few of Jiangsu's TVEs exported until the late 1980s. The difference in export performance between TVEs in Guangdong and those in Jiangsu may be attributed largely to differences in export incentives. Until the formal adoption of the coastal development strategy in 1988, TVEs in Jiangsu faced relatively weak export incentives. Since then, export incentives have improved significantly and an increasing number of TVEs in Jiangsu have become exporters. Exporting is likely to play a role as important in Jiangsu and other coastal areas in the future as it has already in Guangdong. Finally, rural industrialization has occurred at considerable cost to the environment. Although the government has now put in place a regulatory framework to control industrial pollution and to protect farm land, its weak enforcement and the high priority given to jobs and industrial growth do not bode well for the environment.

NOTES

1. The gross output of the five material products sectors—agriculture, industry, construction, transportation, and commerce.
2. The mid-1980s was not the only period when RGVIO grew rapidly. Between 1975 and 1977, Jiangsu's RGVIO expanded at an average annual rate of 46 per cent per year. But this earlier growth started from a smaller base, and also reflected in part the return to commune ownership of some previously elevated handicraft enterprises.
3. Besides this problem, the growth of RGVNAO may be biased upwards for other reasons. Jiangsu's rural social product is the sum of the material products produced by activities with ownership at or below the township level. Although villages are supposed to report the value of gross output in constant prices (1970 prices for years before 1980 and 1980 prices thereafter), fieldwork in Jiangsu suggests that this has not always been done so that some current price data may have been inadvertently mixed in with the constant price data. Another problem is that statistics on non-agricultural activities owned by individuals were not collected before 1984. Thus some of the growth may be a reflection of price increases and improved coverage.
4. *Statistical Yearbook of the Republic of China* (1979), 116 and 420–1.
5. In 1988 Jiangsu's rural labour force or 'workers with rural residence (*xiangcun laodongli*)' numbered 27.37 million, of which perhaps 26.13 million worked in the countryside. See *JSN* 122 and 127, and *JJN* (1989), 3. 27.
6. During 1984–8, employment by industrial TVEs increased at an average annual rate of 11.5 per cent (calculated from data in *JSN* 136 and *JTN* (1991), 114).
7. e.g. during the 1984–8 boom, agricultural employment declined on average by 3.8 per cent per year while employment in TVEs increased at an average annual rate of 10.1 per cent. During the 1989–90 recession, when employment in TVEs declined by 4.4 per cent per year, agricultural employment increased (ibid. 63 and 114).
8. In gross terms, fixed capital (in original prices) employed by TVEs increased from RMB 3.58 billion in 1980 to RMB 24.18 billion in 1988. In net terms, the increase was from RMB 2.90 billion in 1980 to 19.11 in 1988. See *JSN* 136.
9. In 1986, there were 104,292 TVEs in Jiangsu, and of these 2,539 were agricultural enterprises and they produced less than 1 per cent of the gross output produced by TVEs (see *JXQTZ* 2–3). However, the importance of agricultural TVEs was probably greater in earlier years.
10. During the 1989–90 recession, when laid-off workers returned to farming, the share of rural workers in agriculture rose to 62 per cent in 1990.
11. Precise figures for farm workers are not available before the mid-1980s. I estimate that the number of workers employed in farming probably declined from over 17 million in 1978 to around 14 million in 1988.
12. These figures are deduced from data in *JJN* (1989), 3. 27 and *JSN* 122.
13. In 1986, Jiangsu had a cultivated area per capita rural population of 1.34 *mu* (1 *mu* = 0.0667 ha.), substantially higher than that found in the surveyed regions. See Appendix for a description of the surveyed regions.

14. Surveys in other parts of China have found shifts of similar magnitudes, e.g. see *QBLQDZ* 15 and 74.

15. Compared to the two townships surveyed in south Jiangsu, Haitou's per capita distributed rural income in 1987 was 132 *yuan* lower than Hufu's and 365 *yuan* lower than Sigang's and its per capita RGVIO was one-fourth that of Hufu and only about one-tenth that of Sigang.

16. Samuel P. S. Ho, *The Asian Experience in Rural Nonagricultural Development and Its Relevance for China* (Washington, DC: World Bank, 1986), 46–7.

17. Each region is named after its principal city, and this causes some confusion because, in Chinese, both *diji shi* (the region) and municipality are called 'city (*shi*)'. For example, Nanjing *shi* may refer either to the municipality of Nanjing or the much larger region of Nanjing. To avoid confusion, we shall hereafter use city or municipality to refer to the city proper and *shi* to refer to the region. Thus, Nanjing *shi* would include the city of Nanjing and the five *xian* (counties) in the region.

18. Rural population in each region was obtained by deducting the region's non-agricultural population (*fei nongye renkou*) from its total population. (The *JTN* does not provide a precise definition of *fei nongye renkou* except to note that it includes primarily 'the non-agricultural population in cities and towns (*chengzhen fei nongye renkou*)' (see *JTN* (1991), 59).) At the national level, several different definitions of 'rural population' are currently in use. For a discussion of the differences, see Michael F. Martin, 'Defining China's Rural Population', *CQ* 130 (June 1992), 392–401.

19. More specifically, NCR includes Yancheng *shi* and Lianyungang *shi*; NMR includes Xuzhou *shi*, Huaiyin *shi*, Baoying County, Gaoyou County, and Xinghua County; NDR includes Nantong *shi* and Yangzhou *shi* except for Baoyin County, Gaoyou County, Xinghua County, and Yizheng County; SDR includes Suzhou *shi*, Wuxi *shi* except for Yixing County, Changzhou *shi* except for Jintan County and Liyang County, and Zhenjiang *shi* except for Jurong County; and SMR includes Nanjing *shi*, Yizheng County, Jurong County, Jintan County, Liyang County, and Yixing County.

20. See Ch. 4 for more on the relative position of collective activities in rural Jiangsu.

21. However, agricultural (non-farm) collective enterprises were particularly well developed in the north coastal region (NCR), where a larger number of lucrative aqua-culture (particularly prawn) enterprises were developed in the 1980s.

22. Calculated from data in *JXQTZ* 2–3.

23. The presence of these push forces has led Rizwanul Islam to observe that a high level of involvement in rural non-agricultural activities may not necessarily be a sign of prosperity and progress. In some densely populated agricultural areas, peasants turn to rural non-agricultural activities as a last resort, so that increased involvement in non-agricultural activities may be a sign of 'distress adaptation to growing poverty and landlessness' (Rizwanul Islam, 'Non-Farm Employment in Rural Asia: Issues and Evidence', in R. T. Shand (ed.), *Off-Farm Employment in the Development of Rural Asia* (Canberra: National Centre for Development Studies, ANU, 1986), i. 154).

24. See e.g. Samuel P. S. Ho, 'Rural Nonagricultural Development in Asia: Experiences and Issues', in Y. B. Choe and F. C. Lo (eds.), *Rural Industrialization and Non-Farm Activities of Asian Farmers* (Seoul: Korea Rural Economics Institute/Asian and Pacific Development Centre, 1986), 48.

25. This argument is developed in Dennis Anderson and Mark Leiserson, 'Rural Nonfarm Employment in Developing Countries', *Economic Development and Cultural Change*, 28/2 (Jan. 1980), 236–40. They suggest three sources of demand for the products and services of rural industry: (i) rural population's demand for non-food consumer goods, (ii) agriculture's demand for intermediate and capital goods, and (iii) the demand for rural manufactured and handicraft goods that originates outside the rural sector from other regions or abroad. Also see Dennis Anderson, 'Small Industry in Developing Countries: A Discussion of Issues', *World Development*, 10/11 (1982), 913–48.

26. e.g. many county and commune-brigade enterprises developed in south Jiangsu in the 1970s and the early 1980s were located in small towns, and a significant share of their work-force came from the surrounding villages. Surveys of seven 'representative counties' in Jiangsu conducted by Fei Xiaotong and his colleagues found that non-resident day-workers in small towns accounted for about one-third of the county population in the four counties in south Jiangsu and about 15 per cent in the three counties in north Jiangsu (see Philip C. C. Huang, *The Peasant Family and Rural Development in the Yangzi Delta, 1350–1988* (Stanford, Calif.: Stanford University Press, 1990), table 14.1).

27. Since the vigour of rural industry depends in part on the availability of interaction and linkages with urban industry, a spatially decentralized pattern of industrialization is likely to be more conducive to rural industrialization than a highly concentrated pattern. See Samuel P. S. Ho, 'Decentralized Industrialization and Rural Development: Evidence from Taiwan', *Economic Development and Cultural Change*, 28/1 (1979), 77–96, and J. T. Shih, 'Decentralized Industrialization and Rural Nonfarm Employment in Taiwan', *Industry of Free China* (Aug. 1983), 11–20.

28. With 1978 as 100, the cultivated area, grain production, and multiple cropping index in the six villages in 1987 is shown in the table.

	Sigang		Hufu		Haitou	
	Jingxiang	Zhashang	Yangquan	Xiaojian	Haiqian	Haiqi
Agricultural labour	1	7	45	48	108	106
Cultivated area	97	97	77	34	94	98
Grain production	102	97	87	105	176	157
Multiple cropping	127	104	147	186	140	109

The figures are calculated from data in Township-Village Background Data. The agricultural labour figure includes only those who are primarily engaged in agriculture (broadly defined to include cultivation, fishery, sidelines, forestry, and animal husbandry). Part-time workers in agriculture are not included. It should be noted that in Haitou, fishery and sidelines are very important; in Hufu, forestry and sidelines are important, but in the two villages in Sigang the only important agricultural activity is farming. We also learned from our interviews that the amount of labour allocated to farming in the two villages in Haitou declined between 1978 and 1988, and the rise in agricultural labour shown in the above table was caused primarily by the increase in employment in aqua-culture and coastal fishing.

29. Land under tea is counted as forest rather than cultivated land. In the early and mid-1970s, when rural policy was to 'take grain as the key link' and to 'learn from Dazhai', tea lands were converted to dry fields, and dry fields to paddy. The conversion was not entirely successful, and, in the early 1980s, most of these fields were returned to forest, causing a substantial decline in cultivated area.

30. For a detailed discussion of how agriculture performed in China as a whole during this period see Nicholas R. Lardy, 'Overview: Agricultural Reform and the Rural Economy', in US Congress, Joint Economic Committee, *China's Economy Looks Toward The Year 2000*, i (Washington, DC: US Government Printing Office, 1986), 325–35; Frederic M. Surls, 'China's Agriculture in the Eighties', ibid. 336–53; and Kenneth R. Walker, 'Trends in Crop Production', *CQ* 116 (Dec. 1988), 592–633.

31. In the 1970s, Chinese farmers faced several different prices depending on the type of transaction. For goods collected under unified purchase, there was the official procurement (quota) price. Then there was the above-quota price for goods purchased by the state but above the fixed procurement quota. The amount purchased at the above-quota price was set annually. In 1973, a third price, the negotiated price, was introduced to cover the sale of goods (in particular grain) to the state after farmers had met the purchases covered by the other two categories. In the late 1970s, the above-quota price was about 10–30 per cent higher than the quota price, and the negotiated price was about 10–20 per cent higher than the above-quota price. After rural markets were revived in the late 1970s, there was also the market price. The 1979 price reform increased the quota price and the above-quota price. For example, the quota price for grain was increased by 21 per cent and the above-quota price (which had been 30 per cent above the quota price) was raised to 50 per cent above the quota price. Since the procurement quota was fixed, additional production faced not the quota price but the substantially higher negotiated or above-quota prices. Thus the 1979 price reform significantly increased the incentive to produce more. In Jiangsu, the average price (i.e. the weighted average of the official purchase price, the above-quota price, and the negotiated price) of agricultural and sideline products increased by about 23 per cent in 1979 (see *JJN* (1986), 3. 55). Agricultural price reforms were accompanied by reforms of the agricultural marketing system which also opened new opportunities to peasants. For more on China's agricultural marketing system, see Jean C. Oi, 'Peasant Grain Marketing and State Procurement: China's Grain Contracting System', *CQ* 106 (June 1986), 272–90; Joseph Fewsmith, 'Rural Reform in China: Stage Two', *Problems of Communism* (July–Aug. 1985), 48–55, and Andrew

Watson, 'The Reform of Agricultural Marketing in China since 1978', *CQ* 113 (Mar. 1988), 1–28.

32. Deduced from data in *JSN* 168.
33. The best recent discussion of surplus labour in China is found in Jeffrey R. Taylor, 'Rural Employment Trends and the Legacy of Surplus Labour, 1978–86', *CQ* 116 (Dec. 1988), 736–66. The conventional wisdom is that there is a huge surplus of unskilled workers in rural China. However, we have no reliable numerical estimate of the magnitude. In the early 1980s, various Chinese sources repeatedly claimed that roughly one-third of China's agricultural workers or about 100 million workers were in surplus. (For example, this claim was made in interviews I had in 1984 with Chinese officials in Beijing and in Jiangsu. Also see Li Qingzeng, 'Lun woguo nongcun laodongli de guosheng wenti (A discussion of the problem of surplus labour in rural China)', *NJW* 10 (1986), 9.) However, what the Chinese mean by 'surplus' is unclear. For example, in offering their estimates, the Chinese do not always make explicit what assumptions they have made concerning work effort or whether other inputs have been held constant. Therefore it is uncertain that the Chinese actually believe that they can release 100 million workers from agriculture, holding all other inputs constant, without affecting production. One suspects that the most common method used to estimate surplus labour is first to estimate the required labour by multiplying cultivated area by an assumed labour/land norm (where the norm may vary by crop and by region) and then compare the required labour with the actual labour used in agriculture. For example, in Jiangsu, it is widely believed that one full-time worker can cultivate between 6 and 7 *mu* of land. This would imply that, in 1978, Jiangsu, with 69.91 million *mu* of cultivated land, needed between 10 million and 11.7 million full-time workers. Since the actual number of farm workers in Jiangsu in 1978 was about 18 million, this implies that between 6 and 8 million workers were in surplus. This, of course, does not suggest that all 6 or 8 million workers could be removed from farming, other things remaining equal, without affecting output. What it does suggest is that underemployment was prevalent, and that with greater incentive, better organization, and perhaps small increases in other current inputs, a substantial number of workers could leave agriculture without affecting production.
34. *JSN* 26.
35. These and the other figures in this paragraph are from ibid. 134.
36. e.g. only in one of the six villages we surveyed did rural residents of working age work more than 300 standard days in 1987 (see Table 3.6). It is also significant that the average number of standard days worked in the six villages appears to be strongly influenced by employment opportunities outside of farming, i.e. the number of days worked appears to be largely demand driven. The average villager worked more days in a year in Hufu and Sigang, where rural non-agricultural employment opportunities were abundant, than in Haitou, where they were less plentiful. Our village survey also suggests that underemployment in the south was largely seasonal while in the north it was year round.

37. e.g. in 1988, in Taicang County in south Jiangsu, production cost per *mu* for the summer and autumn crops increased by about RMB 50 while receipt per *mu* by only RMB 30 (*JJN* (1989), 2. 33). For the province as a whole, the ratio of rural retail prices of industrial goods to the purchase prices of agricultural and sideline products declined sharply in the late 1980s (see *JSN* 168). Also see Zhongguo nongkeyuan fu chuan diaocha baogaozu, 'Guanyu nongmin zhong liang jijixing wenti (On the problem of peasants' enthusiasm for growing grain)', *NJW* 8 (1986), 30–3, and 'Sowing in Tears', *China News Analysis*, 1323 (Dec. 1986), 1–9.

38. Under the new purchase system, a share of the output is sold to the state under contract at a single procurement price (calculated as a weighted average of the old quota and above quota prices) and the rest is sold on the open market at the market price. Since the new procurement price was lower than the old above quota price, the new arrangement in effect reduced farmers' incentive to produce more. In principle, contractual procurement is supposed to be voluntary, and the expectation was that it would account for about 80 per cent of what the state purchased under the old system. In practice, contractual procurement has operated very much like the old unified procurement. To make matters worse, in 1988, the government used IOUs to pay for some of the contracted grain, a practice never resorted to before (see ibid.). By 1990, because of continued difficulties with the system, contractual procurement was replaced by state procurement (*guojia dinggou*) (see Robert F. Ash, 'The Peasant and the State', *CQ* 127 (Sept. 1991), 495). For more on agricultural pricing, see Terry Sicular, 'Agricultural Planning and Pricing in the Post-Mao Period', *CQ* 116 (Dec. 1988), 671–705.

39. See Nongcun diaocha bangongshi, 'Nongcun gaige yu fazhan zhong de ruogan xin qingkuang (Several new situations in the course of rural reform and development)', *NJW* 4 (1989), 43.

40. e.g. in 1986, Yangquan Village gave each household a subsidy of RMB 5 for every 100 *jin* (1 *jin* = 0.5 kg.) of grain produced. Some villages linked the subsidy to the amount of land cultivated. Some villages also absorbed a portion of the production costs (e.g. the cost of irrigation, ploughing, and insecticide). See Ch. 7 for a more comprehensive discussion of the measures adopted by rural governments to use industry to support agriculture.

41. *JSN* 135.

42. Calculated from data in *JTN* (1989), 363.

43. 'Circular of the Central Committee of the Chinese Communist Party on Rural Work during 1984', *CQ* 101 (Mar. 1985), 133.

44. In 1990 and 1991, the central government increased its investment in agriculture and initiated efforts to rejuvenate grain production. See Zou Jiahua, 'Report on the Implementation of the 1990 Plan for National Economic and Social Development and the Draft 1991 Plan', delivered at the Fourth Session of the Seventh National People's Congress on 26 March 1991, *BR* (22–8 Apr. 1991), 26–33.

45. Nicholas R. Lardy, *Foreign Trade and Economic Reform in China, 1978–1990* (Cambridge: Cambridge University Press, 1992), 126.

46. In fact TVEs in the Pearl River Delta accounted for most of the income

generated by TVEs in Guangdong. In 1986, TVEs in the Pearl River Delta accounted for 70 per cent of all revenue received by TVEs in Guangdong (see Xia Zifen (ed.), *Shanghai xiangzhen qiye jingji, keji fazhan zhanlue he zhengce wenti yanjiu* (Research on the Strategy and Policy for the Economic, Science, and Technology Development of Shanghai's Rural Enterprises) (Shanghai: Shanghai shehui kexue yuan chubanshe, 1988), 219).

47. These are the so-called 'three come-ins and one compensation (*sanlai yibu*)'—specifically, *lailiao jiagong, laiyang jiagong, laijian zhuangbei*, and *buchang maoyi*—i.e. processing of imported materials (for export), customized production of export goods, assembling of imported components (for export), and compensation trade.

48. The '*sanzi* enterprises' are '*hezi qiye* (equity joint venture)', '*hezuo qiye* (co-operative joint venture)', and '*duzi qiye* (wholly foreign owned enterprise)'. Equity joint ventures are enterprises established under the Law of the People's Republic of China on Joint Ventures Using Chinese and Foreign Investment, and co-operative joint ventures are joint projects involving Chinese and foreign partners where the arrangement is not strictly according to China's joint venture law.

49. e.g. between 1978 and mid-1982, of the 1,798 'special export arrangements' in one Pearl River Delta County (Dongguan), 1,391 involved TVEs. In the first four years, TVEs in Dongguan County earned US$82.42 million in processing fees from these arrangements, 82 per cent of all foreign exchange earned in the county. See *Kai chuang xiangzhen qiye xin jumian* (*Start A New Phase For Rural Enterprises*) (Beijing: Zhongguo shedui qiye baoshe and Nong mu yu ye bu shedui qiye guanliju, 1984), 108.

50. Y. Y. Kueh, 'Foreign Investment and Economic Change in China', *CQ* 131 (Sept. 1992), 667.

51. The government used Guangdong and Fujian as test points, where it experimented with the more radical reforms of the open policy. The two southern provinces were the first to acquire the power to manage their economies independently provided their activities did not violate the overall balance of the national plan, and they were also the first to receive special incentives to promote export and the use of foreign capital and technology. The most important reform adopted was the so-called *baogan* (contract) system which allowed local governments at different levels to retain a share of the revenues and foreign exchange earnings they generate. In effect, the contract system gave the provincial and local governments in Guangdong and Fujian a source of revenue and foreign exchange over which they have discretionary control. Thus local governments in the two provinces had strong incentives to increase their revenues and foreign exchange earnings. In addition, the two provinces were given greater authority over their imports and exports and the approval of special export arrangements. In turn, the province passed on some of its increased authorities in these areas to lower-level governments. The government also established four Special Economic Zones in the two provinces, where economic activities involving foreign participation and producing goods primarily for export were given special treatment and tax privileges not available to joint venture projects in other parts

of China. For more on these arrangements, see Samuel P. S. Ho and Ralph Huenemann, *China's Open Door Policy* (Vancouver: University of British Columbia Press, 1984).

52. Sometimes individuals on good terms with cadres in the foreign trade system would obtain export contracts and then transfer them to others for a fee (Enterprise Interview Notes).

53. In Jan. 1981, to give more incentives to exporters, China allowed export earnings to be settled at an internal rate of US$1 = RMB 2.8 instead of the official rate of US$1 = RMB 1.53. The internal settlement rate was discontinued in 1985.

54. Enterprise Interview Notes.

55. For example, special tax preferences, similar to those granted to the SEZs in Guangdong and Fujian, were given to joint ventures located in special economic-technological development zones in the fourteen coastal cities (for details, see *ZJN* (1985), 10. 71–2).

56. The other coastal cities are Dalian, Qinhuangdao, Tianjin, Yantai, Qingdao, Shanghai, Ningbo, Wenzhou, Fuzhou, Guangzhou, Zhanjiang, and Beihai, and the other coastal economic development zones are the Pearl River Delta, the Xiaman-Quanzhou-Zhangzhou Delta, the Liaodong Peninsula, and the Shandong Peninsula.

57. *JJN* (1986), 2. 166.

58. Ibid.

59. 'Direct exports' and 'indirect exports' are not defined. Most exports are 'direct exports'. In 1986, the total exports of rural enterprises in Jiangsu amounted to RMB 1.77 billion, of which RMB 1.26 billion were 'direct exports'. David Zweig suggests that 'indirect exports' may be referring to 'goods exported through other provinces' (see David Zweig, 'Internationalizing China's Countryside: The Political Economy of Exports from Rural Industry', *CQ* 128 (Dec. 1991), 729). Since all goods purchased by the foreign trade system, whether exported in their original state or exported after processing, are reported as 'exports', an alternative interpretation is that 'indirect exports' may be referring to goods purchased by the foreign trade system for processing before they are exported.

60. *JJN* (1987), 4. 66. In 1986, goods worth RMB 7,505 million (valued at actual purchase prices) were purchased in Jiangsu by specialized FTCs, local FTCs, and industry-commercial companies for export either in their original state or after processing (see *JJN* (1987), 3. 82), of which RMB 1,769 million were from rural enterprises (see *JXQTZ* 764–75).

61. For discussions of the reasons why Zhao Ziyang proposed the strategy and the internal debate before and following its adoption, see Dali Yang, 'China Adjusts to the World Economy: The Political Economy of China's Coastal Development Strategy', *PA* 64/1 (1991), 42–64, and Zweig, 'Internationalizing China's Countryside', 716–41. Zhao was apparently influenced by the writings of two economists at the Economic Planning Institute of the State Planning Commission, Wang Jian and Pei Xiaolin. For a summary of the arguments put forward by Wang and Pei, see the following three articles in *Chinese Economic Studies*, 25/1 (1991): Wang Jian, 'The Correct Strategy

for Long-Term Economic Development—Concept of the Development Strategy of Joining the "Great International Cycle"'; Xiao He, 'A Dialogue on the "Great International Cycle"—Interview with Comrades Wang Jian and Pei Xiaolin, Authors of the Theory of the "Great International Cycle"'; and Lin Zheng, 'The Discussion on Chinese Economic Development Strategy'.

62. In fact, in early 1990, the government strengthened the link between the rural economy and the export sector when it announced additional incentives to encourage rural enterprise exports, including priority access to bank credits for export-orientated enterprises (see Zweig, 'Internationalizing China's Countryside', 737).

63. In 1988, many of Jiangsu's local cadres were sent to Guangdong to learn from its experience.

64. *JJN* (1990), 2. 7. For more on the competition between raw material producing and processing regions see Ch. 5.

65. Xia Zifen (ed.), *Shanghai xiangzhen qiye jingji*, 56.

66. For details on the foreign trade contract system introduced in 1987–8, see Lardy, *Foreign Trade and Economic Reform*, 102–4.

67. The incentive to export was also increased by the devaluation of the *yuan*. China experienced strong inflationary pressure in the late 1980s, and to keep its products competitive on the world market in the face of rapidly rising domestic prices, the government devalued the *yuan* in Dec. 1989 by 21.2 per cent and again in Nov. 1990 by 9.6 per cent.

68. *JJN* (1990), 6. 43, 6. 45, and 6. 47.

69. Township-Village Interview Notes and *JJN* (1990), 6. 43 and 6. 45.

70. If the enterprise has no use for the foreign exchange, it may sell it to others through the Bank of China at the swap rate (which in 1988 was about 50–60 per cent higher than the official rate).

71. e.g. Wuxi *shi*, when allocating foreign exchange under its control, gives first priority to enterprises that are already exporting. See *JJN* (1986), 8. 27.

72. In Jan. 1989 alone 326 rural towns were designated as industrial satellite towns. In 1989, besides the industrial satellite towns, nine municipalities under direct provincial jurisdiction and forty county cities were also opened to the outside world. See *JJN* (1990), 6. 35.

73. In 1989, Jiangsu spent US$270 million to import raw materials needed by its export industries (see *JJN* (1990), 4. 135).

74. *JJN* (1987), 3. 82–3, and *JTN* (1991), 239 and 242. In US$, the rise was from US$475 million in 1986 to US$2.04 billion (obtained by dividing the *yuan* figures by the average annual *yuan*–dollar exchange rate). The actual increase in export earnings was probably less since not all the goods purchased by the foreign trade system were always exported.

75. In 1990, Jiangsu's rural enterprises produced industrial output valued at RMB 125.19 billion and sold RMB 10.66 billion to the foreign trade system (see *JTN* (1991), 119, and *Jiangsu Nianjian* (1991), 309). The percentage of output exported in 1990 was three percentage points higher than two years earlier.

76. In the mid- and late 1980s, wages were considerably lower in Jiangsu than in

Guangdong. For example, in the mid-1980s, wages in Guangzhou were over 60 per cent higher than those in Nantong (south Jiangsu), while labour productivity was lower (see Richard Pomfret, 'Jiangsu's New Wave in Foreign Investment', *China Business Review* (Nov.–Dec. 1989), 12).

77. *JJN* (1990), 6. 36, and *Jiangsu Nianjian* (1991), 309 and 366.

78. *BR* (4–10 May 1992), 22, and *BR* (11–17 May 1992), 9–10.

79. The amount invested by the foreign partner in 55 of the 180 joint ventures involving TVEs in Suzhou *shi* in 1989 was less than US$100,000 (*JJN* (1990), 6. 36).

80. Zweig, 'Internationalizing China's Countryside', 718, and 'Learning the Rules of Foreign Trade', *China News Analysis*, 1464 (15 July 1992), 7. About two-thirds of all TVE exports were accounted for by Jiangsu, Guangdong, Shanghai, Zhejiang, and Shandong (see *ZNN* (1988), 331).

81. *JXQTZ* 764–75.

82. 'Learning the Rules of Foreign Trade', 7. About two-thirds of the exports by rural enterprises in Jiangsu in the late 1980s were accounted for by textiles, clothing, and handicraft products.

83. The data are from *ZJN* (1981), 6. 9. It is widely believed that Chinese land statistics are biased downwards. Detailed but incomplete land surveys suggest that actual arable area may be as much as 30 per cent greater than suggested by official statistics (see *CQ* 128 (Dec. 1991), 874). If the downward bias in the land statistics was less in the earlier than the later years, then the decline in cultivated land may be less dramatic than the data suggest.

84. *JTN* (1990), 142.

85. Reported in *Renmin ribao* and cited in Vaclav Smil, *The Bad Earth* (Armonk, NY: M. E. Sharpe, Inc., 1984), 69. Smil also reported a second and lower figure of 29.3 million ha. But either figure suggests an incredible loss of arable land.

86. *QBLQDZ* 36. Although the surveyed villages were supposed to be typical, the results must be interpreted cautiously. See S. Lee Travers, 'Bias in Chinese Economic Statistics: The Case of the Typical Example Investigation', *CQ* 91 (Sept. 1982), 478–85.

87. Smil, *The Bad Earth*, 72.

88. *JJN* (1990), 4. 99.

89. Between 1978 and 1986, the amount of cultivated land occupied by rural enterprises per worker declined from 0.09 *mu* to 0.06 *mu* in 222 administrative villages surveyed in eleven provinces (the decline was from 0.19 *mu* to 0.11 *mu* in the surveyed villages in Jiangsu). But because of the rapid growth of rural industry, the total amount of cultivated land occupied by rural enterprises increased from 152.2 *mu* in 1978 to 506.7 *mu* in 1986. See *QBLQDZ* 36.

90. *XQZFX* 171.

91. Smil, *The Bad Earth*, 106.

92. For a summary of the major environmental problems acknowledged by the Chinese, see ibid.

93. Several major cities, because they were important tourist attractions, were given special priority, e.g. Beijing, Hangzhou, Suzhou, and Guilin.

94. *XQZFX* 12.
95. *ZJN* (1981), 4. 218.
96. Xia Zifen (ed.), *Shanghai xiangzhen qiye jingji*, 52.
97. *Globe & Mail* (23 Feb. 1988).
98. *Kai chuang xiangzhen qiye xin jumian*, 209.
99. Ibid.
100. *JJN* (1990), 4. 110. Moderate and serious were not defined. Nearly all rural industrial enterprises, except for the very small PEs, were surveyed. In 1989, industrial enterprises at the township and village levels numbered 103,841.
101. Yan Yinglong, 'Lun sunan xiangzhen gongye fazhan zhong de huanjing baohu wenti (On the problems of environmental protection in the process of rural industrialization in *sunan*)', *Jiangsu jingji tantao* (*Jiangsu Economic Inquiry*), 6 (1988), 29.
102. *Globe & Mail* (23 Feb. 1988).
103. Xia Zifen (ed.), *Shanghai xiangzhen qiye jingji*, 53. In some industries (e.g. chemical products in daily use, chemical fertilizers, and agricultural chemicals) rural enterprises emitted 40 to 70 times more pollutants that did urban enterprises.
104. Forests were damaged because trees were cut indiscriminately to produce charcoal to fuel the multitude of backyard furnaces that sprang up during the GLF. Of all the questionable policies, probably the most damaging was the disastrous policy of 'taking grain as the key link' that attempted to increase cereal production by converting forests, grasslands, and lakes to crop land (see Smil, *The Bad Earth*, 10–68).
105. See *XQZFX* 52–7.
106. *ZJN* (1984), 5. 158.
107. *JJN* (1990), 4. 99.
108. e.g. if the land required was less than 3 *mu* (1 *mu* = 0.0667 ha.), the approval of the county was needed; if it was between 3 *mu* and 5 *mu* (10 *mu* in some parts of north Jiangsu), the approval of the region (*diji shi*) was needed; and if more than 5 *mu* (10 *mu* in some parts of north Jiangsu), the approval of the province was needed (Township-Village Interview Notes). In 1992, as part of the changes adopted in response to the call for bolder economic reforms, Jiangsu returned the power to approve changes in land use to localities. At present, provincial approval is required only when the change involves 100 *mu* or more of land. This change has apparently caused the decline in farm land to accelerate in some parts of Jiangsu. (Interview with JPASS (Dec. 1992).)
109. *JJN* (1987), 1. 49.
110. The official notice is reproduced in *JJN* (1990), 1. 135–7.
111. These stipulations also applied to neighbourhood (*jiedao*) enterprise. See *XQZFX* 562–4. Some provinces have also issued detailed regulations based on the 1984 law. Jiangsu's detailed regulations, issued in Nov. 1986, can be found in *JJN* (1987), 1. 173–7.
112. The decline in cultivated area is well documented, but less is known about the extent of industrial pollution in rural areas so it is difficult to judge its seriousness.
113. *JJN* (1990), 1. 135. Our impression is that in south Jiangsu, where the

population density is high and the shortage of farm land keenly felt, there is considerable pressure on local cadres to enforce land-use regulations. For example, one township party secretary identified birth control and enforcing the regulations on residential construction as his two most difficult and time-consuming tasks. But apparently this may not be the case in all parts of Jiangsu or in other provinces.

114. e.g. in many townships in south Jiangsu, nearly every rural household is now living in a house constructed or renovated after 1978. For the province as a whole, between 1979 and 1989, 60 per cent of its rural households had constructed new houses or expanded their old houses (*JJN* (1990), 4. 99).

115. In 1992, the authority to establish open areas (*duiwai kaifang qu*) was decentralized to localities, and this created a new source of pressure on farm land as many townships and villages rushed to establish open areas in the hope of attracting foreign investment and gaining the special privileges that accompany it. In Dec. 1992, to end the loss of farm land to open areas, the power to establish open areas was returned to the county level. (Interview with JPASS, Dec. 1992.)

116. e.g. of the 293 electroplating enterprises found in Tianjin's suburbs in 1982, 161 were closed, and the others were required to install equipment to treat waste water (*Kai chuang xiangzhen qiye xin jumian*, 209–10).

117. Yan Yinglong, 'Lun sunan xiangzhen gongye zhong de huanjing baohu wenti', 30. One indication that environmental regulations are enforced more strictly in south Jiangsu is that 70 per cent of the new investments by TVEs in the southern third of the province apparently have complied with the rule of 'three simultaneously' as compared with only 30 per cent in the north (see *JJN* (1986), 4. 99).

118. *JJN* (1989), 8. 18.

119. Ibid.

120. *JJN* (1990), 6. 53.

4

The Rural Non-agricultural Sector
Collective Activities

UNDER the more permissive political environment of the 1980s, 'larger in size and having a higher degree of public ownership (*yida ergong*)' was no longer the preferred form of rural economic organization. Economic activities could now be organized in a variety of ways: individual undertakings, collectives, and various forms of co-operation between collectives, between individuals, and between collectives and individuals. However, in Jiangsu and most other provinces, collective activities have continued to dominate the rural non-agricultural sector, and they are the focus of this and the following chapter. How the collectively owned township-village enterprises (TVEs) functioned in the partially reformed Chinese economy in the 1980s is discussed in Chapter 5. This chapter examines other and more general aspects of the collective component: its ownership composition, its industrial structure, and the system used by township-village governments (TVGs) to manage collective non-agricultural activities. To place the collective component in perspective, the chapter opens with a discussion of its relative position in Jiangsu's rural non-agricultural sector.

The Importance of Collective Activities

One way to gauge the importance of collective activities is by their contribution to rural income. Chinese statistics identify four main sources of rural income: township-village enterprises, economic units under collective and unified management (*jiti tongyi jingying*), 'new economic associations (*xin jingji lianheti*)', and economic activities under family management. Economic units under collective and unified management include all activities (including farming) that are owned and/or managed by collective units below the village level, usually a village small group (formerly the production team) or several such groups. A 'new economic association' (hereafter association) is defined as a stable arrangement that has a regular place of operation, an accounting system, and an income distribution system. Together the activities operated by rural households and associations form the rural private sector, and those under the other two forms of organization are the rural collective sector.

Collective Activities

Table 4.1 presents the gross income generated in rural Jiangsu in 1986 distributed by form of ownership/management and by industry. Reflecting the widespread adoption of the household responsibility system in agriculture, nearly all the gross income from cultivation was from family managed units. However, the collective sector, composed primarily of TVEs, produced nearly 83 per cent of the income from rural non-agricultural

TABLE 4.1. *Gross Income Earned by Rural Units in Jiangsu, Distributed by Management Form, 1986*

	Total	Management form			
		Operated by township and village	Collective unified managed[a]	Association managed	Family managed
Total (RMB million)	83,322	43,828	1,471	588	37,524
% from					
Cultivation	22.7	0.1	12.2	0.3	49.9
Other agriculture[b]	12.8	0.6	21.1	9.5	26.8
Rural non-agriculture	64.5	99.3	66.7	90.1	23.3
Industry	49.5	86.5	41.9	69.0	6.2
Construction	6.1	7.7	5.1	8.2	4.2
Transport	3.0	1.3	4.7	7.5	4.9
Commerce[c]	2.1	1.9	3.4	2.7	2.2
Service	1.2	0.6	1.9	1.5	1.9
Other sources	2.6	1.3	9.7	2.2	3.9

	Total (RMB million)	% from			
		Operated by township and village	Collective unified managed[a]	Association managed	Family managed
Total	83,322	52.6	1.6	0.7	45.0
Cultivation	18,912	0.2	0.9	0.0	98.9
Other agriculture[b]	10,700	2.6	2.7	0.5	94.1
Rural non-agriculture	53,710	80.6	2.2	1.0	16.2
Industry	41,230	92.0	1.4	1.0	5.6
Construction	5,048	66.5	1.4	0.9	31.1
Transport	2,545	22.8	2.6	1.7	72.9
Commerce[c]	1,716	48.8	2.7	0.9	47.6
Service	1,020	26.1	2.5	0.9	70.6
Other	2,151	15.6	6.2	0.6	67.6

[a] The Chinese term is *jiti tongyi jingying*.
[b] Includes forestry, animal husbandry, sidelines, and fishery.
[c] Includes trade and food and drink.

Source: Calculated from data in *JNJSFTZ* 1–10.

activities. Team activities were relatively unimportant because, in the early and mid-1980s, many were privatized and some, through mergers, were elevated to village enterprises. Collective activities were not equally dominant in all non-agricultural sectors. Their domination was most complete in rural industry, where they accounted for over 93 per cent of the income. In the tertiary sector, however, private activities were more important. For example, in services and transport, the private sector accounted for over 70 per cent of the gross income generated.

The distribution of the means of production by ownership provides additional evidence of the continued importance of rural collective non-agricultural activities. In 1986, the collective sector owned 97 per cent of the industrial machinery and equipment in rural areas and 90 per cent of the construction equipment.[1] Transport equipment was the only important category of non-farm equipment owned in large quantities by individuals in rural Jiangsu.

Because TVEs developed earlier and have grown more rapidly in Jiangsu than in other parts of China, the adoption of the household

TABLE 4.2. *Rural Gross Income, Rural Net Income, and Rural Income Distributed to Individuals by Source, 1986 (%)*

	China	Jiangsu
Gross income	100.0	100.0
Township-village enterprises	32.3	52.6
Units under collective and unified management	2.2	1.6
Associations	2.4	0.7
Family operated units	63.1	45.0
Net income	100.0	100.0
By source		
Township-village enterprises	16.8	27.8
Units under collective and unified management	2.2	1.8
Associations	1.6	0.6
Family operated units	79.4	69.8
By allocation		
Taxes	5.9	8.9
Retained by collectives	7.8	10.4
Income distributed to individuals	86.3	80.7
Income distributed to individuals	100.0	100.0
Township-village enterprises	10.0	16.3
Units under collective and unified management	1.9	1.4
Associations	1.4	0.6
Family operated units	86.6	81.8

Sources: Calculated from data in *ZTN* (1988), 286, and *JNJSFTZ* 1–10.

responsibility system diminished the importance of the collective sector less in rural Jiangsu than in rural China as a whole. Table 4.2 shows that, in 1986, collectively generated gross income as a share of total gross income was 54 per cent in rural Jiangsu but only 35 per cent in rural China as a whole. Because Jiangsu's TVEs are concentrated in the industrial sector where the cost of intermediate goods and services used in production as a share of gross income is significantly higher, the relative position of the collective sector is substantially reduced when measured in terms of net income. Furthermore, since not all the income generated by the collective sector is distributed to individual members, the contribution of the collective sector to personal income is lower still. In 1986, in rural Jiangsu, less than 30 per cent of the net income generated and less than one-fifth of the rural income distributed to individuals came from the collective sector. However, these shares are still higher than the national averages. Furthermore, because of its greater size, Jiangsu's TVE sector contributed more in taxes and made more resources available to local governments for development and other purposes (whether measured absolutely or as a share of income) than did the TVE sector in most other parts of rural China.

In the 1980s, even after agriculture was contracted to households and became primarily a household activity, collective activities continued to generate over one-half of the gross income, nearly 30 per cent of the net income, and about 18 per cent of the household income in rural Jiangsu. The main reason for the significantly greater importance of collective activities in rural Jiangsu than in other parts of rural China was the presence of a highly developed rural non-agricultural sector that was dominated by TVEs.

Ownership Composition and Industrial Structure

Ownership Composition

The collective component of the rural non-agricultural sector is composed of enterprises that are owned by either the township, the village, or the team. Table 4.3 presents the employment and gross income of all collectively owned enterprises in Jiangsu. Included among these are a small number of agricultural enterprises engaged in such activities as aquaculture, tea growing and processing, and animal husbandry. Some of these, e.g. tea processing, could just as easily be classified as food processing as agriculture. In any case, these agricultural enterprises comprised less than

TABLE 4.3. *Employment and Gross Income of Collectively Owned Enterprises in Rural Jiangsu, by Industry and by Ownership, 1986*

	Employment (000)				Gross income (RMB million)			
	Total	Township	Village	Team[a]	Total	Township	Village	Team[a]
Total	6,939	3,921	2,835	184	42,425	25,676	15,771	977
% of total								
Agriculture[b]	0.7	0.6	1.0	—	0.6	0.5	0.8	—
Industry	83.4	75.7	94.0	85.7	88.0	83.9	94.7	86.2
Other non-agriculture	15.8	23.7	5.0	14.3	11.4	15.6	4.5	13.8
Construction	11.8	18.9	2.5	4.5	7.4	11.3	1.4	2.8
Transport	2.5	3.4	1.2	4.0	1.4	1.6	1.0	3.9
Commerce[c]	0.7	0.6	0.8	3.3	1.8	1.8	1.5	4.2
Service	0.3	0.2	0.2	1.1	0.3	0.3	0.2	0.6
Other	0.4	0.5	0.3	1.5	0.5	0.5	0.3	2.2

	Total (000)	% from			Total (RMB million)	% from		
		Township	Village	Team[a]		Township	Village	Team[a]
Total	6,939	56.5	40.9	2.6	42,425	60.5	37.2	2.3
Agriculture[b]	52	47.5	52.5	—	243	48.4	51.6	—
Industry	5,790	51.3	46.0	2.7	37,339	57.7	40.0	2.3
Other non-agriculture	1,097	84.6	13.0	2.4	4,843	82.6	14.6	2.8
Construction	822[d]	90.4	8.6	1.0	3,161	91.9	7.2	0.9
Transport	176	76.2	19.6	4.2	607	67.5	26.2	6.3
Commerce[c]	51	44.6	43.6	11.8	747	63.3	31.2	5.5
Service	18	52.8	36.0	11.2	117	62.8	32.1	5.0
Other	31	62.2	28.7	9.0	211	66.2	23.6	10.2

[a] The Chinese term is 'village small group (*cunmin xiaozu*)', formerly the production team. These figures were adjusted to correct apparent arithmetical errors.
[b] Agriculture is defined broadly to include forestry, animal husbandry, and fishery. The enterprises included here are mostly engaged in aqua-culture, tea growing and processing, and other large-scale non-farm activities.
[c] Includes trade, food and drink, etc.
[d] Of these, 501,804 were working on urban construction sites.

Source: Calculated from *JXQTZ* 2–3, 126–65.

1 per cent of all collective enterprises (measured by either employment or gross income) and are therefore of negligible importance. In what follows all collectively owned enterprises at the township, village, and team levels will be considered part of the collective rural non-agricultural sector.

In 1986, Jiangsu reported 21,044 team enterprises, 69,175 village enterprises (VEs), and 35,117 township enterprises (TEs).[2] The average number of workers employed per enterprise was 9 for team enterprises, 41 for VEs, and 112 for TEs. Because they are extremely small, team enterprises, even though they numbered in the thousands, had only a minor impact on either rural employment or rural income. Moreover, their importance declined during the 1980s, when many team enterprises were privatized and others were elevated to village ownership. In 1986, TVEs accounted for about 97 per cent of the employment and of the gross income in the collective sector.

Whether measured by employment or gross income, industry is the most important rural collective activity in Jiangsu, with construction a very distant second. In 1986, of the workers employed by collective enterprises, over 83 per cent were employed by industrial enterprises, nearly 12 per cent by construction enterprises, and less than 5 per cent by enterprises engaged in other activities. Team enterprises were of negligible importance in all sectors other than commerce and services, where they accounted for about 11–12 per cent of employment and 5–6 per cent of the gross income. Of all workers employed by collectively owned industrial enterprises in Jiangsu, team enterprises employed less than 3 per cent, VEs 46 per cent, and TEs 51 per cent. In other words, industrial TVEs provided employment to the vast majority of the non-agricultural workers in rural Jiangsu. In the same year, township enterprises produced 54 per cent of Jiangsu's rural GVIO, village enterprises 38 per cent, and enterprises below the village level (*hezuo gongye* and *geti gongye*) only 7 per cent.[3] By contrast, at the national level, township enterprises accounted for about 40 per cent of the rural GVIO produced in 1986; village enterprises, 36 per cent; and enterprises below the village level, 23 per cent.

The development of the rural non-agricultural sector primarily through the expansion of TVEs is called in China the *sunan* (south Jiangsu) model of rural non-agricultural development, named after that part of Jiangsu where this pattern has been most successful.[4] In this model, rural non-agricultural development occurs primarily through the growth of industrial enterprises that are owned by townships and villages, and the entrepreneurs who seek out the opportunities, initiate the new projects, manage annual production and investment, and allocate annual profits among alternative uses are the township and village cadres. Since the late 1970s, most counties in Jiangsu and many regions in other provinces, have followed the *sunan* model of rural industrialization.

Industrial Structure

It is well known that, because of scale economies in production and distribution and better access to capital, large enterprises in urban locations enjoy important advantages. That rural enterprises, which are small (measured by either employment or capital) and widely dispersed, exist in the presence of large-scale urban industries suggests that there are also technoeconomic influences that enable them to compete in some industries, and that rural enterprises are likely to be concentrated in these industries.

Rural enterprises are found in industries where locational factors favour or protect small enterprises.[5] For example, industries dependent on dispersed raw materials that are costly to transport and where processing substantially reduces the weight of the principal raw material (e.g. agricultural resource industries, sawmills, and stone products) tend to favour production by small establishments in scattered locations. Rural enterprises are also likely to be in industries that serve local markets with bulky or perishable goods that are difficult and costly to transfer (e.g. animal feeds, perishable consumer goods, furniture making and carpentry, and building materials). Service industries that require frequent face-to-face contact between producer and customer (e.g. blacksmithing, welding, metal casting, fabrication and repair of tools and equipment, and various types of repair establishments) also include many small and widely dispersed establishments.

Small enterprises also exist in large numbers in some industries because of the characteristics of the production process. For example, production processes that can be broken into separate parts (e.g. metalworking) permit a high degree of specialization and division of labour, and this often enables small enterprises to exist by specializing in one or two parts of the total process. Large numbers of small enterprises are often found in simple assembly, mixing, or finishing industries where scale economies are relatively unimportant (e.g. toys, textiles, and clothing). These industries are usually labour intensive, produce goods with low transfer costs (so location is not critical), and use simple technology that does not require large initial investment (and therefore entry is easy). The competitiveness of small enterprises is sometimes due to market characteristics. For example, when total demand is small, production runs short, production versatility important, and made-to-order purchases common, small enterprises exist by filling the cracks not occupied by large factories. Although these process and market influences do not have a spatial dimension, they nevertheless provide insights into reasons for the presence of small-scale enterprises in rural areas.

Among less developed countries, regardless of the level of development reached, the bulk of rural industry is accounted for by four broad groups of activities: '(a) food processing; (b) textiles and wearing apparel; (c) wood, including saw-milling, furniture making, and general carpentry; and (d) metal, including blacksmithing, welding, fabrication and the making of tools and equipment.'[6] It is significant that location or process influences that favour or protect small enterprises are present in most if not all these activities. Are these the same activities that dominate Jiangsu's rural industry?

Because rural industry and industrial TVEs are synonymous in Jiangsu, it is the industrial structure of TVEs that needs to be examined. Table 4.4 presents the gross output produced by industrial TVEs in Jiangsu in selected years. Compared to the rural industrial activities found in other developing countries, the industrial structure of TVEs in 1978 showed some similarities as well as some striking differences. In that year, about three-quarters of the GVIO produced by TVEs in Jiangsu was from four sectors: building materials (18 per cent), chemicals (12 per cent), textiles and apparel (13 per cent), and metal products and machinery (34 per cent). There were hardly any wood or wood-related industries, and although Jiangsu was a major producer of agricultural products,[7] less than 3 per

TABLE 4.4. *Gross Output in Current Prices Produced by Industrial Township-Village Enterprises, Old Industrial Classification, Jiangsu, Selected Years*

	1978	1982	1984	1986		
				Total	Township	Village
Gross output (RMB million)	6,257	13,165	22,624	45,900	27,011	18,889
% of total						
Metallurgy	0.8	2.6	3.2	4.9	6.2	3.0
Electric power	0.0	0.0	0.0	0.0	0.0	0.0
Coal	0.1	0.2	0.2	0.3	0.2	0.4
Petroleum	0.1	0.1	0.1	0.1	0.1	0.1
Chemicals	12.0	10.2	10.8	9.9	9.8	10.2
Metal products and machinery	33.6	27.4	29.5	29.5	29.3	29.7
Building materials	18.3	18.7	16.1	15.3	15.6	14.7
Wood and wood products	0.7	0.9	0.8	1.5	1.3	1.8
Food	2.6	3.2	3.5	3.8	3.6	4.1
Textiles and apparel	12.7	23.4	24.2	22.7	25.8	18.3
Leather products	1.8	2.2	1.8	2.2	1.8	2.9
Paper, cultural, educational, and art products	1.7	3.3	2.8	3.9	3.3	4.7
Others	15.6	7.7	7.1	5.9	3.0	10.0

Sources: The 1978, 1982, and 1984 figures are calculated from data provided by JPASS. The 1986 figures are calculated from *JXQTZ* 2–3 and 104–5.

cent of the GVIO produced by its TVEs was in food processing. Since Jiangsu had little forest, the absence of wood and wood-related industries was to be expected. More surprising were the lack of light and food processing industries and the presence of unusually large chemical and metal products and machinery industries. These surprises were largely the legacy of past economic policies and regulations.

In the 1960s and the 1970s, to avoid competition for raw material between rural and state enterprises, communes and brigades were discouraged, if not prohibited, from establishing capacity in agricultural processing industries. This partly explains why food processing was surprisingly undeveloped in the rural collective sector in 1978, and also why the textiles and apparel industry was not more developed in rural areas even though Jiangsu was a major producer of cotton and silk.[8] A contributory reason was the policy in the 1960s and 1970s that emphasized grain production at the expense of non-grain crops which reduced the opportunities for the development of forward linkage agricultural processing industries. Finally, until the late 1970s, China's development strategy assigned a low priority to the consumer goods industries, and this also contributed to the underdevelopment of light industry in general and agricultural processing in particular.

The prominence of chemicals, metal products and machinery, and building materials in 1978 was also in part a legacy of earlier government policy. In the 1960s and the 1970s, rural industry was developed to serve agriculture. The policy encouraged communes and brigades to establish large numbers of enterprises to produce agricultural chemicals, machineries and tools, and construction materials needed to modernize agriculture.

The importance of building materials in Jiangsu's rural industry in 1978 is not unexpected. Of the building materials industries found in rural Jiangsu, the most important was brick-making, which accounted for about 40 per cent of the building materials produced by value, followed in a distant second place by cement and cement products and in third place by the excavation of stones and sand.[9] Since the natural resources used as raw materials in these industries were readily available and widely dispersed and the cost of transporting the final products was high, the production of building materials by small rural enterprises was probably economical.

The metal products and machinery sector of Jiangsu's rural industry was composed of service activities that are commonly found in rural areas in other developing countries (e.g. casting and forging, machinery repairs, the production of metal products used in agriculture) as well as activities rarely found (e.g. the manufacture of metalworking and other industrial machineries and the production of electrical equipment). Furthermore, the activities rarely found in other rural economies accounted for a very large share of the sector in Jiangsu.[10] Had China followed a different strategy of rural development, one that stressed comparative advantage and economic efficiency rather than self-reliance and self-sufficiency, many of the metal products and machinery industries found in rural Jiangsu would not have developed, certainly not to the extent they did. Besides government policy, high and inflexible state prices and a cumbersome distribution system that worked to protect small inefficient plants also encouraged the chemical and the metal products and machinery industries to develop in rural Jiangsu.[11]

In the 1980s, the industrial structure of TVEs changed somewhat in Jiangsu. In particular, the shares of GVIO produced by the chemical, the metal products and machinery, and the building materials industries declined while that by the textiles and apparel industry increased sharply from below 13 per cent in 1978 to around 24 per cent in the mid-1980s. The decline in the chemical and the metal products and machinery industries was an overdue correction of over zealous development in the past. In the early 1980s, local governments closed down some of the most inefficient chemical and metal products and machinery enterprises or forced them to change their output mix. Still, in 1986, metal products and machinery accounted for about 30 per cent of the GVIO produced by Jiangsu's industrial TVEs, much higher than the level generally found in other rural economies. The more interesting change was the sharp rise in the share of rural GVIO produced by the textiles and apparel TVEs, which was largely in response to market forces.

Since the late 1950s, when the central government turned over the management of many of its textile enterprises to local authorities, government control of the textiles and apparel industry has been relatively relaxed. In the 1980s when townships and villages were encouraged to expand their industrial activities, decentralized control and access to local raw materials made it relatively easy for TVEs in Jiangsu to enter the textile industry. However, townships and villages invested not only in traditional activities dependent on local raw materials (e.g. cotton, hemp, and silk spinning and weaving) but also in new activities such as the production of woollen textiles and chemical fibres. In fact, by 1986, these new activities accounted for one-quarter of the output value produced by TVEs engaged in the production of textiles and apparel.[12]

Much of the investments in the non-traditional textile activities were in response to high profits caused by price distortions in a partially reformed economy. One example was the profit opportunities created in the early 1980s by partial price reforms that permitted the sale of most first and second category agricultural goods at the higher negotiated prices except for wool and several other industrial raw materials which were sold at the lower official purchase prices.[13] By keeping the price of wool artificially low while the prices of woollen textiles were rising sharply in response to growing consumer demand, profits were allocated from the early to the final stages of production.[14] This made investment in the woollen textile industry extremely attractive, and township-village governments looking for new industrial investments found it difficult to resist. Price distortions also attracted TVEs to the chemical fibre industry. Even though the cost of producing chemical fibre domestically had declined significantly, the government kept the prices of synthetics and blends at the high levels set originally in the 1960s, when much of the raw material was imported. The high prices made it profitable for TVEs to enter the chemical fibre industry even though they were not low-cost producers.

New TVEs also emerged in the textile industry in the 1980s because they responded to opportunities created by economic reforms, i.e. they 'filled the cracks not occupied by the larger state enterprises. For example, after fulfilling their production targets, textile enterprises in the state sector were permitted to decide what additional products to produce. In making this decision in the more competitive post-reform environment, state enterprises generally emphasized quality and variety. Raw materials used in producing goods outside of the state plan were not supplied by the state but had to be purchased on the market at market prices. To produce goods of greater variety and higher quality, state enterprises needed specialized raw materials, e.g. different types of fibre and blends and different colours, and this created a niche for TVEs. It was not profitable for state enterprises, because of their high set-up costs, to accept small orders of specialized fibres. But TVEs, because of their small size, were well suited to supply special orders involving relatively short (by industry standards) production runs, and they quickly filled this niche.[15]

The industrial structure of Jiangsu's TVEs reflects the heavy industry and the local self-reliance biases of past policies. Thus, compared to rural industries in other developing countries, producer goods industries are overrepresented and light manufacturing and service industries are under-represented. In the 1980s, TVEs developed less according to administrative orders and more in response to market forces, and, reflecting this change, the structure of Jiangsu's rural industry was marginally altered. However, because prices were seriously distorted, TVEs did not always develop in accordance to rural Jiangsu's true comparative advantage.

The Contract Responsibility System

In the late 1970s, TVEs were considered mere extensions of local governments and were operated as such with little attention to efficiency or profits. Because of this, a large number of TVEs were poorly managed and most were either unprofitable or barely profitable. After the historic Third Plenum in December 1978, in an effort to improve economic performance, the government began to experiment with new ways of managing TVEs, and Jiangsu has been at the forefront of these experiments.

In 1979, the Jiangsu government, anxious to improve the performance of its growing TVE sector in an economy that was becoming increasingly market oriented, encouraged its TVEs to adopt the 'several fixes and one reward (*jiding yijiang*)' management system that was then popular among urban enterprises. The system was first tried by township enterprises in Suzhou and quickly spread to other parts of Jiangsu and to village enterprises. A particularly popular version was the 'five fixes and one reward (*wuding yijiang*)' system under which the township assigned the enterprise a fixed number of workers, a fixed amount of capital (fixed and working capital), fixed responsibilities (e.g. fixed output and output value targets), a fixed profit target, and a fixed cost target. If the fixed responsibilities and the profit and cost targets were achieved, the enterprise was rewarded with a share of the above-target profits.[16] To strengthen further the link between responsibility, performance, and reward, some TVEs also adopted the system internally by distributing targets (output, quality, and cost) to workshops, groups, and sometimes even down to individual workers.

In the early 1980s, encouraged by the successful adoption of the production responsibility system in agriculture, Jiangsu began to experiment with contractual arrangements in its rural non-agricultural sector.[17] A small number of TVEs, selected primarily from Suzhou, were contracted out to community members to manage. Despite good results, no attempt was made to implement the contract responsibility system (*chengbao zeren zhi*) on a province-wide basis for fear of political repercussions until 1983, when the Party and the central government gave official sanction to its implementation.[18] The movement of TVEs from the 'several fixes and one reward' system to the contract responsibility system gained further momentum in October 1983 when the CCP Central Committee and the State Council announced their decision to separate government administration from the commune and re-establish township governments.[19] By the end of 1983, most TVEs in Jiangsu had adopted some form of the contract responsibility system.[20]

The adoption of the contract responsibility system was an important development because it signalled a change in the nature of the relationship between the TVE and the TVG, its owner and supervisory unit. The hope was that the contract responsibility system would not only improve management (and therefore enterprise performance) by uniting (*jiehe*) authority (*quan*), responsibility (*ze*), and benefits (*li*), but also reduce

TVGs' discretionary power over TVEs and thus give TVEs greater operational flexibility and autonomy. An hierarchical relationship would be changed to a contractual one in which power, responsibilities, and benefits would be clearly defined. But, as we shall see below, this hope was only partially realized.

Under the contract responsibility system, townships and villages contract the management of their enterprises to individuals, to small groups of individuals, or to enterprise managers representing the enterprises.[21] The length of the contract is usually three years.[22] The contract defines the contractor's power, responsibilities, and rewards, the most important being: (1) a commitment to keep all workers allocated to the enterprise employed, (2) the authority to manage production and marketing, (3) the responsibility of achieving the annual economic targets assigned to the enterprise and the remuneration when they are achieved, and (4) the reward (penalty) for exceeding (not achieving) the targets. That contractors do not have the power to lay off workers reflects of course the high priority given to employment by governments at all levels. Most contracts follow guidelines established by the county or township government, but contractors are usually consulted before the major economic targets (e.g. gross output and profit) are finalized. In any case, the township (village) is fully aware that if terms are made too unfavourable no one would be willing to accept the responsibility and the risk of contracting an enterprise.

Most small (say, fewer than ten workers) TVEs are contracted by individuals, and the most frequently used arrangement is *lirun da baogan*, under which the contractor pays the collective a fixed annual remittance and absorbs all remaining profits or losses. In effect, the contractor is paying a fixed rent for the use of township- or village-owned assets, and thus, in theory, bears all the risk of operating the enterprise. In practice, however, the collective shares some of the risk. The fixed remittance is usually settled quarterly or at the end of the year, and if the enterprise performs poorly during the year, the township or the village is not likely to collect the full fixed remittance since most contractors are poor and have few financial assets of their own. However, consistent poor performance will result in a change in the contractor. In the early 1980s, perhaps one-sixth of Jiangsu's TVEs were under this arrangement.[23] However, during the 1980s, *da baogan* declined in importance as the number of small TVEs decreased. In the late 1980s, only in north Jiangsu where the rural economy was less industrialized and where most VEs were small was *da baogan* still widely employed.[24]

Medium and large TVEs are usually contracted to enterprise managers acting on behalf of the enterprises (the so-called manager contract responsibility system). Two main types of manager contracts are popular in

Jiangsu.[25] Many small-medium (say, 10–30 workers) enterprises have adopted the *lirun baogan, chaoli liuchang, quanjiang quanpei* system under which the enterprise is assigned a profit quota, and if the enterprise meets all its responsibilities and remits the quota profit, it then retains all above-quota profits. In principle, the quota profit must be remitted even if the enterprise is unprofitable, and this is accomplished by reducing wages (i.e. less is distributed to the manager and the workers when wages are settled at year-end). But one suspects that if the loss is substantial, the collective unit (i.e. the township or the village) may find it difficult to collect its quota profit by forcing the enterprise to reduce wages, particularly if it requires wages to fall below the community subsistence level. Thus the system is very much like *lirun da baogan* except that the contractor here is a group rather than an individual. Like *lirun da baogan*, this type of contract declined in importance in the 1980s.

Since the mid-1980s, the vast majority of the medium and large TVEs in Jiangsu (probably all TEs and most VEs in south Jiangsu) have operated under a form of the manager contract responsibility system that has the following characteristics: fixed responsibilities, remuneration tied to performance, above-quota profits shared (*dinge baogan, lianli jichou, chaoli fencheng*). Responsibilities are the economic targets assigned to the enterprise (reflecting the objectives of the local government), and, of these, the most important is profit or a composite economic indicator that includes profit as one major component (wage bill and enterprise reserves are usually the other components). Many localities use a composite economic target rather than just profit because local governments do not want enterprise managers to increase profit at the expense of wage or employment.[26] The salary level of the enterprise manager and the size of the wage fund (or the bonus fund) are linked to how well the enterprise performs during the year relative to the assigned targets. A share of the above-quota profit is retained by the enterprise as a bonus when targets are fulfilled or exceeded, and penalties are levied on the wage fund (or bonus fund) when they are underfulfilled. The enterprise manager determines how the wage and bonus funds are distributed internally within the enterprise. The base wage and bonus of the enterprise manager and of other senior enterprise cadres (usually the party secretary and the chief accountant) are determined by the township or the village, and the method used to determine their size is described in the contract.

Targets are assigned after consultations with both the senior administrative unit (the township or the county) and the enterprise manager.[27] Included among the factors considered before targets are finalized are: the enterprise's performance in previous years, production tasks assigned by the senior unit, market projections for the coming year, and likely changes in final output and raw material prices. Annual targets may be adjusted if economic conditions or government policies change during the year. This flexible approach to target setting is necessary, particularly for the large number of TVEs that are totally market orientated, since it is extremely difficult to project demand, prices, and profit accurately when markets are unstable as they are in China. Otherwise, few would be willing to be contractors. However, the flexibility also means that a hard budget constraint is not always imposed on TVEs.

The following descriptions of contracts used in two of the townships we surveyed give additional details which should help clarify the incentive system embodied in the manager contract responsibility system.

Example 1

In 1988 Zhashang Village (one of sixteen villages in Sigang Township) contracted out all ten of its enterprises using the following method.[28] Four economic targets were assigned to each enterprise: gross output (Q_o), the ratio of sales to gross output, collected revenues from sales

TABLE 4.5. *Rewards and Penalties, 1988, Village Enterprises, Zhashang Village Sigang Township, Jiangsu*

Enterprise performance[a]	W, total wage fund	S, annual salary of enterprise manager
$T = T_o$	W^{*}[b]	$S^* = $ RMB 7,000
$0.8T_o < T < T_o$	$(T/T_o)W^*$	$[1 - 2(T_o - T)/T_o]S^*$
$0.3T_o < T < 0.8T_o$	$(T/T_o)W^*$	$[1 - (T_o - T)/T_o]S^*$
$T > T_o$	$W^* + a(T - T_o)^c$	$S^* + b(T - T_o)^d$
$Q < Q_o$	No effect	S is reduced by RMB 100
$R < R_o$	No effect	S is reduced by RMB 20–30 for every RMB 10,000 not collected
Every unplanned second-order birth in enterprise	No effect	S is reduced by RMB 50
$PR < PR_o$	No effect	S is reduced by RMB 2,000
Each accident with loss exceeding RMB 500	No effect	S is reduced by RMB 50

[a] T, Q, R, PR are defined in the text.
[b] An amount sufficient to pay workers an average monthly salary of RMB 140.
[c] $0 < a < 1$, and varies by enterprise and increases in stages as T rises. For the enterprises in Zhashang, a was between 0.1 and 0.95.
[d] $0 < b < 1$, and increases in stages from 0.02 to 0.065 as T rises.

Source: Zhashang Village, '1988 nian jingji chengbao hetong (Economic responsibility contract, 1988)' (Jan. 1988).

(R_o), and a composite economic indicator (T_o) (the sum of wage bill, contributions to the nine reserves, and actual profit). The enterprise was required to remit to the village (PR_o) one-half of $(T_o - W^*)$, where W^* is the total wage bill available to the enterprise for distribution to workers and staff when T_o is achieved (the other half was to be retained by the enterprise and used as working capital). Those associated with the enterprise would be rewarded or penalized in accordance to the performance of the enterprise. The methods used to determine rewards and penalties are summarized in Table 4.5. There was no ceiling or floor on the size of the total wage bill (W) or the remuneration that enterprise managers could receive (S). Within the parameters stipulated in the contract, the enterprise manager had autonomy over such internal matters as the wages of individual workers and staff,[29] discipline,[30] marketing, and production.

Example 2

In 1988 all twenty-two township enterprises in Sigang operated under the manager contract responsibility system.[31] The contract was between the enterprise manager representing the enterprise and the Sigang Industrial Corporation representing the township. For each enterprise, the contract stipulated its inputs (fixed capital, number of workers, and wage fund), its economic responsibilities (i.e. economic targets such as profit, gross output, sales, and profit remittance), and its management responsibilities (e.g. developing new products, promoting production safety, and implementing good management practices).

The actual performance of each enterprise was assessed against its assigned targets using what is called in China the 100-point assessment (*baifen kaohe*) method, which in turn determined the remuneration of the manager and the size of the bonus fund for workers. The 100 point assessment scheme developed by the Sigang government actually utilized 200 points—100 points to assess economic performance and 100 points to assess management performance (see Table 4.6). The manager's annual remuneration was determined by the total number of points awarded for economic performance and management performance and the value of each point. For the purpose of assigning a RMB value to each point earned for economic performance, Sigang classified its enterprises into four classes (special, first, second, and third) according to size (*guimo*) as measured by capital, labour, and previous contributions to the township. The higher the class, the higher the pay scale for the enterprise manager or the higher the unit value of points awarded for economic performance.[32]

The economic targets were considered hard, or high priority, targets, and were rewarded more generously than the soft management targets. Table 4.6 shows that each hard target point was worth between RMB 50

TABLE 4.6. *The 200-Points Assessment System Employed by Township Enterprises, Sigang Township, Jiangsu, 1988*

A. Economic performance (100 points)

Economic targets	Points for achieving targets	Manager's remuneration for achieving objective by enterprise class (RMB)			
		Special	First	Second	Third
Gross output	10	800	700	600	500
Profit	65	5,200	4,550	3,900	3,250
Remittance	15	1,200	1,050	900	750
Revenue collection	10	800	700	600	500
(Total)	(100)	(8,000)	(7,000)	(6,000)	(5,000)
(Unit value)		(80)	(70)	(60)	(50)

B. Management performance (100 points)

Management targets	Maximum points possible	Maximum increases in manager's remuneration (RMB)
Development of new products	20	200
Superior product awards	20	200
Basic management	20	200
Production safety	15	150
Birth control	15	150
wenming shengchan[a]	5	50
Supporting agriculture	5	50
(Total)	(100)	(1,000)
(Unit value)		(10)

[a] The literal translation is 'civilized production', and it includes such matters as the physical appearance of the enterprise, the morale of workers, etc.

Source: Sigang Township, 'Sigang zhen changzhang renqi mubiao zeren zhi hetong shu (Contract of targets during the factory manager's tenure under the responsibility system, Sigang Township)' (Jan. 1988).

and RMB 80 (depending on the class of the enterprise), while each soft target point awarded was worth only RMB 10. Of the hard targets, profit was the most important. For example, managers were awarded 65 hard target points for achieving the profit target but only 10 points for fulfilling the gross output target. Thus the manager of a special-class enterprise would earn RMB 5,200 for achieving the profit target, but only RMB 800 for the gross output target. When hard targets were under- or overfulfilled, the manager's remuneration would be adjusted upwards or downwards proportionally.[33] For example, if a special-class enterprise achieved 105 per cent of its gross output target, 98 per cent of its profit target, 100 per cent of its remittance target, and 95 per cent of its revenue collection target, the assessed remuneration for its manager for economic performance would be:

$$(1.05)(RMB\ 800) + (.98)(RMB\ 5,200) + (1.0)(RMB\ 1,200) \\ + (.95)(RMB\ 800) = RMB\ 7,896.$$

The management targets were considered soft partly because they were seen as less important, partly because they reflect aspects of enterprise performance that are difficult to quantify, and partly because some of them can be achieved relatively easily.[34] For example, the manager would be awarded points for the development of a new product, for winning an award, when no serious accident occurred in the enterprise during the year, when there was no unplanned birth among its workers, and when all its workers met their grain purchase quotas.

The contract provided the manager with a certain amount of discretionary authority. For example, each manager was entrusted with a small business expense fund on a *baogan* basis.[35] But most importantly, during his or her tenure, the manager had power over income distribution within the enterprise and could therefore reward and penalize staff and workers according to their performance. In other words, wages and bonuses were flexible both upwards and downwards. Both the manager and the workers had a strong interest in ensuring that the profit target was met and exceeded, the manager because his or her salary was tied to profit, and the workers because the wage fund would be reduced proportionally when profit was below the target, and because the size of the bonus fund would be determined by the extent to which profit exceeded the target. More precisely,

Bonus fund = (above target profit)/(1 + profit/wage fund).

In other words, the share of above target profit allocated to the bonus fund would increase as the share of wage fund to total profit increased.[36]

Example 3

TVEs in Hufu Township operated under a system similar to that in Sigang (described in example 2) but with some important differences.[37] Enterprises were contracted to managers for three years but the economic and management targets were determined annually. Like Sigang, Hufu also considered the economic targets hard targets but identified five instead of four targets:[38] gross output, composite profit (*zonghe lirun*, the sum of book (*zhangmian*) profit and contributions to reserves), capital accumulation, reduction in accounts receivable, and sales/output ratio. And, instead of one set of economic targets, Hufu used three: base targets (*jichu zhibiao*), guaranteed targets (*baozheng zhibiao*), and strive-for targets (*zhengqu zhibiao*).[39]

The enterprise's economic performance determined both the manager's salary and the size of the wage fund. The manager's salary was composed of three parts: base salary, floating salary, and bonus. The manager received the base salary for achieving the base targets, and the size of the base salary was determined by the size of the enterprise and its previous performance. A floating salary was added to (or deducted from) the base salary when achievements exceeded the base targets but were below the guaranteed targets (or fell short of the base targets), and its size was determined partly by the absolute difference and partly by the relative difference[40] between actual achievements and the base targets.[41] A bonus was paid to the manager for exceeding the guaranteed targets, and the size of the bonus was determined partly by the absolute difference and partly by the relative difference between actual achievements and the guaranteed targets. The wage fund was tied only to one target, the composite profit. The target wage fund, stipulated in the contract, was paid for achieving the guaranteed composite profit target and would be increased (or decreased) by a predetermined percentage whenever the guaranteed composite profit target was exceeded (or not achieved).

The 100-point assessment system used in Hufu to assess the soft management targets is too complex to be described in detail. Roughly, points were awarded to managers as follows: 22 points to reward production accomplishments (e.g. meeting quality standards and energy consumption standards), 14 points to reward achievements in enterprise management (e.g. establishing a record-keeping system at the workshop level), 10 points to reward safe production, 27 points to reward achievements in financial management (e.g. reduction in inventory), and 27 points for 'constructing spiritual civilization (*jingshen wenming jianshe*)' (e.g. Party work and birth control).

As in Sigang, enterprise managers in Hufu had authority over current production and the power to determine the details of the internal wage system. However, unlike Sigang, Hufu imposed a ceiling on the bonus and on the total salary that a manager could receive. The bonus awarded to the manager could not exceed 30 per cent of the increase in wage fund awarded to the enterprise for exceeding the guaranteed composite profit target, and the manager's total earnings could not be greater than five times the average wage income of workers in the enterprise. However, the manager was also guaranteed a floor income of at least 80 per cent of the base salary.

The contract responsibility system was implemented partly to encourage TVEs to manage their resources more efficiently and to improve their economic performance and partly to change the relationship between TVGs and TVEs from a hierarchical to a contractual one, i.e. to replace the discretionary power that TVGs have over TVEs with a set of rules. The evidence from Jiangsu suggests that while the system has achieved some success in encouraging enterprises to be more efficient, it has not significantly reduced the discretionary power that TVGs have over their enterprises.

The examples presented above illustrate some of the methods used by local governments to unite responsibilities and benefits. Responsibilities include more than just good management or economic performance but also non-economic targets (e.g. family planning and political work). However, unlike the period before 1978, when fulfilling non-economic targets was frequently more important, economic responsibilities have become paramount, and this is clearly indicated in the contracts described above.

If our examples are representative, then the contracts used in Jiangsu in the 1980s were excessively detailed, and the complexities may have reduced their effectiveness. Indeed, some of the enterprise cadres we interviewed admitted that they did not fully understand how their annual remunerations were calculated.[42] This notwithstanding, what is clear from the contracts we examined and from our interviews with enterprise managers is that the intent of the contract responsibility system is to reward good management practices and penalize inefficiency and poor work performance.

Less clear is whether contracts have been scrupulously enforced. Our impression is that enforcement has been patchy. For example, we were told that TVGs did not always pay managers their full bonuses (even when there is no ceiling on income) nor fully penalize them according to the terms of the contract. Erratic enforcement undoubtedly has reduced the incentive impact of the contract system. Nevertheless, the widespread adoption of the contract responsibility system has probably improved enterprise management in rural Jiangsu. Although poor management has not disappeared, the adoption of the contract responsibility system has enabled TVGs to identify the less competent managers and has facilitated the dismissal of the most incompetent ones.[43]

By linking benefits to responsibilities and performance (i.e. the greater the responsibility and the better the performance, the higher the pay), the contract responsibility system has also permitted the rural collective sector to distribute collective income more unequally, particularly to widen the gap between the income paid to the main contractors (enterprise managers and their chief deputies) and to workers. Apparently, in some regions, this has succeeded so well that local governments have had to introduce ways to reduce the gap between the income of workers and that of managers and their deputies.[44] One method has been to impose limits on the size of bonuses,[45] and another has been to restrict the size of managerial income relative to the income of the average worker. Although collective income can now be distributed more unequally in rural China, it is also clear that there are limits to the inequity that the government will tolerate. Of course, setting limits on the income of managers also reduces their incentives to improve enterprise performance.

The hope that the contract responsibility system would reduce TVGs' discretionary power over TVEs and thus give rural enterprises some real autonomy appears to have been misplaced. Judging by what we observed in south Jiangsu in the late 1980s, TVEs have gained little real autonomy since the introduction of the contract responsibility system, and remained effectively under the discretionary control of their TVGs. In fact many of the contracts we examined stated explicitly that enterprises must submit to the supervision of the Party, and of course enterprise managers (the contractors) were appointed by the local Party secretary. We visited fewer TVEs in north Jiangsu, but we were told by senior cadres in the north that TVEs there were probably even more tightly controlled than in the south, perhaps because TVGs in north Jiangsu had fewer enterprises to oversee.[46]

By the late 1980s nearly all the medium and large TVEs in Jiangsu had adopted the manager contract responsibility system. The contracts examined above suggest that TVEs had the greatest autonomy in managing current production (e.g. setting production schedules, deciding on the output mix, and marketing) and in the internal distribution of wages and bonuses.[47] But even in these areas, many decisions needed the approval of the local government/Party branch before they could be implemented. Thus the adoption of the contract responsibility system in the 1980s did not remove TVGs from the daily activities of TVEs.[48] Since investment decisions were made not by the enterprise but by TVGs, TVEs had little or no autonomy over their future.[49] In other words, in the 1980s TVEs were not independent economic entities with lives of their own. Rather, they were a part of a larger collective unit (a village or a township). The many roles they played in that larger unit is one of the topics to be discussed in Chapter 7.

This chapter has focused on three aspects of the collective sector: its ownership structure, its industrial structure, and its management. In most parts of Jiangsu, rural non-agricultural development has followed the *sunan* model, i.e. the sector is dominated by TVEs. This is also the pattern in most other parts of rural China. Because many TVEs were established in the 1960s and the 1970s, the industrial structure of TVEs in 1978 (e.g. the overrepresentation of metal products and machinery and the underrepresentation of light and textile industry) reflected earlier priorities and policies that paid little attention to comparative advantage and economic efficiency. Since 1978, TVEs have become more responsive to market forces, and light and textile industries have become more important. However, because of price distortions, caused in part by partial price reform, market forces have not always given the correct signals to TVEs. As part of rural reform and in an effort to improve enterprise performance, most TVEs have adopted the contract responsibility system. The hope was that by giving enterprises more responsibility, authority, and benefits, enterprise performance would improve. In addition, it was hoped that the arrangement would reduce the discretionary power that TVGs have over TVEs and thus give TVEs more autonomy. While enterprise performance may have improved, enterprise autonomy has not increased. Separating TVGs from TVEs is not a simple matter, and, in view of how closely economics and politics are linked in China, the two may be inseparable.

NOTES

1. *JNJSFTZ* 14–15. In agriculture, collective ownership was less important. When agriculture was decollectivized, some of the collectively owned livestock, farm equipment, and agricultural machinery was distributed among commune workers. Thus, in 1986, 88 per cent of the livestock, 75 per cent of the farm equipment, and 46 per cent of the agricultural machinery were privately owned.
2. These and the other figures in this paragraph are from *JXQTZ* 2–3, 104–5.
3. Calculated from data in *JJN* (1987), 3. 107–12.
4. Strictly speaking, *sunan* encompasses Suzhou *shi*, Wuxi *shi*, and Changzhou *shi*, roughly the Yangzi Delta or the region SDR in Fig. 3.1. For a sample of the Chinese writings on the *sunan* model see Jiangsu sheng zhexue shehui kexue lianhehui (ed.), *Sunan moshi xin tansuo* (*A New Exploration of the Sunan Model*) (Shanghai: Shanghai renmin chubanshe, 1988).
5. The discussion in this and the following paragraph is heavily influenced by ideas discussed in Eugene Staley and Richard Morse, *Modern Small Industry for Developing Countries* (New York: McGraw-Hill, 1965), chs. 5 and 6. Evidence from two Asian economies in support of some of these ideas can be found in Samuel P. S. Ho, *Small-Scale Enterprises in Korea and Taiwan*, World Bank Staff Working Paper No. 384 (Washington, DC: World Bank, 1980).
6. Dennis Anderson and Mark Leiserson, *Rural Enterprise and Nonfarm Employment* (Washington, DC: World Bank, 1978), 25.

7. In 1986, among China's thirty provincial-level administrative units, Jiangsu ranked third in GVAO, second in grain, fourth in oilseeds, and third in meat. See *JJN* (1987), 3. 103.

8. In 1988, Jiangsu produced 14 per cent of China's cotton and 24 per cent of its silkworm cocoons (see *ZTN* (1989), 202–4).

9. Together, the three accounted for about 70 per cent of the value of building materials produced by TVEs in Jiangsu. See *JXQTZ* 208–21.

10. In 1986, even after the closure of many TVEs that produced machineries, industrial and metalworking machineries accounted for one-quarter of the output value produced in the sector.

11. See Christine P. W. Wong, 'Rural Industrialization in the People's Republic of China: Lessons from the Cultural Revolution Decade', in US Congress, Joint Economic Committee, *China Under the Four Modernizations*, part 1 (Washington, DC: US Government Printing Office, 1982), 394–418.

12. *JXQTZ* 208–21.

13. Chinese planners classified agricultural and sideline products into three categories according to their importance. Goods in the first and second categories were among the more important and were under the control of either the central or the provincial government.

14. In early 1984, the price of washed wool in Inner Mongolia was RMB 9,500 a ton, and a ton of washed wool could produce melton (a heavy woollen cloth) worth RMB 21,000 (Enterprise Interview Notes). For a detailed discussion of the post-1978 raw wool market in China, see Andrew Watson, Christopher Findley, and Du Yintang, 'Who Won the "Wool War"?: A Case Study of Rural Product Marketing in China', *CQ* 118 (June 1989), 213–41.

15. Enterprise Interview Notes. Filling the cracks also occurred in other industries. In fact, one TVE we surveyed declared its production strategy to be 'small-scale production, many products, and many special features (*xiao guimo, duo pinzhong, duo tedian*)', i.e. producing what state enterprises were unwilling or unable to produce in small amounts.

16. Mo Yuanren (ed.), *Jiangsu xiangzhen gongye fazhan shi* (*History of the Development of Rural Industry in Jiangsu*) (Nanjing: Nanjing gongxueyuan chubanshe, 1987), 182–3.

17. This paragraph draws heavily from ibid. 230–2.

18. See CCPCC and SC, 'Dangqian nongcun jingji zhengce de ruogan wenti (Some questions concerning current rural economic policies)', *XQZFX* 86.

19. See CCPCC and SC, 'Guanyu shixing zhengshe fenkai jianli xiang zhengfu de tongzhi (Notice concerning the separation of government administration from the commune and the establishment of township governments)', *ZJN* (1984), 9. 9. See Ch. 7 for more on the reorganization of the rural sector.

20. At the end of 1983, about 95 per cent of the industrial TVEs in Jiangsu had adopted some form of the contract responsibility system. See Mo Yuanren, 'Jingying chengbao zeren zhi zengtian le shengji (Management contract responsibility system has increased in vitality)', *JD* 4 (1985), 68.

21. In 1984, of those TVEs in Jiangsu that were contracted out, 74 per cent were contracted to enterprise managers, 10 per cent to groups of individuals, and 16 per cent to individuals (see Mo Yuanren (ed.), *Jiangsu xiangzhen gongye fazhan shi*, 249).

22. Initially contracts were for one year. In 1984, Jiangsu began to experiment with three-year contracts, and by the late 1980s most contracts were for three years. Under the three-year contract, the contractor is not changed for three years but annual targets (output, sales, profits, etc.) may be negotiated yearly.

23. Of the 37,000 industrial TVEs in Jiangsu that operated under the contract responsibility system at the end of 1983, over 16 per cent used the *da baogan* arrangement. See Mo Yuanren, 'Jingying chengbao zeren zhi zengtian le shengji', 70.

24. For example, in 1987, all seven of the VEs in Haiqian (average size, six workers), one of the two villages we surveyed in north Jiangsu, were contracted to individuals or small groups of individuals (Township-Village Interview Notes).

25. See Mo Yuanren (ed.), *Jiangsu xiangzhen gongye fazhan shi*, 236.

26. Township-Village Interview Notes.

27. For example, before setting enterprise targets, a village would consult the township and a township would consult the county.

28. Seven industrial enterprises, one welfare factory (*fuli chang*) for the disabled, a dairy farm, and a service enterprise. The following description is based on Zhashang Village, '1988 nian jingji chengbao hetong (Economic responsibility contract, 1988)' (Jan. 1988).

29. The exceptions were the salaries of the deputy manager and that of the chief accountant, which were set jointly by the manager and the village head.

30. Without the permission of the village, the manager could not lay off workers for economic reasons; individual workers could be dismissed for disciplinary reasons but such action had to be reported.

31. The following description is based on Sigang Township, 'Sigang zhen changzhang renqi mubiao zeren zhi hetong shu (Contract of targets during the factory manager's tenure under the responsibility system, Sigang Township)' (Jan. 1988).

32. For example, each point earned by a special-class enterprise was worth RMB 80 to its manager while that earned by a third-class enterprise was worth only RMB 50. Before 1988, enterprises were not classified by size. The change was made in 1988 in order to widen the earnings of enterprise managers to recognize the fact that the responsibilities of managers and the contributions of managers to the township vary by the size of the enterprise they manage. In 1987, the highest-paid enterprise manager in Sigang earned RMB 6,000 and the lowest earned RMB 4,000. In the same year, the highest-paid enterprise manager in Hufu (a significantly less industrialized township than Sigang) earned RMB 5,600 and the lowest earned RMB 2,200. However, even with the change, the income spread between enterprise managers in Sigang remained smaller than in Hufu.

33. There was the additional penalty of immediate dismissal if profit were less than 65 per cent of the target.

34. However, some are very difficult to achieve, particularly for small TVEs. For example, it is almost impossible for small TVEs to develop new products.

35. If the fund were overspent, 50 per cent of the excess expenses must be

covered by the manager from his or her salary, and if the fund were under-spent, the manager could retain 20 per cent of the surplus.

36. For example, one-half of the above-target profit would be allocated to the bonus fund if total profit equalled wage fund, and two-thirds would be allocated if total profit were one-half of the wage fund.

37. The system developed in Hufu is very complicated and therefore difficult to summarize. Here only its main characteristics are described. The description is based on the following documents: People's Government of Hufu Township, 'Hufu xiang guanyu xiangcun qiye renqi mubiao zeren zhi de yijian (Suggestions concerning the target responsibility system for TVEs during the [contractor's] period of tenure, Hufu Township)', Document No. 3 (1988), People's Government of Hufu Township, 'Hufu zhen 1988 niandu baifen kaohe xize (Detailed standards, 100 point assessment, Hufu Township, 1988)', and Yangquan Village, 'Jingji zeren hetong shu, Yangquan cun (Economic responsibility contract, Yangquan Village)' (1987 and 1988).

38. However, some VEs used only three targets: gross output, net output (value-added), and profit remittance.

39. We were told by the Party secretary of Hufu Township that the practice is to set base targets at levels that are relatively easy to achieve; guaranteed targets at levels to ensure that the targets assigned by the senior administrative unit are fulfilled; and strive-for targets at levels to give enterprises something to shoot for.

40. More precisely, the ratio of (achievement − base target) to (guaranteed target − base target).

41. The amount added (or deducted) also varies by targets.

42. In fact, one manager in Hufu told us that he thought only one person in the township (the one responsible for calculating the final salary settlements) fully understood the contract.

43. Apparently such dismissals are not uncommon. In fact the manager-contractor of one of the enterprises we surveyed was dismissed in 1987 for bad manage-ment practices that resulted in poor profits and a profit remittance far below the target. But, off the record, we were told that personal friction between the manager and the local Party secretary had also contributed to his dismissal.

44. Township and Village Interview Notes.

45. Most have adopted the government regulation that bonuses be limited to no more than 3–4 months of salary. However, compared to SEs, because TVEs can more easily escape the limitations placed on incentives, they have been able to reward their employees, particularly the staff members, more generously.

46. Township-Village Interview Notes.

47. As indicated in the above examples, the average base wage (or the total wage fund for the enterprise) is set not by the manager but by the village or township.

48. This has been noted by many Chinese observers. See, for example, Zou Fengling, 'Xiangzhen qiye chengbao hou de jige wenti (Several problems after the contracting out of rural enterprises)', *JQ* 3 (1986), 63.

49. While some TVEs were permitted to make small investments, they had no effective decision-making power over basic construction since all sizeable

capital expenditures required village or/and township approval. For example, VEs in Yangquan Village could invest up to RMB 5,000 on plant and equipment with village approval. Capital expenditures above RMB 5,000 needed the approval of the township. In Jingxiang Village, VEs could invest up to RMB 500 on their own authority. Larger investments needed village and township approval. Since 1992, TVEs in the more developed areas have been given greater investment decision-making power.

5

Township-Village Enterprises in a Partially Reformed Command Economy

THIS chapter investigates how TVEs managed their development in the 1980s, specifically how they mobilized capital and acquired technology, how they coped with marketing and supply problems, and how they utilized and motivated workers.

Capital and Technology

Among the many problems that TVEs faced in the 1980s, none posed more difficulties than their shortage of capital and their lack of a technological base. How did rural communities raise the capital needed to develop TVEs, and how did they acquire the technology (here defined to include not just technology embodied in machineries and equipment but also disembodied technology, i.e. the skills and knowledge stored in individuals) needed to start production and subsequently to improve product quality and introduce new products.

Sources of Capital

Because TVEs are small and use labour-intensive technology, the amount of fixed capital required to start operation is usually limited. In 1980, for example, the average Jiangsu TVE had slightly more than RMB 38,000 in fixed assets, or about RMB 750 of fixed capital for each of its workers.[1] However, for small rural communities, this still represented a substantial sum. In addition, to begin operation, a TVE needed at least one *yuan* of working capital for every *yuan* of fixed assets.[2] TVEs have received little or no investment from the state, and while enterprises outside a locality have provided TVEs with trade credits and, on occasion, even loans and investments,[3] they have not been a large or dependable source of capital. Furthermore, the capital market in China is undeveloped, and except for some counties in Guangdong and Fujian, foreign investment has been of negligible importance.[4] Thus, for capital, TVEs have had to rely primarily on (i) household savings, (ii) financial institutions (rural credit co-operatives and the Agricultural Bank of China), and (iii) what may be called the community surplus, i.e. the surplus resources controlled by the local community (township, village, or team).[5]

Community Surplus

Before the implementation of rural reforms, the community surplus, then under the control of communes and brigades, was the main source of capital for TVEs. In most regions, the bulk of the community surplus was generated initially by agriculture, and, for this reason, more TVEs were developed in areas where agriculture was relatively advanced, e.g. in south Jiangsu. During the 1970s, an average commune (now township) in south Jiangsu, where agriculture and rural industry were both relatively developed, could generate roughly RMB 1 million in surplus each year.[6] Indeed, in some of the more prosperous and developed counties in south Jiangsu, an annual surplus of this size was generated by many teams.[7] Besides the surplus generated from production, rural communities in China were also able to transform surplus labour into capital for TVEs.[8]

The use of social or duty labour (*yiwu gong*) in rural capital construction was widespread during the Cultural Revolution decade.[9] In addition to the local population, another source of surplus labour was the urban youth sent 'up to the mountains and down to the countryside (*shang shan xia xiang*)' from the late 1960s to the mid-1970s, partly for re-education, partly to avoid open unemployment in urban areas, and partly to supply skills to rural areas.[10] After the implementation of rural reforms in 1978 social labour became less important but did not totally disappear. For example, the sites and buildings of several of the village enterprises we visited were constructed in the mid-1980s with labour mobilized by the village.[11] Another example is the Haitou Prawn Culture Enterprise, a township enterprise that has become an important source of income for the township government as well as for villagers in the region. Investment in this enterprise began in the autumn of 1983 when 7,000 workers were mobilized from the eighteen villages in Haitou to dig prawn-culture ponds.[12] An estimated 240,000 cubic metres of soil were removed forming eight large ponds with a total surface area of 537 *mu* (35.8 ha).

Since the dismantling of the communes in 1984, rural communities in China have not been able to transfer resources (labour and savings) directly from agriculture to TVEs in the manner described above. However, other resources are still at the disposal of rural governments, among which the most important are: (i) extrabudgetary funds, (ii) retained profits of TVEs, and (iii) locally mobilized funds earmarked specifically for TVE development. These are briefly discussed below.

From the late 1970s to the mid-1980s, the development of TVEs was financed mainly by their own profits. Local government finance will be discussed in Chapter 7. Suffice it to say at this point that the most important source of extrabudgetary funds under local government control is the profits remitted by TVEs, and in the 1980s township-village governments (TVGs) used a substantial share of their extrabudgetary funds to develop TVEs. Because TVGs are the owners of TVEs, they also have considerable influence over how TVEs use their retained profits. In south Jiangsu, in the early and mid-1980s, TVEs reinvested 50 to 60 or more per cent of their retained profits.[13] In addition, from time to time, TVEs have been called upon to contribute some of their retained profits as well as their unused machineries and equipment to help establish new TVEs in the same community. Such investments are viewed as internal transfers since the donor and the recipient are owned by the same community.[14] In fact, it is not unusual for a TVE to begin life as a workshop in an established TVE and become independent only after it grows too large to be managed as a workshop or develops conflicts with the mother enterprise.

In the less developed parts of Jiangsu, because few profitable TVEs exist, there is less capital available for the development of TVEs. In some of these counties, the county government has pooled the excess cash of TVEs into a fund to finance short-term loans to TVEs in need of working capital or investment funds.[15] For example, the economic management and guidance station (*jingying guanli zhidao zhan*) in Ganyu County managed one such pool. TVEs in Ganyu began to deposit their excess cash into this pool in 1985, and the pool was opened to individual depositors on an experimental basis in five of Ganyu's townships in 1988.[16] In mid-1988, the pool had over RMB 10 million in deposits, paid a monthly interest of 7 per cent on deposits, and charged monthly interest of 12 per cent to 15 per cent on loans. Both private and collective enterprises could borrow from this fund, and in 1988 the demand for credits exceeded the supply, suggesting that the clearing monthly interest rate was higher than the 12 to 15 per cent charged by the pool.

Household Savings

Rural households have also invested in TVEs. The most common practice has been for newly hired workers to bring small amounts of capital into the enterprise.[17] In the early and mid-1980s, rural households invested in village enterprises in order to ensure factory jobs for family members.[18] For example, to obtain sufficient capital to begin operation, the Sigang Hemp Weaving and Spinning Factory in Jingxiang Village promised a job to any villager who contributed an 'investment of RMB 1,000'. A total of RMB 50,000, more than 10 per cent of the initial investment, was raised in this manner.

These contributions are not equity investments but rather interest-free loans, usually repayable in one or two years. Apparently, to cover un-expected demand for cash, workers often withdrew their contributions prematurely. Consequently, TVEs in Jiangsu have become reluctant to depend on this form of investment, and some enterprises have returned all such contributions to workers.[19] Where it is still in use, it may be as much to keep workers from leaving as to mobilize capital for the enter-prise. In parts of south Jiangsu where the labour-market has become tight, to reduce labour turnover, some TVEs have required workers to deposit trust money (*xinyong jin*), usually several hundred *yuan*, which is returned to the worker (normally without interest) only if he or she remains with the enterprise for a specified period (usually six months or a year).

Workers also lend their wages to the enterprise on a regular basis. In almost all TVEs, workers are paid only a small part of their monthly wage, so that wages are not fully settled until the end of the year. This is possible because workers are also peasants, and therefore do not depend on their wage income for either shelter or food. In effect, TVEs use wages as a source of interest-free loans and rely heavily on them for working capital.

Rural Financial Institutions

In China, rural credit is provided by the Agricultural Bank of China and the rural credit co-operatives under its direct supervision.[20] Since 1979, the value of loans that the Agricultural Bank may make in a region has been linked to the amount of deposits generated in that region, so that as deposits in a region rise so do outstanding loans.[21] This regulation was first implemented at the provincial level and subsequently adopted also at lower administrative levels. Under this system, branch offices of the Agricultural Bank (usually the credit co-operatives) in regions with high per capita income and large bank deposits can make more loans than those in poorer regions. Although capital does move across administrative boundaries, such mobility is limited. Branches of the Agricultural Bank finance projects outside their regions only under unusual circumstances and when they have special permission. The lack of capital mobility is one reason why less developed regions such as Ganyu County have had to develop their own pools of loanable funds.

In the mid-1980s, local governments, following the directives in the 1984 No. 4 Document, encouraged banks and credit co-operatives to help rural communities develop TVEs, and bank loans quickly became an extremely important source of capital for TVEs.[22] Among TVEs in south Jiangsu, the ratio of locally financed investments (i.e. reinvestment of

TVEs plus new investments in TVEs by TVGs) to investments financed by loans declined from 4:1 in 1983 to 1:1 in 1988.[23] For all TVEs in Jiangsu, outstanding bank loans as a percentage of total capital (year-end working capital plus net fixed assets at original prices) increased from 13.7 per cent in 1983 to 25.9 per cent in 1986.[24] It was primarily bank credits that financed the surge in TVE development in the mid-1980s in Jiangsu and elsewhere in China. Borrowing from the bank is undoubtedly the preferred method of raising capital, and it is easy to understand why. Bank loans carry low interest rates,[25] and repayment can be postponed relatively easily.

Since the bank's lending rate is significantly below the market clearing rate, rural credit is allocated administratively. Before making major lending decisions, credit co-operatives generally consult with township governments to seek their views and priorities. The relationship between credit co-operatives and township governments varies in intimacy, but it is generally true that loan applications with local government support are more likely to be approved. However, credit co-operatives do have the final say on loans, and not all proposals supported by township governments are automatically approved. Investment and lending decisions are of course also strongly influenced by the central government's macro policy.[26]

The legal rights of creditors are not well protected by laws in China, so to reduce the risk of default the Agricultural Bank does not make loans to TVEs (except the most established and profitable ones) unless they have a guarantor. The most frequently used guarantor is the community (i.e. the township or the village), usually the township economic commission (*jinglianwei*), or the village economic co-operative (*jingji hezuoshe*). An established and profitable enterprise (either one owned by the community or preferably a state enterprise or a large urban collective enterprise) may also serve as a guarantor, but it is extremely difficult to find one that is willing to take on the responsibility. In other words, TVEs can obtain bank credit only if the community is willing to assume the ultimate responsibility for the loan, so that if a TVE is unable to repay its loan, the debt becomes the responsibility of the community.

When a TVE is unable to repay or service its loan, the common practice is to transfer the debt to another, or to several other enterprises owned by the community. As a method of reducing risk, the guarantor system works reasonably well in south Jiangsu where even a village may operate a dozen or more healthy enterprises. However, in the less developed areas, having the community as the guarantor may not reduce the risk of lending to TVEs by very much since in these areas TVGs have few TVEs, and they are on average small and usually do not earn large enough profits to cover the losses of other enterprises.

An examination of the available data on the sources of funds used by TVEs may help clarify the mechanisms used to mobilize capital and their relative importance. Four sets of data are presented in Table 5.1: (1) the initial fixed capital investment and the subsequent fixed capital investment by sources for four recently formed TVEs (one township enterprise and three village enterprises) in Jiangsu;[27] (2) the fixed capital investment by sources for five established TVEs (four township enterprises and one village enterprise) in Jiangsu from 1985 to 1987;[28] (3) the liability side of the balance sheet of 200 large and successful TVEs (mostly township enterprises) in ten provinces at two points in time (when they were initially formed and in 1986); and (4) the actual investment made by TVEs in China and those in Jiangsu in 1987.

Table 5.1 strongly suggests the importance of community surplus in providing the initial capital for TVEs. Of the liabilities of the 200 large and successful TVEs at the time of founding, nearly 30 per cent consisted of local government capital. Nearly all the initial fixed capital investment of the four recently formed TVEs in Jiangsu came from the local government or were collective assets transferred from other units, usually other TVEs, in the same community. Bank loans and trade credits are the only other important sources of initial capital. The role played by household savings appears to be relatively minor, although the contributions from workers, as a gap filler, can be critical. As TVEs develop, internal funds become more important. Reinvestment of retained profits is one of the two main sources of funds for ongoing TVEs. The other is bank loans. The importance of community surplus and bank loans suggested by the micro data is confirmed by the aggregated data from Jiangsu and from China. In 1987, more than three-quarters of the investment made by TVEs in Jiangsu (and China) was financed either by community surplus or bank loans. The heavy dependence of TVEs on bank credits explains why they were so severely affected when the government tightened credits in 1989–90.

Capital Equipment, Technology, and Skills

The mobilization of savings solves only a part of the investment problem, albeit a large part. Markets in China are highly imperfect, so identifying the correct equipment, locating a supplier, acquiring the engineering, technical, and management know-how, and finding a core of skilled workers to maintain the equipment and to supervise production are common difficulties that all TVEs must face.

In the 1970s and the early 1980s, the usual method used to obtain technical information was to visit an operating enterprise in the same industry, sometimes under false pretences, but usually arranged through

Table 5.1. *Sources of Capital, Township-Village Enterprises*

| | Value (RMB 000) | % financed by | | | | | | |
| | | Community surplus | | | Retained profit | Bank loans | Household[a] savings | Others |
		Subtotal	Local government	Transfer of collective assets				
I. 4 new TVEs in Jiangsu								
Fixed capital investment at founding								
Haitou No. 3 Brick	250.0	76.0	52.0	24.0	—	—	24.0	—
Yixin Telecommunications	172.7	80.0	28.7	51.3	—	20.0	—	—
Sigang Hemp	474.0	50.7	38.0	12.7	—	42.2	7.2	—
Zhashang Wool	178.0	92.7	—	92.7	—	—	7.3	—
Fixed capital investment since founding								
Haitou No. 3 Brick	29.0	100.0	—	—	100.0	—	—	—
Yixin Telecommunications	1,276.1	70.2	—	15.7	54.5	25.1	4.7	—
Sigang Hemp	919.0	45.6	—	—	45.6	54.4	—	—
Zhashang Wool	1,690.4	73.1	—	50.6	22.5	26.9	—	—
II. 5 established TVEs in Jiangsu[b]								
Investment in fixed capital during 1985–7	32,671.2	46.4	—	6.5	39.9	43.1	0.7	9.8

% of total

| | Total | Community surplus | | | Bank loans | Other[c] investment and loans | Household[d] savings | Others |
		Subtotal	Local government	Retained profit				
III. 200 large TVEs								
Total liabilities								
At founding	100.0	29.4	29.4	—	38.2	13.0	6.5	12.9[e]
In mid-1986	100.0	24.2	4.4	19.8	23.4	43.2	4.0	5.2[e]

Table 5.1. *Continued*

% of total

	Total	Community surplus			Bank loans	*Qunzhong jizi*[f]	Outside[g] funding	Others[h]
		Subtotal	Local government	Retained profit				
IV. Actual investment in 1987 by TVEs								
In China	100.0	28.9	7.1	21.8	48.4	4.7	6.8	11.3
In Jiangsu	100.0	32.0	3.8	28.1	44.0	3.9	8.2	11.9

[a] Mostly investments from workers (*jizi*).

[b] Data were collected from 6 relatively mature and well-established TVEs, however one enterprise's data were unusable.

[c] This includes accounts payable to other enterprises as well as loans and investment provided by other enterprises.

[d] This includes wages payable and investments from workers.

[e] Of this, about one-third consists of grants and loans from state agencies, and the rest is unidentified.

[f] 'Funds raised from the masses' (roughly equivalent to household savings).

[g] Funding from outside the locality, including a small amount of foreign investment.

[h] Including state aid (*fuchi jin*).

Sources: The enterprise data in I and II are calculated from Enterprise Background Data. The data for the 200 large TVEs are from Zhou Qiren, 'Fuzhai jingying de hongguan xiaoying—10 sheng 200 jia daxing xiangzhen gongye qiye diaocha de zhengcexing fenxi baogao (The macro-economic effects of debt management—a policy-orientated analytical report on an investigation of 200 large industrial TVEs in 10 provinces)', draft report (May 1987); and reported in William A. Byrd, 'Entrepreneurship, Capital, and Ownership', in William A. Byrd and Lin Qingsong (eds.), *China's Rural Industry: Structure, Development, and Reform* (Oxford: Oxford University Press, 1990). The 1987 actual investment data are from *ZNN* (1988), 322–3 (some of the original figures were adjusted to correct obvious printing errors).

personal contacts and connections.[29] In the 1970s, for skills and engineering and management know-how, TVEs relied heavily on the urban intellectuals sent to the countryside during the Cultural Revolution decade. Another source was skilled workers and technicians who had retired to the countryside. The latter source was particularly important in south Jiangsu, where many skilled workers from Wuxi, Suzhou, Changzhou, and Shanghai had returned to retire. However, because capital goods were centrally allocated in the 1970s, it was almost impossible for TVEs to purchase new machinery and equipment even when they had the funds. Thus what TVEs were able to acquire in the 1970s and early 1980s were mostly old machinery and equipment no longer needed by state and large collective enterprises. The extent to which TVEs have depended on used equipment is indicated by the age distribution of machinery owned by TVEs in Changzhou, one of the most developed regions in Jiangsu, in 1987: 45 per cent manufactured before 1960, 50 per cent before 1980, and only 5 per cent during the 1980s.[30] TVEs in south Jiangsu were able to develop rapidly in the mid- and late 1970s in part because they were close to urban industries and therefore could easily acquire second-hand machinery and equipment and in part because they had access to a large pool of skilled workers who had retired to villages in the region.

TVEs operated in a much less restrictive environment in the 1980s, particularly after 1984. The introduction of economic reforms in urban areas, the adoption of a national development strategy that gave light industry a higher priority, and the recognition of TVEs as an integral part of the economy all made it easier for TVEs to purchase machinery and equipment directly from domestic producers. However, many townships and villages, particularly the less prosperous ones, have continued to equip their enterprises with used machinery from state enterprises.[31] Foreign equipment also became available in the 1980s, particularly in Guangdong where TVEs were able to acquire large amounts of capital equipment through special export arrangements.[32] Occasionally, in the early and mid-1980s, Jiangsu's TVEs were also permitted to import small amounts of machinery and equipment.[33] However, acquiring modern equipment from abroad became a real possibility only after 1987, when China adopted the coastal development strategy, and Jiangsu opened some of its rural areas to foreign investment and allowed localities limited discretionary control over their export earnings. In the first two years after Jiangsu opened its rural areas, its TVEs purchased and installed RMB 493 million worth of advanced (*xianjin*) equipment from abroad.[34]

Since the early 1980s, TVEs have relied primarily on two methods to acquire the technology and know-how needed to begin or upgrade production. When the required skills are relatively simple and therefore can be acquired easily, TVEs have usually sent a few of their workers to enterprises that produce similar products for short periods as trainees. For this type of training, TVEs generally pay a set fee per trainee.[35] The alternative has been to invite engineers, technicians, and skilled workers to the village for up to one or more years to help train workers and to supervise production. Retired engineers, technicians, and skilled workers—the so-called 'high-priced old men (*gaojia laotou*)'—have found this arrangement especially tempting since the attractive wages that TVEs pay are in addition to their retirement benefits, but others have also been attracted to work in rural areas.[36] The movement of urban skills to rural areas was given legitimacy in 1985 when the government announced that technical personnel in urban units may take leaves of absence to work in rural areas without jeopardizing their urban job or residence. Another source of skills has been the pool of rural workers employed by large collective enterprises in urban areas who are now willing to return to their native villages, attracted by the higher wages offered by TVEs, better housing conditions, and the proximity to relatives and friends.

Scientists, engineers, and technicians assigned to work in the more remote parts of China have also found offers from TVEs attractive. For these expatriates, accepting a job with a rural enterprise in their home province or county is often their only way to return home, and TVEs, particularly those in south Jiangsu, have been especially aggressive in recruiting them. The incentives offered include:[37] (1) an increase in base salary by one to three grades and large bonuses so that the total income package is usually more than double what they earned in state enterprises; (2) better housing; (3) assistance in changing their permanent residency (or that of a family member) from a rural to an urban jurisdiction (or at least to a suburb); (4) help in finding employment for family members; (5) pension; and (6) generous annual leaves to visit family in urban areas.

How important is this flow? Our fieldwork and the Chinese literature on TVEs both suggest that it has been important, particularly for TVEs in south Jiangsu, but unfortunately the flow is difficult to quantify. However, some fragments of information from Jiangsu are suggestive. In 1986, Jiangsu's TVEs reported that they employed 67,428 'invited employees or consultants (*pinyong renyuan*)', of which 9,319 had engineering qualifications and 46,469 were skilled workers.[38] In that year, invited engineers accounted for about 14 per cent of the total number of engineers employed by Jiangsu's TVEs. Another interesting statistic is that, of the 6.7 million workers employed by Jiangsu's TVEs in 1986, more than 137,000 (or about 2 per cent of the total) were officially urban residents.[39]

With the rural industry expanding rapidly, TVEs have been under intense market pressure to increase product variety and to improve product quality so they can better compete with other TVEs and with state enterprises. This has made the acquisition of technology and skills more important than ever, and TVEs have found that they can no longer depend on traditional methods (e.g. sending workers out and inviting high-priced old men in) alone to acquire the increasingly sophisticated technology and skills they need.

In 1985, the government adopted the spark plan to help develop appropriate technology and equipment for rural enterprises.[40] Because the spark plan has received little funding, it has had only a limited impact on urban–rural technology transfer. A more important channel for technology transfer is the growth of what is called in China lateral relations (*hengxiang lianxi*) between TVEs and other units. In the mid-1980s, at the beginning of urban economic reform, the government encouraged research institutes, technical schools, and state enterprises to become more market oriented and to develop new sources of revenue by selling their technical skills and know-how. A particularly lucrative market has been TVEs. Since the early 1980s, many urban units have established lateral relations with TVEs to help them improve product quality, develop new products, or install new production processes, usually for a lump-sum payment but sometimes also for a share of the profit.

Of the various lateral relations, probably the most effective for transferring technology to TVEs is the so-called jointly operated TVE (*xiang-cun lianying qiye*). These are TVEs that are operated in co-operation with other (mostly urban) enterprises or units (e.g. technical schools or research institutes). This type of relationship, which generally lasts for an extended period of time, is usually more intimate and more interactive than those described earlier, and is therefore likely to bring the most long-term benefits to TVEs.

Table 5.2 presents the available data on jointly operated TVEs in Jiangsu, and it shows that in 1986 they numbered nearly 2,900 (or about 3 per cent of TVEs), employed 6 per cent of all workers in TVEs, and produced nearly 10 per cent of the gross output produced by all TVEs. Almost 80 per cent of them were in partnership with an urban enterprise (either a state enterprise or a large urban collective enterprise), and slightly more than 12 per cent were associated with other units (e.g. research institutes, vocational colleges, universities, etc.), most of which were undoubtedly also located in urban areas. Less than 10 per cent of the arrangements were between TVEs, usually between a more developed and a less developed TVE. Unlike in Guangdong, where foreign (including Hong Kong and Macao) participation in TVEs is common, only nineteen jointly operated TVEs (less than 1 per cent) in Jiangsu had foreign partners in 1986. However, the number of foreign partners increased sharply after Jiangsu opened its rural areas to foreign investment in 1988. In 1990, Jiangsu claimed 423 joint ventures that involved TVEs and foreign partners.[41]

TABLE 5.2. *Jointly Operated Township-Village Enterprises by Type of Partner and by Type of Relationship, Jiangsu, 1986*

	Jointly operated TVEs		
	Number	Employees (000)	Real gross value of output[a] (RMB million)
Total	2,899	405	5,053
Distributed by type of partner (% of total)			
State enterprise	64.5	72.9	74.4
Urban collective	13.7	11.1	12.3
Other TVE	8.9	6.2	4.6
Foreign enterprise	0.7	0.4	0.3
Other units	12.2	9.4	8.4
Distributed by type of relation (% of total)			
Joint production	24.2	25.2	27.0
Joint marketing	25.4	28.2	26.0
Joint equity	19.6	19.3	20.5
Technology	17.9	17.7	17.5
Other	12.9	9.6	9.0

% of above accounted for by jointly operated TVEs
in Suzhou, Wuxi, Nantong, and Changzhou

	Number	Employees	Real gross value of output[a] (RMB million)
Total	58.2	69.8	78.7
By partner type			
State enterprise	59.5	68.8	76.3
Urban collective	61.2	80.9	90.4
Other TVE	41.5	54.2	75.0
Foreign enterprise	63.2	75.5	90.1
Other unit	58.6	76.2	83.9
By type of relation			
Joint production	56.2	71.6	79.5
Joint marketing	55.0	66.2	77.2
Joint equity	69.4	83.6	89.4
Technology	52.5	55.8	66.8
Other	59.1	74.1	78.7

[a] In 1980 constant prices.

Source: Calculated from data in *JXQTZ* 2–3, 808–35.

TVEs close to urban areas have found it easier to establish and maintain relationships with urban enterprises than those in more isolated areas. One reason is that proximity reduces the transaction costs of such arrangements. Another is that TVEs in areas close to cities have more contacts and connections with urban enterprises since many of the urban workers and technicians were originally from these same areas. It is therefore not surprising that most of the jointly operated TVEs are near the traditional centres of Jiangsu's light industry—Wuxi, Suzhou, Changzhou, and Nantong (see Table 5.2). In Sigang, one of 27 townships in Zhangjiagang (one of 6 counties in Suzhou *shi*), 26 of its 39 TVEs were jointly operated in 1986, and they had economic co-operation arrangements with more than 30 urban units, many more than were found in the other two regions we surveyed.[42]

With an eye to the future, some of the more developed rural regions in Jiangsu have invested heavily in their own technological base by developing research institutes and investing in human capital. In 1992, there were 54 rural research institutes in Wuxi County developing technology for local TVEs in such industries as textiles, metallurgy, and electronics.[43] Funded by TVEs and staffed with researchers lured from universities and state science and technology units, these research institutes worked primarily to improve product quality and to develop new products. Because the state does not assign engineers and technical personnel to TVEs, to ensure a steady supply of technical skills in the future, many rural enterprises have sent promising workers and high school graduates to technical institutes and universities for training, some have established their own technical schools, and a few of the most industrialized counties have opened their own vocational colleges.[44]

Rural communities have had to rely primarily on their own resources to develop TVEs. Because they generated more community surplus, the more developed and prosperous rural communities developed TVEs earlier and at a faster pace. Since the mid-1980s rural communities have also relied heavily on bank loans to finance TVE development. But because the amount of bank loans has been linked to the level of bank deposits in each region, the amount of loanable funds available to finance TVE development has been greater in the more developed than in the less developed localities. In the early 1980s, the primary method used by TVEs to acquire technology and skills was to send workers out for training or to hire retired urban workers and technicians. Since the mid-1980s, urban units have been permitted to sell their technology and know-how to TVEs, and an increasing number of urban units have invested in TVEs, usually by forming jointly operated TVEs. As in the case of capital, TVEs in the more prosperous rural regions have better access to technology and skills, partly because they are closer to urban areas and therefore have better contact with urban units, and partly because they have more money to spend on technology and skills.

Marketing and Procurement

Compared to state enterprises, TVEs depend substantially more on the market in the sense that they produce little for the state plan and receive only a small part of their raw materials from the state run material-supply system. In the mid- and late 1980s, for example, roughly 5 per cent of the gross output produced by TVEs in Jiangsu were included in the state purchase plan.[45] Being dependent on the market in a partially reformed economy has meant that TVEs have had to cope with severe supply and marketing difficulties.

Marketing

To help clarify how TVEs distributed their products in the 1980s, Table 5.3 presents the main channels used by ten TVEs in Jiangsu to market their output in 1987: (i) direct marketing, (ii) planned purchase, (iii) processing, and (iv) joint marketing. Direct marketing refers to goods sold directly to collectively owned commercial units at the township and village levels (*xiangzhen jiti shangye*), to individual peddlers (*geti shangye*), and to end-users. Planned purchase refers to goods produced for, and delivered to, the state for distribution by its supply and com-mercial departments. Processing is when a customer supplies the raw material or the components, in total or in part, to the TVE to process or assemble for a fee. Finally, joint marketing is when a TVE markets its output jointly with a commercial unit. A common arrangement is for the TVE to supply working capital to a retail outlet (either state owned or one operated by the supply and marketing co-operative) in exchange for a special counter at the store to sell its products.[46]

Of the marketing arrangements identified in Table 5.3, only planned purchase is an integral part of the state distribution system. Thus, of the ten TVEs, seven were totally self-reliant in marketing. However, two had purchase quotas from the state. One is Yuhe, which supplied one-third of its 1987 output (frozen prawns) to the Haitou County Foreign Trade Company for sale abroad. Apparently, the county foreign trade company included Yuhe in its plan in order to protect the county's investment in the enterprise.[47] The other TVE with purchase quotas was Zhangjiagang Steel Cable (ZSC), and its relationship with the state sector is sufficiently interesting to describe in detail.

TABLE 5.3. *% of Principal Products Sold by Selected TVEs, by Distribution Channel and by Destination, 1987*

	% of principal products sold			
	Marketed directly	Planned purchase	Processing	Joint marketing
Zhashang Textile	100.00	0.00	0.00	0.00
Zhangjiagang Hemp	100.00	0.00	0.00	0.00
Zhangjiagang Steel Cable	29.04	64.32	0.00	6.63
Zhangjiagang No. 2 Fibre	100.00	0.00	0.00	0.00
Zhangjiagang Radio	90.62	0.00	9.38	0.00
Zhanggong Quarry	100.00	0.00	0.00	0.00
Yixing Telecommunication	100.00	0.00	0.00	0.00
Haitou No. 3 Brick	100.00	0.00	0.00	0.00
Haitou Agriculture Implement	100.00	0.00	0.00	0.00
Yuhe Refrigeration	68.31	31.69	0.00	0.00

	% of principal products sold			
	Within county	To other counties in Jiangsu	To Shanghai	To other provinces[a]
Zhashang Textile	1.99	45.20	10.98	41.98
Zhangjiagang Hemp	44.78	47.82	6.49	0.91
Zhangjiagang Steel Cable	18.69	12.06	5.82	63.43
Zhangjiagang No. 2 Fibre	51.75	25.77	4.33	18.15
Zhangjiagang Radio	2.88	38.21	2.06	56.85
Zhanggong Quarry	5.88	82.81	11.31	0.00
Yixing Copper Material	6.01	19.47	30.16	44.36
Yixing Telecommunication	2.41	88.05	5.26	4.27
Haitou No. 3 Brick	100.00	0.00	0.00	0.00
Haitou Agriculture Implement	0.00	100.00	0.00	0.00
Yuhe Refrigeration	1.65	37.15	0.00	61.20

[a] Including goods that were exported. Two TVEs exported—Yixing Telecommunication exported 3.29 per cent of its output and Yuhe Refrigeration 31.69 per cent.

Source: Calculated from data in Enterprise Background Data.

ZSC is the largest and one of the oldest township enterprises in Sigang Township.[48] In the early 1970s, when the County Material Supply Bureau was unable to acquire sufficient coal to meet local demand, it suggested to

Sigang (then a commune in the county) that its brickyard (a major consumer of coal) produce steel wire for the Bureau to use to exchange for coal. Sigang agreed, and that year the brickyard used some of its profits and purchased used wire-pulling equipment from a state enterprise in Shanghai. Production began in 1975 with steel supplied by the Bureau. This lasted until 1983, when the County Material Supply Bureau terminated the arrangement because it was having difficulties selling the wires, and ZSC was left to fend for itself.

Because steel was allocated by the state, to survive, ZSC had to develop new links with the state sector. Through contacts in Beijing, it gained an introduction to the Ministry of Metallurgical Industry (*yejin gongye bu*), and struck a deal with the Ministry that saved the enterprise. The Ministry appointed ZSC a 'designated supplier of steel cable (*gangsisheng dingdian danwei*)', and in return ZSC agreed to share its profits with the Ministry.[49] Since 1983, ZSC has been included in the Ministry's annual production plan, i.e. each year the Ministry has assigned part of its annual planned output of wire to ZSC and has supplied it with the necessary steel to produce the wire.[50] Without the deal, ZSC would have gone hungry, and Sigang Township would have had to cope with a serious unemployment problem.

The example of ZSC illustrates that while planned purchase may have been relatively unimportant for the TVE sector as a whole, it has been critical for some of the largest and most important TVEs, mostly in heavy industry, that use large amounts of centrally allocated raw materials. Without access to planned purchase, few of these TVEs could have survived, and certainly none could have continued to employ large numbers of rural workers. For these TVEs, profit sharing with the state was the price of survival.

To help TVEs sell their products, local governments do what they can to protect their enterprises by buying locally, but township and county markets are usually too small to be of much help to local enterprises.[51] Even relatively small enterprises depend on customers outside the county. Table 5.3 gives the distribution of sales of eleven TVEs in 1987 by destination. In that year, with one exception, all the TVEs sold a significant proportion of their sales to customers outside the county, and most also sold to customers outside of Jiangsu. The exception was the Haitou No. 3 Brick Factory, which sold 100 per cent of its output within its county. Undoubtedly the reason why Haitou No. 3 did not sell outside of the county was the high cost of transporting bricks. Of course, high transport costs also protected Haitou No. 3 from outside competition.

TVEs have grown rapidly in Jiangsu because they have been able to sell what they produce. One reason for their success is that they have had greater latitude than state enterprises in setting and cutting prices, and have used this flexibility to considerable advantage. However, competitive pricing is not the only reason for TVEs' success. TVEs have also paid close attention to market opportunities, filling gaps in the market and responding quickly when market conditions change. They have also competed aggressively in product quality and in variety. TVEs have described their strategy as follows: 'what others lack, I have; what others have, mine are better; when others are scarce, I am abundant; when others are superior, mine are cheaper (*ren wu wo you, ren you wo you, ren shao wo duo, ren you wo lian*)'.

Since the late 1970s, the TVE sector in Jiangsu has expanded mainly in areas not served or inadequately served by the state sector:[52] for example, (i) goods with strong demand but not included in the state plan (e.g. bricks); (ii) goods in demand but which cannot be included easily in the state plan because the quantity demanded is small, because they are not standardized, or because variety is important; (3) goods produced in insufficient amounts by state enterprises (e.g. textiles). Furthermore, they have been quick to change their product-mix in response to market opportunities. The Chinese attribute the TVEs' ability to change product-mix quickly to the fact that 'small boats can change directions quickly (*chuan xiao tiao tou kuai*)'. In other words, because TVEs possess little fixed capital and usually have only unspecialized tools and equipment, changing product-mix is just a matter of moving workers from producing one labour-intensive good to another.

TVEs also recognize that they operate in a highly competitive market environment, and that timing and quick response are critical to success. With easy entry, only those enterprises that can start or shift production quickly are likely to earn profits. It is not unusual for TVEs (particularly the pioneers) to recover their original investments in one or two years. But since entry drives prices and profits down, latecomers often find it difficult to recover their investments. This explains why, once an opportunity is identified and the decision to invest or to change production is made, township and village cadres work day and night to get the new product in production.

However, with growth, TVEs have lost some of the flexibility they once had when they were very small and relied almost exclusively on labour to produce goods. As more products have come on the market, consumers have become more discriminating and less willing to accept shoddy quality. TVEs have had to adjust to this change by improving product quality. In most cases, significant improvements could be achieved only by adopting a more capital-intensive technology and investing in better and more costly equipment. Thus market competition has pressured townships and villages to increase investments and to make TVEs larger, sometimes through mergers, and more capital intensive. However, this also has made TVEs more vulnerable. When a rural community invests heavily in one enterprise, failure or even a temporary downturn can be very damaging to the community, because the investment tied up in the specialized equipment and machinery cannot be switched easily to other uses, and few TVEs can afford to remain idle for long.

A good example of some of the issues just discussed is provided by the experience of Yixing Telecommunication Material (YTM), established jointly by Dongling Village and Hufu Township in 1979 to produce circuit boards (*fugan ban*) for radios.[53] Circuit boards were selected partly because entry to the industry was easy and partly because they were then in short supply. In 1980, the enterprise purchased and installed a second-hand 800 ton press for RMB 100,000, made arrangements to send ten of its original twenty workers to factories in Changzhou and Shanghai for training, and invited three experts from Shanghai to provide technical and production assistance. Production began in 1981 with sixty-one workers. In its first year of operation, it produced output valued at RMB 618,000 and made a profit of RMB 74,000.

As more enterprises began to produce circuit boards in the early 1980s, YTM found it more difficult to market its products. Customers that had accepted substandard circuit boards before now returned them. Output increased in 1982 but much of it remained unsold and profit declined to a mere RMB 5,200. Thereafter, for several years, YTM worked hard to improve quality by introducing new quality control measures and by refurbishing its old equipment. The next few years were relatively good ones for YTM, partly because of improvements in the quality of its product, partly because it kept its prices low, and partly because China's electronic industry began to take off in 1983. During this period, the enterprise expanded steadily. Its work-force expanded from 68 in 1982 to 130 in 1984 and 230 in 1986, and its gross output value in constant 1980 prices increased from RMB 1.01 million in 1982 to RMB 1.5 million in 1984 and RMB 3.3 million in 1986.

As more TVEs entered the radio component industry, competition intensified in the mid-1980s. In 1986, when a TVE in a nearby village purchased a 1,200 ton press to produce circuit boards, YTM was pressured to respond. The larger press would enable YTM's competitor to produce larger and better quality circuit boards, with greater efficiency in raw material consumption. YTM felt it had no choice but to meet the challenge and invest in a larger press as well. In late 1986 YTM took out a RMB 180,000 bank loan and invested RMB 616,000 in new facilities including a 1,000 ton press. In 1987, when the new equipment came on line, YTM had 238 workers and produced gross output of RMB 4.36 million (in 1980 constant prices).

Shortly thereafter, YTM was hit with two major market shocks. In

1988, along with other radio components, the demand for circuit boards declined sharply, and, at the same time, the prices of raw materials increased. To make matters worse, with the market for radio components in recession, many of YTM's customers were unable to pay for what they had purchased. Since the press could not be used for other purposes, switching to another product was not an option. After a short period of heavy losses, YTM laid off its workers and closed down in late 1988.

Another reason for their marketing success is that TVEs have put considerable resources into marketing. In the 1980s, most TVEs in Jiangsu employed three to five sales representatives (*tuixiao yuan*) who spent most of their time on the road introducing products to end-users and attending ordering and distribution conventions (*dinghuo hui* and *tiaoji hui*) where buyers and sellers meet to make deals and exchange information. TVEs in the Yangzi Delta region alone employed between 150,000 and 200,000 sales representatives.[54] Most worked on commission and were permitted to incur relatively large (by Chinese standards) travel and entertainment expenses. It is significant that marketing personnel employed by the eleven TVEs surveyed in Jiangsu were consistently among the highest-paid employees in the enterprises, sometimes earning even more than the enterprise manager. TVEs were also among the first to use advertising in China, and, in the mid- and late 1980s, many spent hundreds of thousands of *yuan* annually on advertisements. To convince customers that they are reliable producers of quality products, some TVEs have gone to great expense to invite customers to attend appraisal meetings (*jianding hui*) at their plants.

Procurement of Fuel and Raw Materials

In the 1980s, only about 10 per cent of the principal raw materials and fuel needed by Jiangsu's TVEs were included in the state allocation plan.[55] In other words, 90 per cent of the raw materials and fuel used by TVEs were obtained from channels outside of the state plan. Obtaining fuel and raw materials for TVEs has been one of the most important and time-consuming responsibilities of enterprise managers as well as cadres at the county, township, and village levels.

Table 5.4 presents the major sources of raw materials and fuel used by the TVEs surveyed in 1987. Four sources are identified: planned distribution (the state material supply system), materials provided by buyers, resources controlled by the township, and resources purchased from the market. Of the eleven TVEs in the sample, two relied heavily on raw materials supplied, in part or in total, by the state material supply system as part of its planned distribution. In the case of Zhangjiagang Steel Cable (discussed earlier), it received steel from the state because part of

TABLE 5.4. *Principal Raw Materials and Energy (Coal, Electric Power, and Other Fuel) Used by Selected TVEs, by Source, 1987*

	Raw materials (RMB 000)	Source (% of total)			
		Planned distribution	Supplied by buyer	Resources controlled by township	Market
Zhashang Textile	1,556	0.0	0.0	0.0	100.0
Zhangjiagang Hemp	1,726	0.0	0.0	0.0	100.0
Zhangjiagang Steel Cable	11,000	59.1	0.0	0.0	40.9
Zhangjiagang No. 2 Fibre	935	0.0	0.0	0.0	100.0
Zhangjiagang Radio Material	1,670	13.5	20.9	0.0	65.7
Zhanggong Quarry	96	100.0	0.0	0.0	0.0
Yixing Copper Material	11,055	0.0	0.0	0.0	100.0
Yixing Telecommunication	2,896	0.0	0.0	0.0	100.0
Haitou No. 3 Brick	56	0.0	0.0	40.3	59.7
Haitou Agriculture Implements	25	0.0	0.0	0.0	100.0
Yuhe Refrigeration	4,880	0.0	0.0	100.0	0.0

	Energy (RMB 000)	Source (% of total)			
		Planned distribution	Supplied by buyer	Resources controlled by township[a]	Market
Zhashang Textile	n.a.	n.a.	n.a.	n.a.	n.a.
Zhangjiagang Hemp	49	0.0	0.0	0.0	100.0
Zhangjiagang Steel Cable	5,551	0.0	0.0	78.1	21.9
Zhangjiagang No. 2 Fibre	1,397	0.0	0.0	88.4	11.6
Zhangjiagang Radio Material	313	1.3	0.0	76.1	22.6
Zhanggong Quarry	0	0.0	0.0	0.0	0.0
Yixing Copper Material	414	92.6	0.0	7.4	0.0
Yixing Telecommunication	127	0.0	0.0	0.0	100.0
Haitou No. 3 Brick	88	0.0	100.0	0.0	0.0
Haitou Agriculture Implements	2	100.0	0.0	0.0	0.0
Yuhe Refrigeration	641	100.0	0.0	0.0	0.0

[a] Include pooled resources electricity (*jizi dian*).

Source: Calculated from data in Enterprise Background Data.

its output was included in the state plan. The other enterprise was Zhanggong Quarry, and it was included in planned distribution because the raw material it needed (explosive) was a restricted good available only from the government. One enterprise (Zhangjiagang Radio Material) processed raw materials supplied by buyers (*lailiao jiagong*). TVEs are constantly on the look-out for such arrangements since they eliminate both the need to find raw materials and the need to market the end-products.

In the 1980s, most TVEs, including the eleven in Table 5.4, acquired their raw materials primarily from the market. It was relatively easy for TVEs to buy locally produced agricultural raw materials. But, because product markets were imperfect and not fully developed in China, it was more difficult to find and buy raw materials that were not produced locally. TVEs employed a large number of buyers whose job was to cultivate the necessary connections to find and to purchase raw materials. In some regions, permanent markets developed to supply specialized inputs used by rural enterprises.[56] Because TVEs depended on the market, they paid higher prices than did state enterprises for many raw materials. TVEs could afford this because they also charged higher prices for their products. However, some raw materials that were allocated by the state were difficult to find even at high prices, and TVEs often had to use gifts to circumvent government regulations.[57] This has led to the accusation that TVEs promote corruption even though the root cause of corruption is the allocation system (i.e. relying on bureaucrats rather than the market to allocate key resources). To survive in a partially reformed economy, TVEs have had to become adept at circumventing government regulations and policies.[58]

A few of the eleven TVEs in Table 5.4 used resources controlled by the township, reflecting the fact that some of Jiangsu's more industrialized counties and townships have developed elaborate arrangements to acquire raw materials for their TVEs.[59] Some have used barter trade, e.g. exchanging cements, glass, and even automobiles for raw materials.[60] Some have exchanged investments and technology for raw materials. For example, in 1988, Sigang Township signed an agreement to provide a wool-producing base in Inner Mongolia with the equipment and the expertise to establish a wool washing plant. In return the wool-producing base agreed to repay Sigang for the cost of the equipment plus interest in wool, and to sell to Sigang an additional 200 tons of cleaned wool annually for five years at the local price.[61] To gain access to raw materials, some regions in Jiangsu also have signed co-operative agreements with backward but resource rich regions to help develop their resources. Through such arrangements, Wuxi County alone obtained in 1985 760,000+ tons of coal, 70,000+ square metres of lumber, 55,000 tons of pig iron, and 150,000 tons of steel.[62]

Because Jiangsu, particularly its industrialized south, was short of energy in the 1980s, a major problem for TVEs was getting sufficient electric power. TVEs obtained electric power in four ways. Some TVEs were supplied with small amounts of state planned electricity (*guojia jihua dian*) at the state price; most were allowed to purchase limited amounts of negotiated price electricity (*yijia dian*); TVEs that had contributed funds to help construct power stations were able to purchase additional amounts of pooled resources electricity (*jizi dian*) at the negotiated price; and some TVEs also generated electricity for their own consumption. Of these, the least important was self-generated electricity which TVEs used primarily as back-up.[63] Most of the electricity purchased by TVEs in Jiangsu was at the negotiated price, which was substantially higher than the state price (in 1987 it was higher by 60 to 70 per cent). However, the problem was not the high price but that frequently electric power could not be obtained at any price.

The only planned electricity available to villages was the limited amount allocated to agriculture. This effectively restricted village enterprises to industries that required little or no electric power. In fact, many villages, because they could not get sufficient electricity, were forced to sell their enterprises to townships.[64] Townships received some planned electricity for industry, but the amount was usually insufficient to satisfy the needs of TVEs except in relatively undeveloped areas where there were few rural enterprises. It is significant that, of the TVEs listed in Table 5.4, those that relied primarily on planned distribution for their energy supply in 1987 were mostly located in the less industrialized areas (e.g. Haitou in north Jiangsu).

Because of severe shortages of electric power in south Jiangsu, TVEs in that region had a strong interest in assisting the state to develop new energy sources. Many helped by contributing funds for the construction of new power plants. For example, Zhangjiagang No. 2 Chemical Fibre contributed RMB 1.4 million in instalments to help construct the Jianbi Power Plant.[65] In return, it was given the right to buy up to 10,000 kWh per day from Jianbi for 20 years at the negotiated price. In China, this type of electricity is called 'pooled resources electricity (*jizi dian*)', and is included in Table 5.4 as 'resources controlled by the township'. A considerable amount of capital has been raised in this manner from TVEs in south Jiangsu anxious to break the energy bottleneck that has constrained their growth.

Price flexibility, a quick response to changes in market conditions, and aggressive marketing are some of the reasons why TVEs have been successful. Price distortions in a partially reformed planned economy and failures (past and present) of state enterprises to respond to consumer needs have created numerous profit opportunities in China, and TVEs, because they operate outside the state plan, have been able to respond to these opportunities more quickly and in ways that state enterprises could not. TVEs have also been adaptable and innovative in finding sources of energy and raw materials outside the state material supply system. With the help of local government, they have developed an elaborate network of connections and special arrangements to acquire energy and raw materials.

Labour-Market and Wage Determination

Rural communities developed TVEs in order to move peasants off the land and into factories, an objective that has been achieved with considerable success. This section (1) describes how rural factory jobs were allocated among peasants and the characteristics of those who moved into TVE jobs, (2) discusses how wages were determined in the collective sector, and (3) examines one aspect of labour mobility in rural areas, i.e. the extent to which TVEs used workers from outside the locality.

Allocation of TVE Jobs and Hiring Practices

To a significant extent, township-village cadres have controlled the hiring of workers by TVEs.[66] Because TVEs are community owned, community members have first priority for TVE jobs. In the late 1970s, the desire to leave the land was very strong in rural Jiangsu, particularly in the more densely populated south, and factory jobs were especially coveted. Because so many wanted factory jobs, many gained employment in TVEs by using the back door (*zou houmen*), i.e. they used personal connections. In many rural communities in Jiangsu, the majority of workers selected to enter TVEs before 1981 were relatives and friends of rural cadres.

Because TVEs were desperately short of capital in the early 1980s, TVGs offered factory jobs to anyone in the community who was willing to contribute capital. Depending on the capital requirements of the TVE, a job could be had for as little as several hundred *yuan*. But jobs in some TVEs required contributions of several thousand *yuan*. To develop its enterprises, a township would also turn to its villages for capital, and allocated jobs to the villages in proportion to the size of their contributions.[67] Until the mid-1980s, contributing capital was the surest way for rural households to secure jobs at TVEs for family members.[68] As the number of TVEs increased, local governments gave TVE jobs first to those who contributed capital and then allocated the remaining jobs among households in the community more or less equally among teams or households, apparently in an effort to give every household in the village an opportunity to earn non-agricultural income.[69]

By the mid-1980s, in many rural communities in south Jiangsu, nearly all those who wanted them had factory jobs. Thus, in these communities, local workers became increasingly selective about what jobs they would accept, and many were unwilling to do the dirty and the heavy work. Some TVEs even found it difficult to fill what were once considered desirable jobs. Once the local supply of labour was exhausted, TVEs in south Jiangsu began to turn to other townships, counties, and provinces for workers, and gradually a rural labour-market developed. Workers from outside the community were usually employed on contract, usually for a season or for a year.

The hiring practice used in the townships we visited in the late 1980s was as follows. The local government determined, after consultation with the enterprise, the number of workers that a TVE was to employ during the year. The enterprise did the hiring, but usually had to submit the names of those hired to the local government for examination and approval (*shenpi*). Most TVEs consulted with the local government before hiring new workers. Some TVEs selected new workers through examinations, but special priority was given to members of local households in economic difficulties and to retired servicemen. Many rural communities used the allocation of TVE jobs to help poorer households and to equalize household income within the community. Before a TVE could hire outside workers, it must have the approval of the labour bureau at both township and county level.

Enterprise managers may dismiss workers for disciplinary reasons but must report such cases to the local government.[70] TVEs can and often do go under, and when they do workers are laid off. In other words, workers in TVEs do not have the same job security as workers in state enterprises, and this is probably the most important difference between a job in a TVE and one in a state enterprise. The use of trust money (*xinyong jin*) in parts of south Jiangsu suggests that workers do move from one enterprise to another, but apparently this does not occur frequently. TVGs also have the authority to reallocate local workers among enterprises under their jurisdiction.[71] Some villages review the allocation of labour at the beginning of each year, and if needed workers are transferred from one village enterprise to another. Those unwilling to move may be dismissed.

There is considerable evidence that the number of workers employed by TVEs is frequently far greater than they need. Table 5.5 presents the distribution of the employees of eleven TVEs by job classification, and it shows that 73 per cent of the workers were engaged in production and 27 per cent involved in support work. The ratio of production workers to total workers was especially low in Sigang, the most developed and the most densely populated of the three surveyed townships. Not only did TVEs appear to use more administrative and technical personnel than

TABLE 5.5. *Distribution of All Employees by Job Classification, Selected TVEs, 1987*

| | Total | % of total[a] | | Production workers | | | Non-production workers | Agricultural workshop |
		Administrative and technical	Buyer and seller	Total	Regular	Others[b]		
11 TVEs in Jiangsu	4,874	12.3	2.8	73.2	63.0	10.2	9.9	1.8
Sigang								
Zhashang Textile	305	6.6	6.6	86.9	86.9	0.0	0.0	0.0
Zhangjiagang Hemp	204	8.8	2.5	82.4	77.5	4.9	4.9	1.5
Zhangjiagang Steel Cable	2,067	14.5	2.1	63.0	43.9	19.1	18.3	2.1
Zhangjiagang No. 2 Fibre	575	12.7	2.1	73.2	63.3	9.9	10.3	1.7
Zhangjiagang Radio Material	236	11.0	4.2	80.1	78.8	1.3	4.2	0.4
Hufu								
Zhanggong Quarry	541	2.4	0.4	87.4	87.4	0.0	4.4	5.4
Yixing Copper Material	441	9.5	4.5	85.9	85.9	0.0	0.0	0.0
Yixing Telecommunication	245	33.5	6.9	59.6	59.6	0.0	0.0	0.0
Haitou								
Haitou No. 3 Brick	135	5.9	0.7	93.3	93.3	0.0	0.0	0.0
Haitou Agricultural Implements	65	9.2	3.1	86.2	33.9	52.3	1.5	0.0
Yuhe Refrigeration	60	21.7	6.7	71.7	71.7	0.0	0.0	0.0

[a] The average number of employees in 1987.
[b] Apprentices and temporary workers.

Source: Calculated from data in Enterprise Background Data.

were needed, but many workers were engaged in activities that were not directly related to the enterprise.

TVGs in Jiangsu have used jobs in TVEs to reduce the income differentials between agriculture and non-agricultural activities and to reduce disguised unemployment in agriculture. Such practices are apparently widespread among TVEs, particularly in south Jiangsu, which is more developed and more prosperous and therefore better able to afford it. One senior county official we interviewed asserted that as many as one-third of the workers employed by TVEs in his county were 'unnecessary'.[72] It would appear that underemployment in some parts of rural Jiangsu has been transferred from agriculture to the TVE sector.

Table 5.5 shows one common method used by TVEs to absorb more workers from agriculture and to equalize earnings between farming and non-agricultural work. In some parts of Jiangsu, primarily in the more prosperous south, the most important agricultural tasks (e.g. ploughing, irrigation) are performed on behalf of all households in the village by a team of specialists who are placed in agricultural workshops (*chejian*) in the village enterprises and receive wages as enterprise workers.[73]

Another method is to integrate agricultural and sideline activities with factory activities. The Zhangjiagang No. 2 Chemical Fibre Factory, a township enterprise in Sigang, is a good illustration of this practice. In early 1988, Zhangjiagang No. 2 Chemical Fibre reported a total workforce of about 596 workers, of which 370 were directly involved in the production of chemical fibre, 86 were engaged in activities that were indirectly related to the production of chemical fibre (e.g. office workers, quality control workers, warehouse workers, etc.), and the remaining 140 workers were engaged in what can only be described as sideline activities: vegetable farming, animal husbandry, beancurd making, tailoring, etc.[74] Apparently the practice of integrating agricultural and sideline activities with factory activities was fairly common in south Jiangsu. It was a way of reducing the income differential between agriculture and non-agricultural activities, but it was also a way of reducing the TVE's income tax.

Township-village governments have used TVEs to absorb underemployed workers in agriculture. Initially, rural cadres distributed TVE jobs as favours to friends and as rewards. Subsequently, as more jobs became available, they were distributed more or less equally among households in the community. Only after the supply of local workers was exhausted, did TVEs hire workers from outside the community, first from neighbouring villages and townships and then from other counties and provinces. Agricultural workers were frequently placed in TVEs as a way of reducing the income differential between agriculture and non-agricultural activities. Such practices have transferred some of the rural underemployed in Jiangsu from agriculture to the TVE sector.

Characteristics of TVE Workers

Table 5.6 presents the age–education characteristics of workers employed by eleven TVEs in Jiangsu, and it suggests that it was the younger and better-educated rural workers who were absorbed by TVEs. In 1987, 71 per cent of the workers in the eleven TVEs we surveyed were below the age of 35 and 72 per cent had at least an elementary school education. This is not unexpected since in other developing countries it was also the young and better educated who first left farming for factory work. The young are less fixed in their ways and thus more likely to accept the discipline needed to work in factories. They are also better educated and therefore easier to train. Because the sample is so small, our findings cannot be used to represent Jiangsu as a whole, even though we suspect that throughout Jiangsu it has been the young and better-educated workers who have moved from farming to TVEs.

Unfortunately, an age breakdown for all workers employed by Jiangsu's TVEs is not available.[75] However, a breakdown by education is available. The Jiangsu Bureau of Rural Enterprises reported that, in 1986, 58 per cent of the workers employed by TVEs in the province had more than six years of education (44 per cent had between seven and nine years and 14 per cent had more than nine years of education). This is a smaller percentage than found in the eleven TVEs we surveyed but still substantially higher than the 38 per cent found in the provincial labour force as a whole. In fact the percentage of workers with more than six years of

TABLE 5.6. *Employees by Age and by Years of Education, Selected TVEs*

	% of total by age			
	Below 20	20–34	35–50	50+
Jiangsu labour force, 1982	17	44	27	12
11 TVEs in Jiangsu, 1987	17	54	24	5
Zhashang Textile	23	56	20	*
Zhangjiagang Hemp	26	40	22	12
Zhangjiagang Steel Cable	19	52	23	6
Zhangjiagang No. 2 Fibre	29	42	29	0
Zhangjiagang Radio Material	14	68	18	0
Zhanggong Quarry	9	74	16	2
Yixing Copper Material	6	55	34	5
Yixing Telecommunication	16	53	18	12
Haitou No. 3 Brick	15	59	19	7
Haitou Agricultural Implements	0	66	30	4
Yuhe Refrigeration	3	57	40	0

TABLE 5.6.   *Continued*

	% of total by years of education[a]			
	0–6	7–9	10–12	12+
Jiangsu labour force, 1982				
Total	62	27	11	1
Agricultural labour force	74	21	5	*
Manufacturing labour force	41	39	19	1
All TVEs in Jiangsu, 1986	42	44	14	*
11 TVEs in Jiangsu, 1987	28	50	22	*
Zhashang Textile	27	65	8	0
Zhangjiagang Hemp	36	47	16	0
Zhangjiagang Steel Cable	34	45	20	1
Zhangjiagang No. 2 Fibre	10	31	59	0
Zhangjiagang Radio Material	8	80	12	0
Zhanggong Quarry	29	58	12	0
Yixing Copper Material	32	55	13	0
Yixing Telecommunication	21	55	24	1
Haitou No. 3 Brick	56	37	7	0
Haitou Agricultural Implements	42	54	2	2
Yuhe Refrigeration	5	67	29	0

* less than 1 per cent.

[a] Illiterates and those with elementary school education are placed in the 0–6 years category, those with junior middle school (*chuzhong*) education in the 7–9 category, those with senior middle school (*gaozhong*) and special or technical secondary school (*zhongzhuang*) education in the 10–12 category, and those with senior technical school (*dazhuan*) education in the 12+ category. However, it should be noted that during the Cultural Revolution decade China experimented with a 5-year elementary school system, and that in the countryside, because of a shortage of teachers, *gaozhong* is sometimes only 2 years.

Sources: The 1982 data are census figures and are from Du Wenzhen and Gu Jirui (eds.), *Zhongguo renkou (Jiangsu fence) (China's Population, Jiangsu)* (Beijing: Zhongguo caizheng jingji chubanshe, 1987), 237 and 250. The 1986 data for Jiangsu's TVEs are from *JXQTZ* 862–3. The other data are from Enterprise Background Data.

education is about the same in TVEs (58 per cent) as in all manufacturing enterprises (59 per cent). In terms of education, the main differences between TVE workers and urban manufacturing workers are (1) more urban workers have had higher education, and (2) the quality of education received by urban manufacturing workers was probably better since urban schools are significantly better than rural schools.

Wage Determination

One reason often cited to explain the competitiveness of TVEs is that labour costs (wages and benefits) are lower in TVEs than in state enterprises. Another is that workers in TVEs do not 'eat from one large pot (*chi daguofan*)', but are paid according to the dictum, 'more work, more pay'. This is in contrast to state enterprise workers who are paid more or less the same regardless of whether they work, or how hard they work.[76] In other words, the income of workers is more closely tied to their performance in TVEs than in state enterprises. Indeed, most of the TVEs we surveyed employed incentive schemes that linked wages to performance, but the link in some cases was tenuous at best. Although 'eating from one large pot' may be less of a problem among rural collectives than among state enterprises, the practice has not entirely disappeared.

In China, the annual wage is composed of three components: base wage (including piece wage and overtime), bonus (including 'floating wage'), and subsidies (benefits). As in hiring, TVGs in Jiangsu have considerable influence over wages.[77] The principal wage parameters (e.g. the average base wage (or the wage fund) and guidelines for bonuses) are set by the township industrial corporation at levels reflecting the economic conditions and income level of the region.[78] Using these wage parameters, TVEs then determine the wages of individual workers. One important reason why local governments play a direct role in wage determination is to prevent a widening of the gap between agricultural and non-agricultural earnings.

In those parts of rural Jiangsu where non-agricultural activities are particularly well developed, TVGs are under considerable pressure from county and provincial governments to prevent additional resources from leaving agriculture. This they do partly by keeping TVE wages from rising too rapidly and partly by using TVE earnings to subsidize agriculture.[79] Using income from industry to subsidize agricultural earnings is justified partly on the ground that, since TVEs are owned by the entire community, all community members should share in their earnings.[80]

The base wage is paid when workers fulfil the targets (usually output, quality, and raw material consumption) assigned to them individually or to the workshop as a whole. Targets are set at levels that are not difficult to achieve. The base wage of individual workers may vary by seniority and by the type of job performed. A slightly higher wage is paid for working on Sundays, but not all TVEs follow this practice. When performance can be precisely measured, a piece wage scheme is usually substituted for the base wage.[81] And, of course, the piece wage links reward directly to performance. Piece wage workers usually do not get a higher piece rate for working on Sundays.

There are two types of bonuses—the floating wage (sometimes in the form of a monthly bonus) and the year-end bonus. The size of the floating wage is tied to individual performance, and is usually awarded to workers who exceed their production quotas.[82] The incentive is usually symmetric in the sense that pre-determined amounts are also deducted from the base wage if the monthly output falls below the target. In many TVEs, workers are also given bonuses for economizing on material inputs (raw material, energy, etc.) and for good attendance.

Many TVEs also distribute an annual bonus at the end of the year if the annual profit target is exceeded.[83] The annual bonus is also symmetric in that a penalty is levied against the wage fund if the profit target is not achieved. The size of the annual bonus varies by categories of workers, but workers within each category usually receive the same bonus. Thus the annual bonus is like a profit-sharing labour contract that links workers' income to enterprise performance. Having a share contract means that TVE workers bear part of the risk of business fluctuations but it also provides them with more stable employment since it allows TVEs to lay off fewer workers during economic downtowns.[84] Following the practice of state enterprises, some TVEs have also introduced food and other subsidies, and of course these benefits are distributed equally to all workers. Because year-end bonuses and benefits are rewarded more or less equally, their use tends to equalize income within the community.

An example will help clarify the wage system employed by many TVEs in Jiangsu and how it has evolved over the years. The Zhangjiagang Steel Cable Factory, one of the leading township enterprises in south Jiangsu[85] (with over 2,000 workers, it is also one of the largest), has altered its monthly wage system four times since 1971: in 1975, 1979, 1983, and 1986. The characteristics of the original wage system and the subsequent changes are summarized in Table 5.7. Between 1971 and 1974, workers at Zhangjiagang Steel Cable received a base wage and an overtime wage. Since everyone worked seven days a week, the overtime wage was effectively a part of the base wage package. The political environment in the 1970s discouraged the use of material incentives so bonuses were not distributed, and in any case workers at that time received only a fraction of their monthly earnings as the bulk was transferred to the team for distribution by work points at the end of the year.

The first wage adjustment came in 1975, when the base wage was increased and a performance bonus was introduced. Subsequent changes in 1979, 1983, and 1986 greatly increased the importance of the performance bonus. A food subsidy was introduced in the early 1980s and a medical subsidy in 1986. The principal impacts of these changes were to reduce the importance of the base wage and increase that of the monthly performance bonus.[86] Together, base and overtime wages accounted for

TABLE 5.7. *Composition of Monthly Wage of Workers, Zhangjiagang Steel Cable Factory, Selected Years (RMB per month)*

Year	Total	Base wage[a]	Overtime	Average bonus[b]	Subsidies
Male					
1971	32.94	28.00	4.94	0.00	0.00
1975	52.35	36.00	6.35	10.00	0.00
1979	65.88	43.25	7.63	15.00	0.00
1983	91.86	54.71	9.65	20.00	7.50
1986	127.55	74.00	13.05	30.00	10.50
Female					
1971	25.88	22.00	3.88	0.00	0.00
1975	40.94	28.00	4.94	8.00	0.00
1979	52.35	36.00	6.35	10.00	0.00
1983	73.14	40.50	7.14	18.00	7.50
1986	109.48	58.75	10.23	30.00	10.50

Year	% of total				
	Total	Base wage	Overtime	Average bonus	Subsidies
Male					
1971	100.0	85.0	15.0	0.0	0.0
1975	100.0	68.8	12.1	19.1	0.0
1979	100.0	65.6	11.6	22.8	0.0
1983	100.0	59.6	10.5	21.8	8.2
1986	100.0	58.0	10.2	23.5	8.2
Female					
1971	100.0	85.0	15.0	0.0	0.0
1975	100.0	68.4	12.1	19.6	0.0
1979	100.0	68.8	12.1	19.1	0.0
1983	100.0	55.4	9.8	24.6	10.3
1986	100.0	53.7	9.3	27.4	9.6

[a] Between 1975 and 1982, workers in their first year received a lower base wage than that received by the other workers.
[b] Does not include the annual bonus.

Source: Zhangjiagang Steel Cable Factory, *Zhangjiagang shi gang sheng chang lishi* (*History of Zhangjiagang Steel Cable Factory*) (Dec. 1987).

100 per cent of the total monthly wage in 1971, 80 per cent in 1975, and between 65 per cent and 70 per cent since 1983. In the same period, the monthly bonus increased from 0 in 1971 to about 24 per cent of the monthly wage in 1986. In addition to the monthly bonus, workers also received a year-end bonus, which averaged around RMB 250 in the late

1980s. When the year-end bonus is included, the share of the base wage and overtime wage in the total annual wage was only about 45 per cent in 1987 (see Table 5.8).

The average annual wage and its composition for workers and for staff members for selected TVEs in the three surveyed townships are presented in Tables 5.8 and 5.9. Alan Gelb, in his study of wages in rural enterprises, found that 'the community serves as a powerful reference point for the acceptable level of pay in the firm'.[87] In other words, the average wage of workers in TVEs is closely linked with the general prosperity of the village or township. The data in Tables 5.8 and 5.9 are generally consistent with Gelb's conclusion. For example, workers in the five TVEs in Sigang, the most prosperous of the three townships, earned an average income (RMB 1,893) that was substantially higher than that in Haitou (RMB 723), the least prosperous of the three.[88]

Staff members earned higher wages than workers, usually higher by 30 to 80 per cent, and bonuses accounted for a larger share of their income. Because TVEs are totally market orientated, their success depends largely on the product development and marketing skills of their staff. Presumably, the higher income and the larger bonuses earned by staff members are reflections of their greater contribution to the enterprise. TVEs can also reward their middle management in ways not reflected in the wage data (nor available to SEs). For example, one TVE we visited rewarded its section chiefs by providing each with the use of a new car. Another rewarded two of its senior engineers (who had retired from a SE in Shanghai) by providing them with the use of apartments it had purchased in Shanghai.

Of the eleven TVEs, the five in Sigang gave a much higher share of their wages in bonuses and benefits than did the others.[89] On average, the base wage accounted for about 45 per cent of the annual income earned by workers in the five TVEs in Sigang but over 70 per cent in the TVEs (other than Yuhe) in Hufu and Haitou. Year-end bonuses and benefits may also be thought of as a dividend, paid to workers for being members of the community. Since more or less the same amount is given to all workers, these payments help to equalize income in the community. Sigang, being a relatively developed and prosperous township, could pay a higher community dividend than the other two less prosperous townships.

It should be noted that in 1987 the Haitou No. 3 Brick Factory and the Haitou Agricultural Implement Factory still employed the wage system that was abandoned by most TVEs in south Jiangsu in the mid-1980s. Little or no bonus payments were distributed, and, more importantly, the base wage and the bonus were distributed more or less equally to all workers. For example, Haitou Agricultural Implement employed 50

TABLE 5.8. *Average Wage (RMB per year) and Composition of Wage of Workers[a] in 11 Jiangsu TVEs, 1987*

	Average total wage	Average base and piece wage	Composition (%) of total wage			
			Base and piece wage	Overtime	Bonus	Subsidies and others
Sigang	1,893	850	44.9	6.2	33.0	15.9
Zhashang Textile	1,413	1,081	76.5	0.0	23.5	0.0
Zhangjiagang Hemp	1,565	664	42.4	9.6	36.8	11.2
Zhangjiagang Cable	2,240	904	40.4	4.6	38.1	17.0
Zhangjiagang No. 2 Fibre	1,286	672	52.3	17.7	9.4	20.6
Zhangjiagang Radio Material	1,320	677	51.3	6.7	25.3	16.8
Hufu	1,314	1,027	78.2	14.4	6.4	1.1
Zhanggong Quarry	1,538	1,158	75.3	24.3	0.4	0.0
Yixing Copper Material	1,123	971	86.5	0.0	13.2	0.3
Yixing Telecommunication	1,005	703	70.0	1.6	19.3	9.2
Haitou	723	563	77.9	6.3	8.6	7.2
Haitou No. 3 Brick	486	444	91.2	0.0	8.8	0.0
Haitou Agricultural Implements	709	709	100.0	0.0	0.0	0.0
Yuhe Refrigeration	1,434[b]	720	50.2	16.7	13.9[b]	19.1

[a] All workers except administrative, technical, and sales staff.

[b] Because Yuhe was a Sino-Japanese joint equity venture, its wage system was governed by China's joint equity venture laws. Due to a dispute between Yuhe and the Haitou FTC, the enterprise was still uncertain at the time of our investigation of the size of its 1987 profits. Consequently, the enterprise distributed the same bonus (RMB 200) to all its employees with the promise that differential bonuses would be distributed once the profit picture becomes clear.

Source: Calculated from data in Enterprise Background Data.

TABLE 5.9. *Average Wage (RMB per year) and Composition of Wage of Staff Members[a] in 11 Jiangsu TVEs, 1987*

	Average total wage	Average base and piece wage	Composition (%) of total wage			
			Base and piece wage	Overtime	Bonus	Subsidies and others
Sigang						
Zhashang Textile	2,575	945	36.7	6.6	44.4	12.4
Zhangjiagang Hemp	2,457	1,232	50.2	0.0	44.0	5.9
Zhangjiagang Cable	2,287	1,061	46.4	7.0	44.8	1.7
Zhangjiagang No. 2 Fibre	2,960	991	33.5	6.3	47.7	12.5
Zhangjiagang Radio Material	1,584	678	42.8	12.6	26.8	17.8
	1,559	736	47.2	6.9	26.7	19.2
Hufu						
Zhanggong Quarry	1,556	1,060	68.2	0.0	12.4	19.5
Yixing Copper Material	2,900	2,700	93.1	0.0	6.9	0.0
Yixing Telecommunication	1,983	1,108	55.9	0.0	8.1	36.0
	1,085	782	72.1	0.0	19.4	8.5
Haitou						
Haitou No. 3 Brick	1,575	1,062	67.4	10.1	9.5	13.0
Haitou Agricultural Implements	1,156	967	83.7	0.0	16.4	0.0
Yuhe Refrigeration	1,140	1,140	100.0	0.0	0.0	0.0
	2,003[b]	1,076	53.7	15.8	10.0[b]	20.5

[a] Administrative, technical, and sales staff.
[b] See n. b, Table 5.8.

Source: Calculated from data in Enterprise Background Data.

production workers (excluding apprentices) in 1987, and of these 28 were paid RMB 750 and 22 RMB 687.[90] The wage system at No. 3 Brick was even more equal: all 126 of its production workers received the same total income (wage + bonus) of RMB 486 in 1987. Thus, apparently, 'eating from one large pot' is still very much the practice in some TVEs, mainly those in the less developed regions in north Jiangsu.

To examine income differentials within TVEs, the employees in eleven TVEs are grouped according to their total payroll earnings in 1987. The results are presented in Table 5.10. Zhanggong Quarry, which relied totally on a piece rate system and paid the second-highest average wage among the eleven TVEs surveyed, had by far the widest dispersion of income levels. With this exception, incomes of employees of the other TVEs in Hufu and Haitou were concentrated in a relatively narrow income range. For example, 96 per cent of the employees of Yixi Telecommunication Materials earned incomes in the range of RMB 800–1,200, and about 96 per cent of the employees of Yixing Copper Material earned income either in the range of RMB 800–1,000 (39.0 per cent) or RMB 1,200–1,600 (56.5 per cent). Income is more dispersed in the five TVEs in Sigang, particularly in the largest TVE in the sample (Zhangjiagang Steel Cable). Still, in two enterprises (Zhashang Textile and Zhangjiagang No. 2 Chemical Fibre), over 50 per cent of their employees earned a total income between RMB 1,200 and RMB 1,600. Income dispersion within a region, however, may be considerably wider, since the fortunes of individual enterprises and of smaller communities often vary significantly within a region.

Our survey indicates that wages in TVEs are not always linked to work or performance, and that intra-enterprise income differentials are surprisingly narrow. Discussions with rural cadres suggest that a larger survey of TVEs is likely to yield similar results. A widely held belief among Chinese familiar with state enterprises and TVEs is that workers in TVEs generally work longer hours and harder than do workers in state enterprises. The question is why, when wages are not strongly linked to performance and effort? Perhaps the answer is that workers in TVEs do not have job security (or what is called in China the 'iron rice bowl') to the same degree as workers in SEs. Most are contract workers, hired for the year or the season, and the contracts of unsatisfactory workers are simply not renewed. Furthermore, workers also know that their jobs will not be protected for long if their enterprise is unsuccessful. This knowledge undoubtedly also helps to maintain work discipline in TVEs.

Labour Mobility

In the 1960s and 1970s, local governments in Jiangsu kept rural workers from leaving the land by imposing fines on those (mostly craftsmen) who

TABLE 5.10. *Workers and Staff Members Distributed by Total Earnings in 1987, as a Percentage of All Employees*

	RMB 0–400	RMB 400–800	RMB 800–1,000	RMB 1,000–1,200	RMB 1,200–1,600	RMB 1,600–2,000	RMB 2,000–3,000	RMB 3,000+
Sigang	1.1	2.1	4.0	11.4	24.8	25.9	29.2	1.5
Zhashang Textile	0.0	9.2	12.2	5.4	59.5	4.5	8.6	0.6
Zhangjiagang Hemp	4.3	9.7	10.9	5.1	16.0	45.9	8.2	0.0
Zhangjiagang Steel Cable	0.0	0.0	1.8	4.5	12.2	33.1	46.0	2.4
Zhangjiagang No. 2 Fibre	0.0	0.7	1.9	29.8	54.6	11.3	1.7	0.0
Zhangjiagang Radio Material	9.8	4.7	8.4	37.8	21.1	10.9	6.9	0.4
Hufu	7.1	4.9	27.6	13.4	28.4	6.2	9.2	3.3
Zhanggong Quarry	15.9	11.1	8.1	10.4	17.9	13.1	19.8	3.7
Yixing Copper Material	0.0	0.0	39.0	0.0	56.5	0.0	0.0	4.5
Yixing Telecommunication	0.0	0.0	50.8	45.4	0.0	1.7	2.1	0.0
Haitou	0.0	0.0	1.2	3.5	16.9	3.9	4.2	0.0
Haitou No. 3 Brick	0.0	70.4	0.7	0.0	5.9	0.0	0.0	0.0
Haitou Agricultural Implements	0.0	93.3	3.1	9.2	0.0	0.0	0.0	0.0
Yuhe Refrigeration	0.0	87.7	0.0	5.0	60.0	16.7	18.3	0.0

Source: Calculated from data in Enterprise Background Data.

worked outside their villages.[91] Despite these penalties, rural workers continued to ply their trade outside their villages because of the higher income they were able to earn. Some communes also made arrangements for their members to work in nearby enterprises as temporary workers.[92] However, the wages earned were credited to the production teams and workers received only work points. These examples notwithstanding, labour mobility in rural Jiangsu was severely restricted before 1978.

Conditions changed gradually after 1978. The adoption of the household responsibility system released many workers from the land to seek non-agricultural employment, and of course many were absorbed by local TVEs. In those parts of Jiangsu where rural non-agricultural growth was rapid, labour shortages soon developed, and TVEs in these regions began to look to the less developed regions for workers. With fewer restrictions on travel and on outside employment, workers from the less developed regions also travelled to other regions in pursuit of employment. Table 5.11, which presents the composition of workers in eleven TVEs by their

TABLE 5.11. *Permanent Residence of Employees in Sample TVEs in Jiangsu, 1987*

	% of total employees from			
	Home township (village)	Other townships (villages) in home county	Other counties in Jiangsu	Other provinces
Sigang				
Zhashang Textile[a]	66	18	13	3
Zhangjiagang Hemp[a]	68	11	14	8
Zhangjiagang Cable	90	3	5	2
Zhangjiagang No. 2 Fibre	98	2	0	0
Zhangjiagang Radio Material	99	1	0	0
Hufu				
Zhanggong Quarry[a]	73	11	5	11
Yixing Copper Material	100	0	0	0
Yixing Telecommunication	100	0	0	0
Haitou				
Haitou No. 3 Brick[a]	37	63	0	0
Haitou Agricultural Implements	96	4	0	0
Yuhe Refrigeration	90	8	2	0

[a] Village enterprise.

Source: Enterprise Background Data.

place of permanent residence, suggests that rural labour has become considerably more mobile in rural Jiangsu since 1978.

Outside workers are those who are not members of the community that owns the TVE (i.e. for township enterprises, they are from another township; for village enterprises, another village). For convenience, we shall call the labour flow between different townships (villages) in the same county intra-regional flow, and that between different counties and different provinces inter-regional flow. While the bulk of the labour employed by TVEs in Jiangsu is local, i.e. from within the community, Table 5.11 indicates that intra-regional labour flow is quite common and that inter-regional flow is becoming important.

Because villages are much smaller than townships in terms of area and population, many of the larger village enterprises rely heavily on workers from neighbouring villages within easy commuting distance. This of course shows up in Table 5.11 as intra-regional flow, and explains why intra-regional labour flow is so much more important for village enterprises than for township enterprises. Not surprisingly, inter-regional labour flow is more prominent in Sigang and Hufu, where rural non-agricultural development started earlier than in Haitou, and where the labour-market is also tighter. What is somewhat surprising is that inter-regional flow has not been greater, considering the substantial regional wage differentials that have existed in Jiangsu.

The personal characteristics of fifteen outside workers in one village enterprise (Zhanggong Quarry) in Hufu Township are presented in Table 5.12. Most of the outside workers were young (the average age was 36), had attended elementary school (only one was illiterate), and were married. Except for one, the family of all the others had remained at home,[93] and all were from households whose principal source of income and employment at home was farming and animal husbandry. They worked as common labour both at home and at the quarry. The usual routine is for migrant workers to work in the quarry for eleven months (from mid-January to mid-December) and to spend one month at home. All had found their jobs at the quarry by themselves, and most had worked in the quarry for more than one year. The higher income at Zhanggong Quarry was clearly what attracted these workers to Hufu. Like other TVEs, Zhanggong Quarry paid the same wage to local and outside workers, and on average outside workers were able to earn nearly three times more than they did at home. A substantial share of the earnings of the outside workers was remitted home, and such remittances accounted on average for nearly one-half of their total household income. Because these findings are based on a small sample of workers from only one enterprise, they should not be used to draw conclusions about outside workers in general. Nevertheless the findings are suggestive and

TABLE 5.12. *Personal Characteristics of 15 Migrant Workers, Zhanggong Quarry, Hufu, Jiangsu, 1986 Year-end*

Code	Age	Family size	Number of family labour	Work at home	Monthly income at home (RMB)	Monthly income at quarry (RMB)	1986 family income (RMB)	Migrant worker's remittances as % of family income
1	25	3	2	Digging clay	100	200	3,000	0.66
2	26	5	5	Brick kiln	100	200	5,000	0.26
3	33	5	4	Temporary worker	100	250	4,000	0.63
4	38	5	4	Digging clay	100	250	4,000	0.50
5	36	4	2	Making bricks	100	150	2,500	0.60
6	33	3	2	Brick kiln	100	250	4,000	0.50
7	41	3	2	Digging clay	90	300	4,000	0.50
8	32	6	6	Farming	30	200	5,000	0.32
9	39	5	2	Farming	30	180	2,000	0.65
10	32	3	2	Village quarry	50	120	2,000	0.45
11	37	3	2	Village quarry	50	170	2,600	0.38
12	38	4	2	Village quarry	30	250	4,000	0.50
13	34	3	2	Village quarry	50	170	2,600	0.38
14	50	6	6	Village enterprise	50	150	5,000	0.20
15	45	6	5	Village enterprise	50	150	7,000	0.71[a]
Average	36	4.3	3.2		69	199	3,780	0.48

[a] Five members of this family worked at Zhanggong Quarry, and together they remitted home a total of RMB 5,000.

Source: Survey of Migrant Workers, Zhanggong Quarry. The survey was conducted in July 1987.

consistent with what we know about migrant workers in other less developed countries.

Outside workers were used not only to fill TVE jobs unwanted by local
workers, but also to do work that had previously been done by local
workers now employed by TVEs. To gauge their total impact on the
community, we collected information on outside workers (defined as a
worker who works in a location other than his or her place of permanent
residence) in three townships. The principal findings are summarized in
Table 5.13, and they suggest some interesting regional differences, particularly that (1) inter-regional flow is much more important, and (2)
seasonal variations (as measured by the coefficient of variation) much less
prominent in the two more prosperous townships in the south than in
Haitou. For example, Haitou had no inter-regional flow, and its coefficient
of variation of the number of outside workers by month was almost three
times that of Hufu. These regional differences are elaborated briefly
below.

The three townships in Table 5.13 were among the most developed in
their respective regions, and therefore all attracted outside workers. To
put the size of the labour influx in perspective, the number of local
workers in the three regions in 1987 was 17,998 in Sigang, 13,012 in Hufu,
and 15,609 in Haitou. In other words, outside workers represented between 4 and 8 per cent of the local labour force.[94] There was also some
labour outflow. For example, about as many workers left Haitou to work
outside as there were outside workers in Haitou. In 1986, about 700
Haitou villagers worked outside of the township; roughly 400 worked as
urban construction workers for eight months each year and another 300
worked in Shandong Province for about five months each year as contractors of prawn ponds.[95]

Outside workers worked primarily as common labour (e.g. digging
clay, firing bricks and tiles, and as quarry and construction workers). The
share of outside workers who worked as common labour was 100 per cent
in Haitou, 61 per cent in Hufu, and 53 per cent in Sigang. In other words,
in all three regions, outside workers did primarily the heavy and the dirty
work that local workers were unwilling to do now that they had better
alternatives.

Haitou, the least developed of the three townships, showed the greatest
seasonal variations in its use of outside workers. The peak months for
farm activities in and around Haitou, which is in a two-crop agricultural
region, are June–July and September, and the slack period is the winter
months. The evidence suggests, therefore, that most of Haitou's outside
workers were farmers from the poorer neighbouring townships seeking
extra income during the slack season. Haitou had few TVEs, and most
were quite small, so there were few opportunities for year-round work in
Haitou for outside workers.

TABLE 5.13. *Outside Workers by Month and Origin in Three Jiangsu Townships: Sigang, Hufu, and Haitou, 1987*

Month	Sigang	Hufu	Haitou
January	n.a.	520	860
February	n.a.	618	900
March	n.a.	739	1,160
April	n.a.	1,231	1,000
May	n.a.	1,279	900
June	n.a.	1,273	120
July	1,076[a]	1,273	0
August	n.a.	1,273	300
September	n.a.	1,273	0
October	n.a.	1,087	250
November	n.a.	1,087	600
December	n.a.	1,087	950
Annual average	n.a.	1,062	587
Standard deviation	n.a.	266	409
Coefficient of variation	n.a.[b]	0.25[b]	0.70[b]
% of outside workers in April 1987 from	Sigang[a]	Hufu	Haitou
Other townships in home county	17	26	100
Other counties in Jiangsu	63	7	0
Other provinces	20	67[c]	0

[a] The data are for July 1986 and include only workers in TVEs. However, few outside workers were employed by team or private enterprises. The data were collected by Zhangjiagang Bureau of Agricultural Work.

[b] Monthly data for 6 villages, 2 in each of these 3 townships, were also collected, and the coefficients of variation for outside workers in these villages were calculated to be: Sigang: Jingxiang, 0.27; Zhashang, 0.16; Hufu: Yangquan, 0.52; Xiaojian, 0.16; Haitou: Haiqian, 0.80; Haiqi, 1.21.

[c] Most are from Anhui and Zhejiang, the provinces immediately adjacent to Hufu.

Sources: Calculated from data provided by the Zhangjiagang Bureau of Agricultural Work and from Township-Village Background Data.

Conditions were different in the two townships in south Jiangsu. The employment of outside workers in Sigang and Hufu showed considerably less monthly variation.[96] In other words, a large number of outside workers in these regions were employed year round, returning to their native villages only for the Spring Festival (lunar New Year). However, because a large number of the outside workers were tea-pickers in Hufu, the number of its outside workers did show more seasonal variation. The tea-picking season lasts about six months each year, usually from mid-April to mid-September, and this is clearly reflected in the larger number of outside workers employed between April and September. In those villages that had sizeable tea plantations, e.g. Yangquan, the seasonal variation was very pronounced (see Table 5.13, n. b). The number of outside workers in villages in Sigang, where TVEs were extremely well developed, showed the least monthly variation as nearly all of them were employed as contract workers on a year-round basis. But because outside workers cannot settle permanently in Sigang, the turnover is high. Of the 1,076 outside workers in Sigang in July 1986, 56.7 per cent were new workers who were in their first year of employment and 31.5 per cent were in their second year. Only 4.1 per cent had worked in Sigang for more than three consecutive years.

A common destination of migrant workers in Jiangsu is rural enterprises in *sunan* (the Yangzi Delta).[97] Apparently it is very easy for rural enterprises in *sunan* to attract workers from other regions. Once a rural enterprise in *sunan* announces that it is hiring outside workers, the news travels quickly through friends and relatives to north Jiangsu and neighbouring provinces (e.g. Anhui, Zhejiang, and Henan). Those interested travel to *sunan* and try out for the jobs. New workers start out as apprentices (usually for several months), and are promoted to contract workers once they have proven themselves. Formal recruiting is necessary only when an enterprise needs a large number of outside workers, particularly young female workers (e.g. as tea-pickers or textile workers). In such cases, the village or the township provides the enterprise with letters of introduction to local governments in surplus labour regions to assure them that workers will be looked after and to enlist their help in screening out potential trouble-makers. Recruited workers are sometimes provided with transportation from their native villages to the enterprise.

Given the sizeable wage differentials between counties, particularly between counties in the north and those in the south, why were inter-regional labour flows not greater?[98] One possible reason is that while temporary movement was possible in the 1980s, migration in the sense of settling permanently in a different village was still very difficult.[99] Furthermore, local governments in south Jiangsu were often reluctant to allow TVEs to hire more temporary outside workers. For example, in several villages we visited where labour shortages were quite severe, we were told that they do not plan to hire more outside workers but instead will expand production in the future by investing in more capital-intensive technology, i.e. substituting capital for labour.

Local governments in south Jiangsu offered various reasons for their reluctance to accept more outside workers, e.g. an already tight market for consumer goods, the increased social tension from the presence of large numbers of outsiders, and the shortage of land to accommodate more newcomers.[100] But the principal reason why local governments and TVEs have been reluctant to use more outside workers is their desire to ensure that 'fertilized water does not fall on fields cultivated by outsiders (*feishui bu luo wairen tian*)'.

It is important to remember that TVEs are more than just production units. Local governments use them to distribute collective benefits to community members, and most communities do not want to share these benefits with outsiders. In other words, community members and their local governments are reluctant to share with outsiders the community dividend that is distributed as part of TVE wages. The alternative would be to discriminate against outside workers and not pay them the community dividend, but this would appear too much like labour exploitation.

What is good for a local community may not, of course, necessarily be good for the national economy. The reluctance of local communities to use more outside workers and the restrictions on inter-regional and rural–urban labour mobility have produced a highly fragmented labour-market in Jiangsu and in China. Because of this, neither state enterprises nor TVEs face an unlimited supply of labour at subsistence wages even though labour exists in great abundance and many workers are under-utilized in the countryside. In an earlier chapter, it was noted that the real average wage in TVEs has increased rapidly in Jiangsu, despite the presence of surplus labour in the province. Unless labour becomes more mobile and the labour-market becomes less fragmented, the pressure for rural wages to rise will persist in many parts of rural Jiangsu, and, of course, rising wages will reduce industrial profits (and investments) and lead to the premature substitution of capital for labour.

Township-village enterprises in Jiangsu are market orientated and operate with little assistance from senior governments. To prosper and grow in a partially reformed command economy without state assistance, TVEs have had to be resourceful, flexible, and competitive. Because they are outside the state sector, TVEs are less encumbered by government bureaucracy and have a greater degree of operational autonomy than state enterprises, and TVEs have used their autonomy effectively both in marketing and in procuring scarce resources. The evidence suggests that TVEs (at least in Jiangsu) often employ more workers than they need, pay wages at levels closely linked to the community average, and have relatively equal intra-enterprise income distribution. It appears that township-village governments in Jiangsu have used jobs in TVEs to shift disguised unemployment from agriculture to industry and as a community dividend to help keep income distribution within the community relatively equal. Because each local government has only a small number of enterprises and because they are its main source of revenue, TVEs also face a relatively hard budget constraint when compared to state enterprises. In other words, TVEs cannot afford to be an 'iron rice bowl' for their workers nor can they remain unprofitable for long. It is precisely because they are market orientated and face a relatively hard budget constraint that TVEs have had to be competitive, self-reliant, and responsive to opportunities.

NOTES

1. *JSN* 136.
2. In 1980 the ratio of working capital to fixed assets was 1.23:1.
3. In 1986, out of a total 87,060 industrial TVEs, there were 568 jointly operated TVEs (*xiangcun lianying qiye*) that involved equity investment from other enterprises (see *JXQTZ* 808). In Sigang, one of three townships where in-depth fieldwork was conducted, eleven of its thirty-nine TVEs in 1986 had equity investment provided by enterprises from outside the township. Under these arrangements, profits are shared among the investors.
4. Almost all foreign investments have come from overseas Chinese investors. For example, of the RMB 500 million invested in TVEs in Quanzhou (Fujian) in 1987, about half came from overseas Chinese. See Shen Liren *et al.*, *Xiangzhen qiye yu guoying qiye bijiao yanjiu* (*Comparative Studies of Rural Enterprises and State Enterprises*) (Beijing: Zhongquo jingji chubanshe, 1991), 61.
5. The following discussion is based on evidence drawn primarily from Jiangsu, but the main sources of capital are more or less the same in other parts of China, e.g. see Chen Wenhong, 'Zhongguo xiangzhen gongye de wenti (The problems of rural industry in China)', *Guangjiaojing*, 160 (Jan. 1986), 68–71.
6. Shen Liren *et al.*, *Xiangzhen qiye yu guoying qiye bijiao yanjiu*, 33.
7. For example, in Changshu (Suzhou *shi*), surplus generated at the team level in 1977 was in excess of RMB 33 million, or more then RMB 1 million per production team. During 1977–8, relying heavily on its community surplus, Changshu established over 1,000 new TVEs (see ibid.).
8. That surplus labour is a potential source of capital is of course an old idea discussed by, among others, Ragnar Nurkse (see his *Problems of Capital Formation in Underdeveloped Countries* (Oxford: Basil Blackwell & Mott, Ltd., 1953)). For a more recent discussion of the economics of transforming agricultural surplus labour into industrial capital in the Chinese context, see American Rural Small-Scale Industry Delegation, *Rural Small-Scale Industry in the People's Republic of China* (Berkeley, Calif.: University of California Press, 1977), ch. 4.
9. Workers mobilized for capital projects were given work points by their production teams. Perhaps over 100 million people were mobilized each winter to construct water conservancy and land improvement projects during the 1970s. For example, Thomas Rawski estimates that rural workers contributed about 5 billion workdays of labour during the winter construction

campaign in 1975 (see his *Economic Growth and Employment in China* (Oxford: Oxford University Press, 1979), 115).

10. At the end of 1976, rural Jiangsu had absorbed over one-half million educated urban youth (see *Jiangsu sheng guomin jingji tongji ziliao* (*Jiangsu Provincial National Economic Statistical Material*) (1976), 217). One of the villages we surveyed used its rusticated youth to enlarge its tea plantation, and on the pretext of providing work for the rusticated youth it also requested and received permission to open a smeltery to extract metal from scraps. The profits from the smeltery were then used as working capital for the tea plantation. When we visited the village in the late 1980s, the tea plantation had developed into its most important and profitable enterprise.

11. The information in the rest of this paragraph is from Enterprise Interview Notes.

12. Workers were paid RMB 0.4 for each cubic metre of soil removed in IOUs, which were never fully settled although the township government did sell walking tractors at a discount to some of the IOU holders. Large-scale projects using surplus labour in the slack season to develop non-farm activities occurred in other parts of Jiangsu as well. For example, Hufu Township, one of the other townships we surveyed, mobilized workers from all 22 of its villages in 1979 to dig a 1 km. canal to enable more villages to open rock quarries. Each village was given a quota of the digging, and the entire project was completed in about a month.

13. e.g. TVEs in Yanqiao (Wuxi County) reinvested on average 63 per cent of their retained profits during 1978–86 (Shen Liren *et al.*, *Xiangzhen qiye yu guoying qiye bijiao yanjiu*, 51).

14. Of the TVEs we visited, most started with financial assistance from local governments or from the more prosperous TVEs in the community, and a few received assistance from both.

15. Township-Village Interview Notes.

16. We were told that while the Agricultural Bank did not approve of this experiment, it had not openly opposed it.

17. Some TVEs have also experimented with issuing shares to their workers or to residents in the community. This experiment has more to do with separating TVEs from the control of TVGs to stop the predatory fiscal practices of some local governments. As yet, few TVEs in Jiangsu have issued shares. For examples of this experiment, see Song Dahai, 'Xiangzhen qiye shixing zichan gufenzhi de diaocha (An investigation of the implementation of the capital stock-share system in rural enterprises)', *NJW* 1 (1987), 35–8, and Liu Huazhen, 'Tantan xiangcun jiti qiye de gufenzhi wenti (A discussion of the stock-share system for rural collective enterprises)', *NJW* 1 (1987), 29–31.

18. Of the eleven industrial TVEs we surveyed, five accumulated some of their capital from individuals in this fashion between 1975 and 1987.

19. Enterprise Interview Notes.

20. For all practical purposes, the credit co-operatives are branch offices of the Agricultural Bank. One senior county cadre we interviewed remarked that 'there is no credit co-operative in the countryside because the existing ones

belong to the state'. Besides the Agricultural Bank, the Provincial Bureau of Rural Enterprises through its county branches also makes loans (interest free and usually repayable in 3–5 years) to TVEs. The programme began in the late 1970s when the state distributed grants to less developed localities to encourage TVE development. In 1981 the grants were replaced by interest-free loans (see Samuel Ho, *The Asian Experience in Rural Nonagricultural Development and its Relevance for China* (Washington, DC: World Bank, 1986), 91–2). These loans are targeted for specific industries and selected geographic (usually the poorer) areas. As loans are repaid, the funds are recycled to new projects. Over the years, the state has added fresh funds so the revolving fund has gradually become larger. Between 1979 and 1980, the Jiangsu provincial government established a RMB 50 million fund to assist TVE development in its less developed regions (primarily in the north), from which interest-free loans were made to TVEs (Mo Yuanren (ed.), *Jiangsu xiangzhen gongye fazhan shi* (*History of the Development of Rural Industry in Jiangsu*) (Nanjing: Nanjing gongxueyuan chubanshe, 1987), 154 and 175). For more on rural finance in China, see World Bank, *China, Rural Finance: A Sector Study* (Washington, DC: World Bank, 1982) and On-Kit Tam, 'Rural Finances in China', *CQ* 113 (Mar. 1988), 60–76.

21. The difference between loans and deposits in a region in the base year determines the level of lending in that region in the next year. Thus, if deposits rise by a certain amount in a year, outstanding loans may rise by the same amount in the next year. Subsequent modifications to credit planning (e.g. differential re-deposit rates) have weakened the link some-what, but it is still the case that the amount of outstanding loans in a region is closely tied to the level of deposits in that region. This description of rural credit planning is based on the discussion in William A. Byrd, 'Entrepreneurship, Capital, and Ownership', in William A. Byrd and Lin Qingsong (eds.), *China's Rural Industry: Structure, Development and Reform* (Oxford: Oxford University Press, 1990), 200–1.

22. See CCPCC and SC, 'Guanyu kaichuang shedui qiye xin jumian de baogao (Report concerning the initiation of a new phase in commune-brigade enterprises)', No. 4 Document (1 Mar. 1984), in *XQZFX* 111–28.

23. Shen Liren *et al.*, *Xiangzhen qiye yu guoying qiye bijiao yanjiu*, 53–4.

24. Calculated from data in *JSN* 136.

25. The interest rates paid by TVEs for bank loans are higher than those paid by state enterprises but are still lower than the shadow interest rate. In 1988, TVEs paid a monthly rate of 9.6 per cent for equipment loans and 7.2 per cent for working capital loans (comparable rates for state enterprises were 2.4 per cent and 6.6 per cent). These interest rates were, of course, much lower than the 12 to 15 per cent charged by the Ganyu County government.

26. In 1986, to gain greater control over TVE investments, the government announced that investment projects have to be examined not only by local branches of the Agricultural Bank and the Bureau of Rural Enterprises but also by higher levels of government. In Jiangsu, investments in TVEs below RMB 1 million require the approval of the county, those between RMB 1 million and RMB 5 million require the approval of the region (*shi*), those

between RMB 5 million and RMB 30 million require the approval of the province, and those above RMB 30 million require the approval of the State Planning Commission (*JJN* (1987), 1. 115).

27. One was established in 1981, one in 1984, and two in 1985.

28. Pre-1985 data were also collected, but these enterprises were so closely linked with other TVEs in their community that their finances in earlier years could not be disentangled.

29. Competition has made enterprises increasingly reluctant to share information. To gain access to basic technical information about equipment, specifications, etc., several villages we surveyed sent representatives to visit enterprises in other regions and made the arrangement by pretending to be potential buyers.

30. Shen Liren *et al.*, *Xiangzhen qiye yu guoying qiye bijiao yanjiu*, 133.

31. e.g. in 1987 Haitou Township signed an agreement with the No. 2 Automobile Factory under which it would supply No. 2 Auto with 1,000 tons of fish and prawns a year at 10 per cent discount, and in return No. 2 Auto would sell used machinery to Haitou on credit.

32. e.g. from 1978 to mid-1982, TVEs in one Pearl River Delta county (Dongguan) acquired over 35,000 sets of machinery and equipment through various processing arrangements with Hong Kong firms. See *Kai chuang xiangzhen qiye xin jumian* (*Start A New Phase For Rural Enterprises*) (Beijing: Zhongguo shedui qiye baoshe and Nong mu yu ye bu shedui qiye guanliju, 1984), 109.

33. e.g. Jiangsu's TVEs imported about US$41 million worth of equipment in 1984 (Mo Yuanren (ed.), *Jiangsu xiangzhen gongye fazhan shi*, 271).

34. Included in this sum is US$103 million in foreign exchange. Advanced is not defined. See *JJN* (1990), 6. 36.

35. Apparently this type of training is sometimes provided without charge (for example, when the enterprise manager is from the village or has a particularly close relationship with someone in the village) so the only cost to the TVE is the subsistence cost of its workers while away from the village.

36. Many of the TVEs we surveyed, at one time or another, had used high-priced old men to improve their operation. For example, in 1978, the Haitou Agricultural Implements Factory hired a craftsman from Shandong for RMB 200 per month (at a time when the annual per capita national income (*guomin shouru*) in Jiangsu was only RMB 282) to help it use an old furnace to cast moulds from scrap iron. In fact, many of the more than 200 engineers and technicians attracted to Ganyu (Haitou is one of thirty townships in Ganyu) in the mid- and late 1980s were these high-priced old men (Township-Village Interview Notes).

37. Shen Liren *et al.*, *Xiangzhen qiye yu guoying qiye bijiao yanjiu*, 103.

38. These and the other figures in this paragraph are from *JXQTZ* 862–3. The 1986 national survey of rural labour conditions found that 9.4 per cent of the 7,793 outsiders working in the 222 surveyed villages were engineers, technicians, and retired master craftsmen (see *QBLQDZ* 18).

39. An urban resident is one whose registered permanent residence (*hukou*) is in an urban area and is therefore eligible to buy grain at the official price.

40. Among the specific goals the SSTC adopted for implementation under the spark plan during the 7th Five Year Plan period (1986–90) were to develop 100 categories of complete sets of equipment suitable for use by rural enterprises, to organize the large-scale production of this equipment, and to supply 500 model rural enterprises with new technology and new management and quality control systems. See CCPCC and SC, 'Guanyu 1986 nian nongcun gongzuo de bushu (Concerning the deployment of rural work in 1986)', Document No. 1 (1 Jan. 1986), in *XQZFX* 164. Among Jiangsu's goals for the 7th Five Year Plan period was the development of 10–20 sets of affordable equipment suitable for rural areas (see *JJN* (1987), 10. 1).

41. *Jiangsu Nianjian* (*Jiangsu Almanac*) (Nanjing: Nanjing daxue chubanshe, 1991), 309.

42. Township-Village Interview Notes. Of the 26, 11 involved equity capital investment, 3 were compensation trade arrangements, 8 involved processing, and 4 were producing under licensing arrangements.

43. *BR* (17–23 Aug. 1992), 9.

44. Zhangjiagang in south Jiangsu was the first county in China to establish its own vocational college (it graduated its first class in 1987). In 1986, the college had five departments (electronics, textiles, construction, machinery, and enterprise management) and a total enrolment of about 300 students (Township-Village Interview Notes). Since the early 1990s, TVEs have been able to recruit some of their engineers directly from universities. When this occurs, the TVE compensates the university for the cost of training, an amount that is subject to negotiation.

45. Shen Liren *et al.*, *Xiangzhen qiye yu guoying qiye bijiao yanjiu*, 139. In Shanghai, where TVEs were more integrated with the state sector, about one-third of the gross output produced by TVEs in 1985 was directly or indirectly (e.g. goods produced on behalf of state enterprises with purchase quotas) for the state plan. See Xia Zifen (ed.), *Shanghai xiangzhen qiye jingji keji fazhan zhanlue he zhengce wenti yanjiu* (*Research on Problems Concerning the Strategy and Policy for the Economic and Technological Development of Rural Enterprises in Shanghai*) (Shanghai: Shanghai shehui kexue yuan chubanshe, 1988), 183.

46. Shen Liren *et al.*, *Xiangzhen qiye yu guoying qiye bijiao yanjiu*, 120. Another arrangement is for TVEs and commercial units to pool their resources and construct new retail outlets.

47. Yuhe is a Sino-Japanese joint equity venture formed in 1985 with 25 per cent of the investment from the county government, 50 per cent from the township, and 25 per cent from the Japanese. The joint venture agreement called for the Japanese partner to take full responsibility for the export marketing of the frozen prawns, and the original expectation was that most of the prawns would be exported. However, as of 1988, the Japanese partner allegedly had not fulfilled his side of the bargain.

48. ZSC employed 2,000 workers in 1987.

49. In 1987, the Ministry received RMB 500 in profits for every ton of steel it supplied to ZSC.

50. At first, ZSC received 1.08 tons of steel for every ton of wire and cable it produced for the Ministry. Periodically, the input norm was reduced, and, by 1987, it was 1.04 tons of steel for every ton of wire. The Ministry supplied steel to ZSC at the low state-allocation price but also bought the wires produced by ZSC at the low state-purchase price. Steel obtained as planned distribution accounted for 63 per cent of the steel acquired by ZSC in 1987.

51. However, protective practices by local governments can be very effective in some cases. One of the TVEs we surveyed lost nearly all its market because of such protective practices. For many years, Haitou Agricultural Implements produced street drains under a processing arrangement with Xuzhou. In 1983, as part of its reform, Jiangsu reorganized its administrative divisions, and in the shuffle Haitou was moved from Xuzhou's jurisdiction to Lianyungang's. Because it wanted to give its business to an enterprise in its jurisdiction, Xuzhou terminated its processing arrangement with Haitou Agricultural Implements, and in one stroke the enterprise lost four-fifths of its market.

52. Shen Liren *et al.*, *Xiangzhen qiye yu guoying qiye bijiao yanjiu*, 114.

53. Because at that time many considered TVEs unreliable, 'Yixing' was included in the name to give the impression that the enterprise was owned by Yixing County. The description in this and the following three paragraphs is based on Enterprise Interview Notes, Enterprise Background Data, and Enterprise Background Papers.

54. Shen Liren *et al.*, *Xiangzhen qiye yu guoying qiye bijiao yanjiu*, 119.

55. Ibid. 139.

56. e.g. in 1986 Wenzhou *shi* had 150 specialized markets that supplied inputs used by rural enterprises in the region.

57. Some have resisted the use of gifts to open doors or to close deals. For example, in one of the TVEs we surveyed, the younger cadres and some of the older cadres held very different views about the use of gifts to achieve commercial success. The younger cadres considered gifts a normal and necessary part of business, but, to some of the older cadres, who came up the ranks during the more idealistic 1950s and 1960s, the practice smacked of bribery and corruption. When the younger cadres in the enterprise wanted to make a particularly expensive gift (a colour TV) in order to seal a deal, the dispute between the two groups became so heated that the township had to step in. The disagreement over gifts was a major reason for the eventual break-up of the enterprise into several smaller enterprises.

58. Economic crimes and corruption apparently have become an indispensable part of doing business in China. See Chen Wan, 'Rural Reform in Southern Jiangsu Province', *China Spring Digest* (July/Aug. 1987), 15–16, and Jean-Louis Rocca, 'Corruption and Its Shadow: An Anthropological View of Corruption in China', *CQ* 130 (June 1992), 408–16.

59. Competition from local governments and from TVEs have made it difficult for government commercial departments to fulfil some of their annual procurement quotas, even though they have procurement priorities. For a discussion of this problem in one market, see Andrew Watson, Christopher Findley, and Du Yintang, 'Who Won the "Wool War"?: A Case Study of

Rural Product Marketing in China', *CQ* 118 (June 1989), 213–41.

60. Shen Liren *et al.*, *Xiangzhen qiye yu guoying qiye bijiao yanjiu*, 115.

61. In 1988, the price of wool was about 15 per cent lower in Inner Mongolia than in Jiangsu.

62. Mo Yuanren (ed.), *Jiangsu xiangzhen gongye fazhan shi*, 288.

63. Among the TVEs we surveyed, several had the capacity to generate power. For example, 4.2 per cent of the power Yixing Copper Material consumed in 1987 was self-generated.

64. One of the village enterprises we surveyed was sold to Sigang Township in 1982 for just this reason.

65. TVEs in Sigang contributed a total of RMB 8.0 million to the construction of Jianbi Power.

66. This and the following two paragraphs are based on Township-Village Interview Notes and Township-Village Background Papers.

67. For example, when Hufu township needed additional capital to purchase equipment for its copper material enterprise in 1979, it turned to its twenty-two brigades (villages) for financial contributions. Sixteen villages responded and jobs at the enterprise were allocated to these villages in proportion to the size of their contributions.

68. Between 1981 and 1985, in Jingxiang, a village of 371 households and one of the villages we surveyed, 135 of its members gained employment in village enterprises and 163 in township enterprises, and most did so by contributing capital.

69. e.g. Zhanggong Quarry, a village enterprise in Hufu township, divided its jobs among its production teams according to their population shares. Each team then determined which of its members would work at the quarry. Fei Xiaotong also found that many of the TVEs he visited in south Jiangsu recruited one worker from each peasant household in the village (see Fei Xiaotong, 'Xiao chengzhen zaitansuo (A further inquiry into small towns)', in Jiangsu sheng xiao chengzhen yanjiu ketizu (ed.), *Xiao chengzhen da wenti (Small Towns, Big Issues)* (Jiangsu renmin chubanshe, 1984), 51.

70. Sometimes, the dismissed worker, if he or she is a member of the community, is sent to work in a less desirable enterprise, but in most cases no provision is made for dismissed workers.

71. The reallocation involves only local workers since outside workers are employed from year to year on a contract basis.

72. Township-Village Interview Notes. In theory, TVEs have the right to lay off surplus workers, but in practice they do not. The political pressure to keep as many workers employed as possible is just too great.

73. How agriculture is organized is discussed in greater detail in Ch. 7.

74. Enterprise Interview Notes.

75. In a survey of 1,172 workers, selected on a stratified random basis, from forty-nine rural enterprises in four counties (one each from Jiangsu, Guangdong, Anhui, and Jiangxi provinces), only 14 per cent of the workers were over the age of 39. (See Alan Gelb, 'Workers' Incomes, Incentives and Attitudes', in Byrd and Lin Qingsong (eds.), *China's Rural Industry*, 284.)

76. An important component of urban reform has been wage reform. However, the reform has been only partially successful.

77. What follows describes wage determination in Jiangsu. Apparently, the market plays a much greater role in setting wages in TVEs in some parts of China, e.g. Guangdong (see Meng Xin, 'The Rural Labor Market', in Byrd and Lin Qingsong (eds.), *China's Rural Industry*, 299–322).

78. However, major changes in the wage system must be approved by the township economic commission.

79. Before the introduction of the household responsibility system (*da baogan*), wages of workers in township enterprises were transferred to production teams for distribution. For example, in the 1970s, workers at the Zhangjiagang Steel Cable Factory received only 20 per cent of their monthly earnings. The remaining 80 per cent were transferred to the production teams, and workers were credited with work points. At the end of the year, the production teams distributed income to both agricultural and factory workers according to the number of work points they had accumulated during the year. By using work points to distribute income, the team was effectively transferring a share of the workers' earnings to other team members. This method was gradually abandoned in Jiangsu.

80. Villagers feel strongly that they all have a right to share in the prosperity brought about by rural industrialization. This can be illustrated by an example from Zhashang Village. Some years ago, when a county enterprise was built on village land, forty-seven villagers from Zhashang were given employment by the enterprise as required by government policy. The Chinese call these workers 'land labour (*tudi gong*)'. Zhashang's land labour continued to live in the village, but their work unit (*danwei*) was changed from their production teams to the county enterprise, a switch that was much envied by other villagers. In 1978, Zhashang began to establish its own enterprises and the village quickly became very prosperous. (By the mid-1980s, every production team in the village operated an enterprise.) In 1983, the land labour requested that they be permitted to return to their production teams since they could earn more working in the team enterprises. The village turned down the request but decided that since they were still village members they were entitled to an annual subsidy of RMB 200. This arrangement lasted for a year or two, but in early 1985, when another team enterprise was about to be established, the land labour again raised their demand to return to their production teams. By 1985, Zhashang's enterprises paid wages that were RMB 30–40 higher per month than the wage paid by the county enterprise. This time the pressure on the village was so intense that it relented and permitted the land labour to return but only on the condition that they wait three months after returning to their teams before taking employment. In the end 32 of the 47 land labour returned to their teams. The 15 who did not were either near retirement or had been promoted to middle-level cadres.

81. e.g. Zhanggong Quarry paid its workers strictly by piece rate. Since its workers were used primarily to load and to transport stones, the output of its workers (tons of stones moved) could be easily measured.

82. Like the base wage, the floating wage (or penalty) is settled at the end of the year.

83. e.g. in the Zhangjiagang Hemp Textile Factory 30 per cent of the above-

target profit may be distributed to staff and workers as bonus (30 per cent for the staff and 70 per cent for the workers).

84. In many ways, the wage system employed by TVEs resembles the scheme proposed by Martin Weitzman in his book, *The Share Economy: Conquering Stagflation* (Cambridge, Mass.: Harvard University Press, 1986).

85. That Premier Zhao Ziyang visited the factory in 1984 suggests that Zhangjiagang Steel Cable was considered a model TVE.

86. Throughout the entire period, the base and overtime wages of female workers remained about 78 per cent of those of male workers. However, starting in 1986, female workers received the same average monthly bonuses as male workers.

87. Gelb, 'Workers' Incomes, Incentives and Attitudes', 294.

88. It should be noted that Yuhe Refrigeration, a Sino-Japanese joint equity venture, was governed by special regulations and its wage package was therefore substantially superior to those in other TVEs in Haitou.

89. So also did Yuhe Refrigeration. As was noted earlier, Yuhe is a Sino-Japanese joint equity venture and is therefore a special case.

90. The figures in this paragraph are from Enterprise Background Data.

91. Sometimes grain was also withheld from those who worked away from the village. This paragraph draws heavily on Township-Village Background Papers.

92. For example, before 1982, Xiaojian Village regularly sent its members to work in a nearby town as temporary workers. At the peak, about 70–80 workers (out of a labour force of about 500 at the time) were sent each year. With the development of enterprises in the village after 1978, the labour outflow declined.

93. The exception was a man of 45 who had a family of six. One member of the family was left at home to look after the responsibility land, while the other five came to Hufu to work at the quarry.

94. This was roughly the order of magnitude of outsiders working in the more developed counties in south Jiangsu in the late 1980s. For example, in mid-1987, 33,662 outsiders worked in Wuxi *xian*, one of the counties in Wuxi *shi*, accounting for 6.8 per cent of the county's rural labour force (*JJN* (1989), 8. 19). A 1986 survey of 222 administrative villages in eleven provinces found the number of outsiders working in the villages to be about 4 per cent of the total labour force (6.4 per cent in villages surveyed in Jiangsu), and about 45 per cent of the outsiders were seasonal workers (see *QBLQDZ* 15 and 18). Since 1992, the number of outside workers in parts of *sunan* has increased significantly.

95. Because Haitou was a pioneer in the development of prawn culture, many of its people were highly skilled in aqua-culture. In the Jiangsu villages included in the 1986 national labour survey, 946 residents left to work outside the villages (either for a season or the year) while 655 outsiders came in to work in the villages. Of course villages in less developed areas tend to experience more labour outflow than inflow. For example, the 1986 national labour survey found five times more outflows than inflows in the villages surveyed in Hebei. See *QBLQDZ* 77–9 and 148–50.

96. Although monthly data for Sigang are not available, the data for two of its villages show little monthly variation. See n. b, Table 5.13.
97. The 1986 national labour survey found that 86 per cent of the outsiders in the villages surveyed in Jiangsu were employed by industrial enterprises (*QBLQDZ* 79).
98. In some parts of China, inter-regional labour mobility was much greater. For example, in the late 1980s, about a million Hunanese were working in Guangdong, mainly in the Pearl River Delta (interview with the Hunan Provincial Commission of Economic Structural Reform, 30 Apr. 1990).
99. In the regions we visited, only a few migrant workers who had married local residents were permitted to settle permanently, i.e. move their permanent household registration (*hukou*) to the village.
100. The worry about land may be motivated by a concern that once migrant workers have spent time in the village and have contributed to its development, they may feel they have a claim on community property.

6

The Rural Non-agricultural Sector
Private Activities

IF the *sunan* (south Jiangsu) model represents the collective development of the rural non-agricultural sector, then the alternative is represented by the so-called Wenzhou model,[1] named after the region on the southern coast of Zhejiang Province where rural non-agricultural development has been fuelled primarily by the rapid growth of small private industries owned and operated either by individual households or groups of individuals.[2] In 1985, Wenzhou had over 133,000 private enterprises, and they produced 67.9 per cent of Wenzhou's RGVIO.[3] In total some 830,000 rural residents were involved in non-agricultural activities in Wenzhou.[4] In the Wenzhou model, local governments do not control private enterprises but do exert indirect influence on them through licensing and other regulations, and the energy behind rural non-agricultural development is private entrepreneurship operating in response to market forces, rather than rural cadres.[5]

Most, but not all, regions in Jiangsu have attempted to follow the *sunan* model. The collective development of rural industry is most pervasive in south Jiangsu where RGVIO is not only very substantial but most of it (95 per cent in 1986) is produced by TVEs.[6] However, in the north, not only is RGVIO significantly smaller but the share produced by TVEs is also less. In 1986, 20 per cent of the RGVIO produced in north Jiangsu was accounted for by industrial activities below the village level. Compared to the south, north Jiangsu is less developed, less industrialized, and less urbanized. Thus TVEs, which depend on lateral linkages with urban enterprises for information, technology, capital equipment, and skilled labour, find the environment in the north much less hospitable and conducive to development than in the south.

Repeated failures to establish viable and profitable TVEs have forced many regions in north Jiangsu to explore alternatives to the *sunan* model, and some of the most successful cases of rural non-agricultural development in north Jiangsu have been attained not through the expansion of TVEs but rather through the growth of the private sector. One example is Ganyu, a coastal county in north Jiangsu, where private activities have contributed in a major way to rural non-agricultural development. More than one-third of Ganyu's relatively robust RGVIO was produced in the private sector, and in Haitou, the most prosperous township in Ganyu and one of the three townships we surveyed, the private sector produced more than 60 per cent of the RGVIO (about the same share as in Wenzhou). Thus, even in Jiangsu, the home of the *sunan* model, the Wenzhou model has relevance.

This chapter examines the development of private non-agricultural activities in rural Jiangsu, particularly since the mid-1980s, and is divided into three parts. An overview is first presented. The second section then discusses in greater detail two important components of the private sector, the so-called 'specialized households' and the 'new economic associations'. Finally, private enterprises that use hired labour, particularly the small number of large private enterprises, i.e. those that use eight or more hired workers, are examined.

Overview

The initial stimulus for the development of private activities came in the late 1970s when the central government encouraged rural households to supplement their income by engaging in sideline activities. When the household responsibility system was introduced in the late 1970s and early 1980s and farm activities were contracted to individual households, some non-farm collective activities were also contracted out. Subsequently, these and other small-scale non-farm collective activities were privatized.[7] Local governments in Jiangsu at first did little to support the development of private activities, partly because most were initially committed to the *sunan* model and therefore concentrated their energy and resources on developing TVEs.[8] The 'three fears (*san pa*)'—the fear that policy might change, the fear of public criticism,[9] and the fear of risk—also worked to hold private initiatives in check.[10]

The pace of development quickened in 1984. After several years of trying to imitate south Jiangsu, local governments in the north concluded that, unlike their counterparts in the south, they lacked the resources (capital, technology, skilled labour, and adequate transportation and communication infrastructure) to develop TVEs on a large scale. This realization came at a time when the provincial government, as part of its effort to implement the rural policy of the central government, was urging the counties to pay greater attention to the development of team enterprises, specialized households, and various forms of private enterprises and to the development of non-agricultural activities other than industry.[11] Thereafter, Jiangsu turned to a more balanced approach to rural non-agricultural development, one that was more broadly based and depended less on collective rural industrial activities. Since the mid-1980s, some

TVEs have subcontracted work to private enterprises and others have provided private enterprises with raw materials and waste materials for processing.[12]

To govern the development and operation of private industrial and commercial activities in rural areas, the State Council promulgated numerous regulations in 1984.[13] Local governments were formally notified that private individuals may engage in those 'industrial, handicraft, commercial, food and beverage, service, repair, transportation, and house renovation undertakings that can be suitably handled by rural residents as well as other occupations permitted by the state'.[14] The right of private enterprises to hire workers was also reaffirmed. To develop skills lost in past decades when private activities had been discouraged, skilled workers who had retired to the countryside were permitted to engage in private businesses without fear of losing their retirement benefits as long as they passed on their special skills to apprentices or could restore or develop special or famous-brand products.[15] Rural residents were also permitted to set up stalls or operate services in market towns provided they supplied their own food rations. Goods that could be freely marketed included third-category agricultural and sideline products and third-category industrial products.[16] First- and second-category agricultural and sideline products could also be marketed once farmers had satisfied the state's unified and quota purchase plans. Peasants were also permitted to purchase and own, either individually or jointly, tractors, motor vehicles, and boats for the purpose of transporting either goods or passengers, and such activities would not be limited either with respect to distance or administrative boundaries. The promulgation of these regulations helped to ease the three fears and gave further encouragement to local governments to support private activities.

One consequence of official support for private activities is that since 1984 the state has collected and published statistics on private enterprises. Under its mandate to administer rural enterprises, the Bureau of Rural Enterprises (*xiangzhen qiye guanliju*) collects information on two loosely defined categories of private enterprises: joint household enterprises (*lianhu qiye*) and individual enterprises (*geti qiye*). Both may employ hired workers (as contract workers for a season or a year, or as temporary workers on a day-to-day basis) but are organized differently. A joint household enterprise has contributions (cash, labour, equipment, or know-how) from several rural households and is therefore similar to a partnership while an individual enterprise is owned and operated by a single individual or household. Some of these enterprises operate only seasonally or on a part-time basis, and most are informally organized. The official data probably do not capture all private activities, as data are collected only from private enterprises that are registered. But not all private activities are registered, and the regulations are unclear as to when a household sideline becomes sufficiently important to make registration necessary. Thus one suspects that some household sidelines and part-time activities are not recorded.

To complicate matters further, some TVEs are actually private enterprises in disguise.[17] Because of tax and other preferential treatments given to TVEs, private enterprises have many reasons to wear the 'red hat'—i.e. declare themselves collective. And, of course, once a private enterprise puts on the 'red hat' it is counted statistically as a collective enterprise. The quality of the official statistics is also questionable. We found numerous internal inconsistencies in the data. This is hardly surprising since most private enterprises are tiny and few of them keep accurate records. Despite these shortcomings, the official data collected by the Bureau of Rural Enterprises (see Table 6.1) provides the best and most comprehensive picture of the private sector.

Although underdeveloped by comparison to the collective sector, private rural non-agricultural activities are not insignificant in Jiangsu. Table 6.1 shows that, in 1986, there were over 640,000 private enterprises engaged in various non-agricultural activities in Jiangsu's villages, and

TABLE 6.1. *Number, Employment, Gross Income, and Average Size of Joint Household (*Lianhu*) Enterprises and of Individual (*Geti*) Enterprises by Industry, Rural Jiangsu, 1986*

	Number of enterprises (000)		Year-end employment (000)		Gross income (RMB million)		Average size (workers/ enterprise)	
	Lianhu	Geti	Lianhu	Geti	Lianhu	Geti	Lianhu	Geti
Total	45	595[a]	259	1,168[a]	994	3,671[a]	5.7	2.0
	% of total							
Industry	61.2	41.9	64.1	48.8	73.2	52.0	6.0	2.3
Construction	12.9	10.0	20.9	11.6	12.2	8.3	9.3	2.3
Transportation	13.8	19.1	6.5	14.6	7.6	17.4	2.7	1.5
Trade	5.4	14.6	3.4	11.7	3.7	12.9	3.6	1.6
Food and drink	1.6	3.1	1.4	3.3	1.1	3.3	4.9	2.1
Service	2.1	5.8	1.3	4.9	0.8	3.0	3.6	1.7
Other	3.0	5.6	2.3	5.1	1.4	3.2	4.4	1.8

[a] Individual enterprises with more than 7 workers numbered 5,766, employed 55,213 workers, and earned gross income of RMB 224.2 million.

Source: Calculated from data in *JXQTZ* 104–5. The original figures were adjusted to correct apparent arithmetical errors.

they employed 1.43 million part-time and full-time workers and generated over RMB 4.6 billion in gross income. By contrast, in the same year, rural collective enterprises employed 6.94 million workers and earned over RMB 42 billion in gross income.[18] Most private enterprises were individually owned and operated. For every joint household enterprise in rural Jiangsu, there were over thirteen individual enterprises. Within the private sector, individual enterprises accounted for 80 per cent of the workers and 78 per cent of the gross income.

The data in Table 6.1 show that most private enterprises in rural Jiangsu are very small. Even though, in 1986, joint household enterprises were on the average nearly three times the size of individual enterprises, the average still employed fewer than six (family and hired) workers. Less than 5 per cent of the workers employed by individual enterprises worked in enterprises with more than seven workers. That private enterprises are tiny is not surprising since the employment of workers by private individuals remains a sensitive political issue in China, where the accepted doctrine is that capitalists exploit workers and must therefore be placed under strict regulations.

Slightly more than half of those engaged by private non-agricultural enterprises in 1986 were in industry, another quarter in construction and transportation, and the rest scattered among the retail trade, food and drink, services, and other miscellaneous activities. In the three townships we surveyed, workers in private industry did mostly handicraft work, processing, and simple assembling using relatively primitive technology, and there is no reason to believe that they were not engaged in similar types of activities in other parts of Jiangsu. In other words, the small rural private enterprises in today's Jiangsu are not that different from those found in most developing countries or in pre-1949 China. Probably the single most important difference is that Jiangsu's private rural non-agricultural activities rely much more heavily on mechanical power today than they did in pre-1949 China, when power was almost entirely human or animal. For example, rural transport in Jiangsu today relies much more on walking tractors, small trucks, and motor barges than on human and animal power.[19]

The vast majority of the private non-agricultural activities in rural Jiangsu exist to serve local needs or produce non-traded goods, such as commerce, services, and service-type industries. Thus, not surprisingly, the distribution of private rural non-agricultural employment by region roughly parallels that of the rural population (see Table 6.2). In 1986, 48.7 per cent of Jiangsu's rural population resided in the south (17.4 per cent in SDR, 20.4 per cent in NDR, and 10.8 per cent is SMR) and 51.3 per cent in the north (17.6 per cent in NCR and 33.7 in NMR). In the same year, the distribution of privately employed rural non-agricultural

TABLE 6.2. *Distribution of Employment and Gross Income of Private Enterprises by Region, Rural Jiangsu, 1986 (% of provincial total)*

Region[b]	Employment		Gross income[a]	
	All enterprises	Industrial enterprises	All enterprises	Industrial enterprises
Geti and *lianhu* enterprises				
South	40.6	44.8	53.2	61.1
SDR	16.7	19.3	29.1	34.5
NDR	13.4	18.3	13.4	18.1
SMR	10.5	7.2	10.7	8.5
North	59.4	55.3	46.8	38.8
NCR	17.6	15.4	16.2	16.2
NMR	41.8	39.9	30.6	22.6
Geti enterprises				
South	41.4	43.2	53.9	61.2
SDR	16.2	17.1	28.5	32.8
NDR	13.8	18.9	14.0	19.7
SMR	11.4	7.2	11.4	8.7
North	58.6	56.7	46.1	38.8
NCR	18.3	16.5	17.3	18.0
NMR	40.3	40.2	28.8	20.8
Lianhu enterprises				
South	37.1	49.7	50.2	60.8
SDR	18.6	26.4	31.1	39.0
NDR	11.9	16.3	11.2	13.9
SMR	6.6	7.0	7.9	7.9
North	62.8	50.2	49.7	39.2
NCR	14.2	11.5	12.3	11.7
NMR	48.6	38.7	37.4	27.5

[a] No gross income was reported from Lianyungang. Some gross income figures used to construct this table were estimated either from gross output value figures, when they were available, or from employment figures.
[b] The regions are defined in Ch. 3.

Source: Calculated from data in *JXQTZ* 126–61.

workers was 40.6 per cent in the south (16.7 per cent in SDR, 13.4 per cent in NDR, and 10.5 per cent in SMR) and 59.4 per cent in the north (17.6 per cent in NCR and 41.8 per cent in NMR), and the distribution of privately employed rural industrial workers followed that of rural population even more closely.[20]

Obviously not all private non-agricultural activities in rural Jiangsu serve local needs. Indeed, the most successful and the most rapidly expanding private ventures are not those that primarily serve local needs but rather those that produce for the larger regional, national, or international markets. But in Jiangsu, with a few exceptions, only a small number of private enterprises in any given region produce for the larger outside market. One such exception is Haitou Township, where we were told that 'almost every rural household' produced either fishnets or semi-finished products for the fishnet industry, and most of the finished nets were sold to buyers from outside the region.[21] In this region the organization of the private sector is very similar to that found in Wenzhou:[22] i.e. rural households in each locality specialize in the production of one commodity (e.g. buttons,[23] plastic flowers, rubber elastics, hardware, electric components), with some households producing the semi-finished products and others specializing in the final finishing. The industry is held together by a handful of local entrepreneurs who supply part of the working capital and the raw materials. The region becomes known for the specialized product, and outside buyers come to the region to purchase the product. In other words, a specialized market develops in the region.

Table 6.2 suggests that privately employed rural non-agricultural workers in the southern region of SDR earned significantly more income than those in other parts of Jiangsu, particularly in the NMR. SDR, with 16.7 per cent of the private non-agricultural employment in rural Jiangsu, accounted for over 29 per cent of the gross income generated by private rural non-agricultural activities. In contrast, NMR had nearly 42 per cent of the employment but accounted for less than 31 per cent of the gross income. These large regional differences may be explained by differences in economic conditions. SDR is the most developed and prosperous region in Jiangsu and NMR is the least prosperous. Given SDR's higher per capita income, rural demand for goods and services is much stronger there and is therefore better able to support a higher level of private economic activity. The other major difference is that land is relatively more abundant in north Jiangsu than in the south, particularly in SDR. Thus, the demands of farm work are much lighter in SDR than in NMR so that the average rural worker can devote more days per year to non-agricultural activities, including self-employed activities, in SDR than in NMR. In other words, private rural non-agricultural activities are more likely to be part-time or seasonal occupations in NMR and full-time occupations in SDR.

Despite fairly rapid development since the mid-1980s, private rural non-agricultural activities in rural Jiangsu, though no longer insignificant in strength, are still relatively weak in comparison to collective activities. Most private activities employ few workers other than family members, and most exist to serve local needs. However, as in other parts of China, a small number of relatively large private enterprises have emerged in rural Jiangsu that produce for the larger regional, national, or international markets.

Specialized Households and Associations

To move the rural economy away from self-sufficient and semi-self-sufficient (*ziji banziji*) production to large-scale commodity production (*da guimo shangpin shengchan*), the government has encouraged rural households to become more commercialized and to integrate agricultural production more closely with industrial and commercial activities. One consequence of this policy has been the emergence of the so-called specialized households (*zhuanyehu*)[24] and new economic associations (*xin jingji lianheti*). Over time, they have become the most active and commercialized components of the private sector.[25]

Organization

In principle, a specialized household is one that (1) allocates at least 60 per cent of the labour time of its main workers to the specialized activity, (2) markets at least 80 per cent of the output from the specialized activity, (3) earns at least 60 per cent of its gross household income from the specialized activity, and (4) has per capita earnings from the specialized activity that are at least twice the local average or RMB 700.[26] But, in Jiangsu, as in other parts of China, many are classified as specialized households that do not meet all of these conditions. However, they are substantially more specialized than the average rural household. A household may specialize in any of a wide variety of activities, including cultivation, animal husbandry, fishing, handicrafts, industry, transportation, and various services. In 1986, about one-quarter of the registered individual enterprises in rural Jiangsu were specialized households engaged in non-agricultural activity.[27]

In the autumn of 1978, the Central Committee urged rural areas to break down the artificial boundary that had separated agricultural production from non-agricultural activities and encouraged rural households to co-operate and combine agricultural production with related industrial and commercial activities. The objective was to integrate agricultural production, processing, and marketing (including the transport of agricultural products to market), and also to integrate, more broadly, industries that supplied or serviced agriculture (e.g. feed and fertilizer), agriculture (broadly defined to include cultivation, animal husbandry, fishery, forestry, and sidelines), and agricultural processing and marketing (including transportation).[28]

In response, various co-operative ventures, called 'new economic associations (*xin jingji lianheti*)', hereafter 'associations', developed among households. These associations are jointly owned by their members, who also participate as workers, though many employ hired workers as well. Some arrangements developed almost immediately after the introduction of the household responsibility system, in part because some collective activities could not be readily divided into units small enough to be contracted out to individual households. Many who participated in associations were formerly specialized households who decided to pool resources and operate jointly.[29]

At its most basic, an association involves only the pooling of labour. But, more often, it involves the pooling of capital, technology, and raw material as well.[30] Participants are mostly rural households, and most of them also cultivate responsibility land, so in effect they have one foot in agriculture and one foot in the non-agricultural sector. In a small number of cases, collective and state enterprises are also members of associations.[31] Usually an association includes only households from the same production team, but many also involve households from different teams and even from different counties. Income is distributed among association members according to the amount of labour contributed, capital shares, or a combination of the two (sometimes skills, technology, labour, raw materials, and capital are all converted into shares).

To be included in the official statistics as an association, the co-operative arrangement must be a stable one, have a regular place of operation,[32] use an accounting system, have a well-defined income distribution system, and if the arrangement is seasonal it must operate for at least three months each year.[33] In 1986 about one-third of the registered joint household enterprises in Jiangsu were designated associations.[34] Although most associations are engaged in industrial and commercial activities, some are primarily involved in agricultural production.

The story of the fishing association headed by Chen Zhoulu illustrates how a typical association developed in Haitou Township in north Jiangsu.[35] Chen was a member of what was once a fishing production team in Haiqian Village. Because the team did not have enough boats to contract out to all its members when the household responsibility system was introduced to the village in the winter of 1983, Chen and three other households decided to form an association and go on their own. The four households raised RMB 4,000 among themselves and borrowed RMB 2,000 from friends and relatives, and used the money to build a boat suitable for coastal fishing. In its first year of operation (1984), the association earned more than RMB 18,000 in gross income which, after deducting operating expenses, was sufficient to repay the RMB 2,000 loan and provide an income of RMB 1,800 to each of the four participating households.

In 1985, four other households from the village that had jointly renovated an old boat joined the association, bringing total memberships to eight households.[36] The addition of the second boat substantially enhanced production efficiency, since it was easier to pull nets with two boats. The association expanded a second time in February 1988, when a third boat and two new households were added, bringing total membership to ten households and total value of fixed capital to RMB 65,000. The main reason for the second expansion was to provide employment to Chen's oldest son, who joined the association as a crew member of the third boat.

After deducting operating expenses, fees, and taxes, net income earned by the association was distributed to each household as follows: 60 per cent as wages (the captain and the engineer of each boat received wages that were twice that of crew members) and 40 per cent according to capital shares. In 1987, each of the eight households then in the association received (in wages and profits) RMB 5,700—ranging from a low of RMB 4,600 to a high of RMB 6,200—well above the average net income of RMB 3,918 earned by those households in Haitou Township that participated in associations.

Industrial Composition and Pace of Development

In 1984, the first year for which provincial data are available, rural Jiangsu reported over 289,000 specialized households and over 18,000 associations (Table 6.3). Of the specialized households, 37 per cent were in agriculture (20 per cent in farming and 17 per cent in non-farm activities), and the rest in non-agricultural activities, mostly in transportation (21 per cent), industry (16 per cent), and commerce (13 per cent).[37] A small number of associations were also in agriculture, but most were in industry (44.5 per cent) and transportation (25.5 per cent).[38] Associations generally employed substantially more workers than did specialized households.[39] On average, in 1984, each association produced about RMB 16,339 in gross output and each specialized household earned gross income of RMB 5,227 (both in 1980 prices). Specialized households and associations employed, in addition to family members, over 86,000 hired workers.[40]

Between 1984 and 1988, the number of specialized households declined by over 67,000, from 289,413 to 222,236. At the lowest point, in 1986, specialized households numbered 200,722. Most of the decline occurred among specialized households in agriculture, particularly those in cul-

TABLE 6.3. *Specialized Households and New Economic Associations, Distributed by Industry, 1984 and 1988, Rural Jiangsu*

	New economic associations				Specialized households		
	Number	Number employed	Gross output (RMB million current prices)	Average gross output (RMB 1980 prices)[a]	Number	Gross household income (RMB million)	Average gross household income (RMB 1980 prices)[a]
1984 total	18,265	117,992	299.4	16,339	289,413	1,660.5	5,227
	Distributed by industry (% of total)				Distributed by industry (% of total)		
Cultivation	3.7	3.3	2.1	7,822	20.0	15.5	3,567
Other agriculture[b]	8.5	7.4	6.1	9,889	16.9	15.5	4,239
Industry	44.5	56.5	60.1	22,756	15.7	22.9	8,305
Transportation	25.5	12.6	13.6	8,598	21.1	22.3	5,734
Construction	5.0	8.8	7.9	25,007	7.0	5.5	4,247
Commerce[c]	8.5	6.3	6.4	11,979	12.7	12.7	5,419
Other	4.3	5.1	3.8	14,218	6.6	5.5	4,519
1988 total	13,479	112,390	776.4	46,515	222,236	3,691.8	12,696
	Distributed by industry (% of total)				Distributed by industry (% of total)		
Cultivation	0.4	0.4	0.2	19,733	7.6	3.4	3,802
Other agriculture[b]	15.0	10.5	11.6	22,850	18.4	15.5	7,172
Industry	58.9	68.3	68.1	57,449	16.8	34.6	29,513
Transportation	11.5	5.7	6.8	27,185	31.8	23.5	9,714
Construction	5.7	9.6	6.6	52,126	5.9	5.1	11,354
Commerce[c]	6.4	3.4	4.6	33,264	15.0	13.8	12,159
Other	2.1	2.0	2.0	43,677	4.5	4.1	12,052

[a] Deflated by the implicit gross output deflator for each industry in Jiangsu.
[b] Includes forestry, animal husbandry, and fishery.
[c] Includes trade, food and drink, and services.

Sources: Calculated from data in *JNZXTZ* 6–7 and 454–5 and *JJN* (1989), 3. 34–5.

tivation. Nearly 58,000 rural households in Jiangsu specialized in cultivation in 1984 but less than 17,000 did so in 1988.[41] Rural households specializing in cultivation accounted for one-fifth of all specialized households in Jiangsu in 1984 but less than 8 per cent in 1988. A small number of associations were also involved in farming in 1984, but by 1988 they had all but disappeared.[42] Of the few associations that remained in cultivation in 1988, the average gross output (in 1980 prices) produced was RMB 19,733, about 2.5 times that in 1984. However, between 1984 and 1988, the real gross household income earned by the average specialized household in cultivation increased only slightly, from RMB 3,567 to RMB 3,802.

Probably the single most important reason for the decline in the number of specialized households and associations in cultivation was that, after 1984, farming became much less profitable (see Chapter 3).[43] With more and better economic opportunities emerging in the rapidly expanding rural non-agricultural sector, few rural households wanted to farm full-time when more could be earned by farming part-time and working part-time or full-time off the land. A contributing factor was that land consolidation in Jiangsu did not occur to the extent needed to support a large number of specialized farm households. Although farming was not very profitable and a decreasing number of rural households wanted to farm full-time, few were willing to give up their economic safety net by transferring all or most of their contract land to specialized agricultural households to cultivate.[44] At least in Jiangsu, the expectation that farming will be done increasingly by specialized households has not materialized.[45]

What is more surprising is that, between 1984 and 1988, the number of specialized households and of associations in non-agricultural activities also declined, the former by nearly 18,000 units and the latter by about 5,000.[46] The decline occurred in all industries but one. The number of specialized transport households increased from 61,000 in 1984 to nearly 71,000 in 1988. In the 1980s, a large number of rural households in Jiangsu invested in 'walking' tractors and used them to transport goods in the countryside, and apparently many of them were able to earn enough from this activity to be designated specialized transport households.[47] However, those specialized households and associations that remained apparently enjoyed a significantly higher level of economic activity, as measured by either real average gross output or real average gross household income (see Table 6.3).

While the number of specialized households and associations has declined, the number of hired workers employed by them has increased.[48] In other words, either more specialized households and associations were using hired labour or those using hired labour were employing more of them or, more likely, a combination of the two possibilities. This suggests that the decline in the number of specialized households and associations was part of a consolidation process that weeded out the less able from the large number of rural residents who went into business for themselves in the early and mid-1980s, nearly all without previous experience and most without sufficient capital and adequate market information.

The strong growth of TVEs in Jiangsu in the 1980s also contributed to the decline in the number of specialized non-agricultural households and non-agricultural associations. In China, rural households turn to private activities usually only as a last resort when better and safer economic opportunities are not available. The rapid expansion of TVEs in rural Jiangsu during 1984–7 created a large number of jobs in the collective sector and thus made it less necessary for rural households to turn to private non-agricultural activities to supplement their income.

Size, Capital Intensity, and Earnings

Tables 6.4 and 6.5 present more detailed information on Jiangsu's specialized households and associations. A comparison of the data in the two tables suggests that except for size (as measured by employment) specialized households and associations in the same industry were not very different in other ways. Associations generally employed several times more workers than did specialized households, partly because they employed more hired workers, but partly because, with several families involved in each association, more family workers were involved in production. Of the non-agricultural activities, brick-making, weaving and embroidering, and construction were organized in units with the largest average size (in terms of total employment), but even these were not very large. For example, the average brick-making association employed fewer than fifteen workers. However, because some specialized households and associations used the putting-out system to organize production, the total employment impact may be considerably larger than these figures suggest.[49]

The amount of capital invested in the average private rural enterprise in Jiangsu was very small. In 1985, the average rural association in Jiangsu had fixed capital of less than RMB 6,500 while the average specialized household had about RMB 2,300, indicating that both employed simple labour-intensive technology. If the capital structure of private enterprises in Jiangsu is similar to that found in other parts of China, then the average rural association in rural Jiangsu probably had about RMB 12,000 in total capital and the average specialized household RMB 4,200.[50] The main sources of capital were the accumulated savings of owners and loans from friends, relatives, and private moneylenders.[51] Banks provided relatively little capital to rural private enterprises.[52]

TABLE 6.4. *Average Size, Capital–Labour Ratio, and Per Capita Income of Specialized Households in Jiangsu by Industry, 1985*

	Average number of workers in specialized activity		Fixed capital[a] per worker (RMB)	Per capita		
				Gross household income (RMB)	Net household income[b] (RMB)	Disposable income[c] (RMB)
	Total	Hired				
Total	2.19	0.45	1,065	1,968	1,043	948
Cultivation	2.18	0.10	567	1,245	860	806
Other agriculture[d]	1.84	0.11	944	1,735	1,022	963
Industry	3.43	1.56	558	3,139	1,182	1,016
Feed	2.11	0.40	1,176	1,478	875	812
Food	2.34	0.36	732	2,011	907	803
Brick	5.43	3.72	486	3,905	1,209	1,006
Weaving and embroidery	2.75	0.74	280	2,518	1,048	962
Metalworking	3.18	1.37	910	4,235	1,555	1,281
Other	3.60	1.73	510	3,454	1,274	1,068
Transportation	2.46	0.86	2,632	1,835	1,095	986
Motor vehicle	1.75	0.29	7,612	3,943	2,070	1,732
Boat	1.71	0.10	2,955	1,965	1,155	1,045
'Walking' tractor	1.32	0.06	2,411	1,529	944	863
Construction	2.47	0.90	358	1,575	964	910
Trade	1.74	0.12	733	2,043	974	862
Food and drink	2.40	0.41	716	2,175	992	900
Services A[e]	2.19	0.46	713	1,650	1,012	945
Services B[f]	2.26	0.51	530	1,611	1,009	944
Other	2.02	0.35	495	1,612	947	886

[a] At original prices and probably unadjusted for depreciation.
[b] Gross household income minus production expenses (including wages paid to hired workers).
[c] Net household income minus taxes and payments to collectives.
[d] Includes forestry, animal husbandry, and fishery.
[e] Service activities related to the production of material products.
[f] Service activities unrelated to the production of material products.
Source: Calculated from data in *JNZXTZ* 2–5.

Associations generally employed more fixed capital per worker than did specialized households, but the difference was not so large as to suggest that they used different technologies. For example, in industry as a whole, the average fixed capital per worker was RMB 558 among specialized households and RMB 674 among associations. The most capital-intensive activity was transportation, where the fixed capital–labour ratio averaged RMB 2,632 for households specializing in this activity and RMB 2,440 for associations. The most common transport

TABLE 6.5. *Average Size, Capital–Labour Ratio, and Per Capita Income of New Economic Associations in Jiangsu by Industry, 1985*

	Workers per enterprise		Fixed capital[a] per worker (RMB)	Gross income per worker (RMB)	Wage per hired worker (RMB)	Net income[b] per partner member (RMB)	Disposable income[c] per partner member (RMB)
	Total	Hired					
Total	7.66	2.42	839	3,654	650	1,907	1,459
Cultivation	5.30	0.62	492	1,544	366	890	789
Other agriculture	5.63	0.51	843	3,525	734	1,870	1,672
Industry	9.51	3.66	674	3,740	625	1,916	1,377
Feed	4.12	0.43	1,338	1,788	667	958	832
Food	7.13	1.76	1,101	4,993	560	1,927	1,237
Brick	14.67	6.85	685	2,398	506	1,659	1,343
Weaving and embroidery	13.41	7.56	423	3,175	427	2,349	1,413
Metalworking	8.14	2.92	644	4,930	715	2,350	1,601
Other	9.33	3.46	625	3,918	692	1,902	1,374
Transportation	3.32	0.25	2,440	3,776	1,145	2,262	1,921
Construction	16.57	5.66	368	3,375	700	1,908	1,479
Trade	4.54	0.81	1,162	5,497	564	1,758	1,211
Food and drink	4.96	0.96	664	3,556	701	1,452	1,112
Services A[d]	7.03	2.31	733	3,217	1,190	1,458	1,107
Services B[e]	6.78	2.14	749	2,771	496	1,393	1,070
Other	7.32	1.80	750	3,203	669	1,535	1,217

[a] At original prices and probably unadjusted for depreciation.
[b] Gross income minus production expenses (including wages paid to hired workers).
[c] Net income minus taxes and retained earnings.
[d] Services related to the production of material products.
[e] Services unrelated to the production of material products.

Source: Calculated from data in *JNZXTZ* 452–3.

equipment in rural Jiangsu was the walking tractor, and given its price in the mid-1980s, the statistics in Tables 6.4 and 6.5 suggest that on average each specialized transport household owned slightly more than the equivalent of one tractor in fixed capital while each transport association owned the equivalent of between two and three tractors. What we observed in the three surveyed townships is consistent with the statistical evidence just presented.

Average earnings of specialized households and of associations are significantly above the average for all rural households in Jiangsu. In 1985, the average per capita gross income for all rural households in Jiangsu was RMB 681.7 and the average per capita net income was RMB 493.[53] In the same year, the average per capita income earned by urban households in Jiangsu was RMB 831.[54] When comparing these figures with the income data in Tables 6.4 and 6.5, it is important to remember that, while the data on specialized households include all household earnings, the data on associations include only the earnings generated by the association. Furthermore, the association data are not per capita income figures but income per participating member in the association.

The average per capita gross household income earned by specialized households was nearly three times that earned by all rural households. But because specialized households also had substantially higher production costs, the average per capita net income of specialized households was only about twice that of all rural households. Even after deducting taxes and contributions to the collective, the average per capita (disposable) income of specialized households was still 1.9 times the average per capita net income of all rural households. For comparison purposes, the income data for those in associations must first be converted from income per participating member to per capita income. If we assume that in 1985 each person who participated as a member of an association supported on average two people,[55] then the average per capita income of those involved in rural associations was RMB 730 (RMB 1,459/2), which is lower than the average per capita disposable income of specialized households but still 50 per cent higher than the average per capita net income of all rural households. However, the above estimate understates the average per capita income of association members since earnings from sources other than the association are excluded.

The data in Table 6.5 also allow us to compare the earnings of hired workers and that of association members. The income differential was particularly large in weaving and embroidering where the average association member earned 5.5 times that of the average hired worker. But, on average, most association members earned between two to three times more than hired workers. This is not a particularly large differential considering that the earnings of association members include not only wages but also returns to capital and to entrepreneurship (or risk-taking).[56]

Per capita gross income of specialized households varied considerably by industry. To take an extreme example, the average per capita gross income of households specializing in motor vehicle transportation was over three times that of households specializing in cultivation. However, once production costs are deducted, the differences in average per capita earnings by industry are much reduced. The substantially higher earnings of households that specialized in motor vehicle transportation can be explained largely by the fact that it was the most capital-intensive activity by far. The amount of fixed capital available to workers in these households was seven times greater than what was available to workers in the average specialized household. Similar variations by industry are observed in the earnings of associations.

The above discussion deals with average earnings, but averages tell only a part of the story. Table 6.6 presents the distribution of specialized households by gross household income and by net household income (both in current prices). In 1984, shortly after many rural residents first went into business for themselves, the average gross household income of specialized households was RMB 5,530, the medium was below RMB 5,000, and about 70 per cent of the households had gross income at or below the average, i.e. the distribution was skewed to the lower incomes. But since the average gross household income for all rural households in 1984 was only RMB 2,876, most specialized households still earned considerably more gross income than did the average rural household. Unfortunately we do not have net household income figures for 1984, but since specialized households have a lower ratio of net to gross income, we can assume that the gap between the net income earned by specialized households and that by all rural households was less than that for gross income.

What is more interesting is that, since 1984, the gap between the average gross income earned by specialized households and that by all rural households has widened considerably. The average gross income of specialized households exceeded that of all rural households by RMB 2,654 in 1984, RMB 5,120 in 1985, and RMB 11,933 in 1988.[57] Because specialized households used more raw materials and because raw material cost increased sharply in the late 1980s,[58] the differential between the average net household income of specialized households and that of all rural households increased much less, from RMB 2,138 in 1985 to RMB 3,844 in 1988.

A part of the increase was due to inflation, i.e. the gap widened less in real than in nominal terms. But two other factors also contributed to the increase. First, after the initial rush to establish private undertakings,

TABLE 6.6. *Specialized Households in Rural Jiangsu Distributed by Gross Household Income and by Net Household Income,[a] 1984, 1985, and 1988*

	1984	1985	1988
Total number of specialized households	289,413	210,946	222,236
Average gross household income (RMB)	5,530	8,104	16,612
Average net household income (RMB)	n.a.	4,294	7,240
Number of households distributed by gross household income (% of total)			
Total	100.0	100.0	100.0
Below RMB 5,000	57.8	41.0	14.9
RMB 5,000–6,000	13.8	14.6 ⎱ 25.3	
RMB 6,000–7,000	7.9	9.9 ⎰	
RMB 7,000–8,000	6.5	7.8 ⎱ 15.9	
RMB 8,000–9,000	3.7	5.6 ⎰	
RMB 9,000–10,000	2.4	4.6	
RMB 10,000–15,000	4.6	8.3	17.9
RMB 15,000–20,000 ⎫			9.2
RMB 20,000–50,000 ⎪			11.8
RMB 50,000–100,000 ⎬	3.4	8.2	3.4
Above RMB 100,000 ⎭			1.6
Number of households distributed by net household income (% of total)			
Total	n.a.	100.0	100.0
Below RMB 2,000	n.a.	16.0	6.8
RMB 2,000–4,000	n.a.	45.1	24.1
RMB 4,000–6,000	n.a.	22.9	27.0
RMB 6,000–8,000	n.a.	8.8	17.3
RMB 8,000–10,000	n.a.	3.6	10.1
RMB 10,000–15,000	n.a.	2.3	8.7
RMB 15,000–20,000	n.a.	0.7	3.1
Above RMB 20,000	n.a.	0.5	3.0

[a] Gross household income minus production expenses (including wages paid to hired workers).

Sources: *JJN* (1989), 3. 35, and *JNZXTZ* 2, 3, and 8.

consolidation weeded out the less able, i.e. those with earnings substantially below the average, and many were undoubtedly among the 78,000 households that dropped out of the specialized household category between 1984 and 1985.[59] The departure of the less able would have had the effect of raising the average income of specialized households. Secondly, those specialized households with better than average entrepreneurial and production skills flourished in the new economic environment, and many of them began to make large profits. As more households moved into the higher income categories (i.e. the upper tail of the distribution), the distribution of specialized households by income became less skewed, and this would also have helped to raise the average. The percentage of specialized households with gross income above RMB 10,000 was 8 per cent in 1984, 16.5 per cent in 1985, and nearly 44 per cent in 1988. When measured in terms of net income, the shift to higher income was less dramatic but still impressive. Less than 4 per cent of the specialized households in Jiangsu earned more than RMB 10,000 in net income in 1985 but nearly 15 per cent did so in 1988.

Among private activities in rural Jiangsu, some of the most dynamic and commercialized in the 1980s were the so-called specialized households and associations. A sizeable number of them were engaged in agriculture, but most were involved in industry, transportation, and services. During the decade, the number of specialized households and associations declined, partly because of stiff competition and partly because of improved employment opportunities in the collective sector. However, because entrepreneurship, skills, and hard work were better rewarded than in the past, the differential between the average income of specialized households and that of all rural households widened considerably during the 1980s. Indeed, in the late 1980s, many of rural Jiangsu's *nouveau riche*, the so-called 10,000 *yuan* households (*wanyuanhu*), were specialized households.

Large Rural Private Enterprises

Relative Importance

Most rural private enterprises provide employment only for the owner-proprietors and family members. In 1985, only 13.5 per cent of the specialized households (28,534 out of 210,946) and 35 per cent of the associations (7,061 out of 20,158) in rural Jiangsu used hired labour, and together they employed 128,390 hired workers. If only industrial enterprises are considered, then the percentage of those with hired labour is substantially higher—35 per cent for households specializing in industry and nearly 50 per cent for industrial associations. To put these numbers in perspective, in the same year Wenzhou *shi* in Zhejiang Province (the region in China where rural private activities are reportedly most developed), with a population less than one-tenth that of Jiangsu's, had 13,121 specialized households and associations that used hired labour and they employed 42,263 hired workers.[60] Thus, relative to its population and size, Jiangsu is still some distance behind Wenzhou in the development of private enterprises. However, Jiangsu is ahead of most other provinces.[61]

Table 6.7 presents the distribution of rural private enterprises in Jiangsu that used hired labour in 1985 by size and by industry, and the figures show that most used only a few hired workers—the average was 3.6 (3.2 for specialized households and 5.4 for associations).[62] But the figures also show a small number of large private enterprises, i.e. those with eight or more hired workers. About 8 per cent of the specialized households that used hired labour in 1985 employed eight or more hired workers, and 2 per cent employed sixteen or more hired workers. However, these large units absorbed one-third of all hired workers employed by specialized households. Of the associations that used hired labour, about 23 per cent used eight or more hired workers and 9 per cent used sixteen or more hired workers. Associations that employed eight or more hired workers accounted for 58 per cent of all hired workers employed by associations. The data in Table 6.7 also show that in 1985 there were over 1,000 private industrial enterprises (490 specialized households and 568 associations) in rural Jiangsu that employed sixteen or more hired workers, and together they employed over 24,000 hired workers. These are interesting, and somewhat unexpected, statistics since the 1984 State Council Document No. 26 clearly states that the maximum number of workers private industrial and commercial enterprises may employ is seven.[63] In the late 1980s, at least in Jiangsu, large private enterprises existed primarily in rural areas.[64]

Private Entrepreneurs

These large private enterprises are of interest because to have expanded beyond using family workers and a handful of helpers suggests that they no longer serve just the local market but have successfully penetrated the larger regional (and even possibly the national or the international) market. In other words, they include the most competitive and successful private enterprises, and running them would be some of the most successful private entrepreneurs in rural Jiangsu. Who are they, and where did they come from? How did they accumulate the capital and obtain the labour? To start the enterprise, what difficulties did they have to overcome? Should we expect more of these large enterprises to develop?

The findings of a 1988 survey of large rural private enterprises shed light on some of these questions. First, a few words about the 97 enterprises that were surveyed.[65] The selected enterprises were from 11 provinces and included (i) individual enterprises that employed 8 or more hired workers, (ii) individual enterprises that employed fewer than 8 hired workers but had total capital in excess of RMB 50,000, and (iii)

TABLE 6.7. *Specialized Households and New Economic Associations with Hired Workers Distributed by Industry and Number of Hired Workers, Rural Jiangsu, 1985*

| | Specialized households | | | | | | | New economic associations | | | | | | |
| | Total[a] | | Percentage distributed by number of hired workers | | | | | Total[b] | | Percentage distributed by number of hired workers | | | |
	Number	%	1–3	4–7	8–15	16+		Number	%	1–3	4–7	8–15	16+
Number of household	28,534	100	76.1	16.5	5.4	2.1		7,061	100	53.2	24.3	13.4	9.2
Cultivation	722	2.5	87.1	8.6	2.5	1.8		31	0.4	67.7	6.5	12.9	12.9
Other agriculture[c]	2,199	7.7	90.5	7.4	2.0	0.2		275	3.9	70.2	18.9	6.9	4.0
Industry	12,903	45.2	64.2	23.5	8.6	3.8		5,262	74.5	47.7	26.9	14.6	10.8
Transport	3,086	10.8	94.6	4.9	0.4	0.2		471	6.7	92.1	5.3	1.1	1.5
Construction	3,464	12.1	72.6	19.9	5.6	1.9		389	5.5	37.3	26.5	28.8	7.5
Commerce[d]	4,677	16.4	87.5	10.1	2.2	0.2		445	6.3	76.2	17.1	3.6	3.1
Other	1,481	5.2	86.8	8.9	3.7	0.6		188	2.7	60.1	22.3	10.1	7.5
Number of hired workers	90,288	100	41.3	25.3	17.0	16.5		38,102	100	20.9	20.9	21.2	37.0
Cultivation	1,955	2.2	49.9	15.5	9.8	24.9		159	0.4	24.5	5.7	25.8	44.0
Other agriculture[c]	4,210	4.7	70.7	18.1	8.7	2.5		1,028	2.7	30.6	25.9	20.7	22.8
Industry	54,367	60.2	30.3	27.5	20.4	21.9		30,761	80.7	18.2	21.3	20.6	39.9
Transport	4,863	5.4	82.0	13.0	2.6	2.3		868	2.3	75.1	13.0	5.1	6.8
Construction	11,417	12.6	38.1	28.4	17.4	16.1		2,760	7.2	16.9	17.6	38.3	27.2
Commerce[d]	9,790	10.8	65.0	22.9	10.1	2.0		1,583	4.1	39.6	21.5	10.6	28.3
Other	3,686	4.1	59.1	19.4	16.1	5.3		943	2.5	28.1	21.3	22.7	27.9

[a] Incomplete as data excludes Wujin County, Liyang County, and the suburban districts of Wuxi City and Changzhou City.
[b] Incomplete as data excludes Wuxi County, Yixing County, Wujin County, Liyang County, and the suburban districts of Wuxi City and Changzhou City.
[c] Includes forestry, animal husbandry, and fishery.
[d] Includes trade, food and drink, and services.

Source: Calculated from data in *JNZXTZ* 9 and 456.

joint household enterprises (or associations) that used twice as many hired workers as family workers and had total capital in excess of RMB 50,000. Of the 97 private enterprises, 46 belonged to category (i), 14 to category (ii), and 37 to category (iii). Of the 37 joint household enterprises, 32 were associations involving between 2 and 5 households, and 5 involving 6 or more households. Seventy-one of the enterprises surveyed started from scratch in the sense that they had no predecessor. The others were either developed from household activities, team enterprises that had been privatized, or started as joint household enterprises and then bought out by a single proprietor. A few were in the service sector, but 89 of the 97 were in industry or construction. On average, these enterprises had been in operation for 3.4 years, employed 22.8 hired workers (26.8 if family workers are included), had RMB 144,000 in total capital (RMB 78,000 in fixed capital), and earned RMB 247,000 in gross income in 1987.[66]

Entrepreneurship is the ability to see an economic opportunity (e.g. a new product, a new technique, a new use for existing resources, or a new way to organize production or distribution), to seize it by assembling the capital, labour, and materials needed to launch a new enterprise, and to get the new enterprise going.[67] This is not an easy task in any economy but is particularly difficult in a partially reformed command economy, such as China's, where information is highly imperfect and resources are still largely controlled and allocated by a clumsy bureaucracy. To be successful in such an environment, the entrepreneur must be energetic, bold (*danda* or a willingness to take personal risks), and have a wide network of friends and contacts (*guanxi*) and the ability to make contacts and cut deals (*da jiaodao*).[68] *Guanxi* and the ability to make contacts are particularly important since, in doing business in China, everything from getting the operating licence to finding raw materials and capital to transporting the end-products to market, depends on personal connections.

Given these traits, it is not surprising that the survey found the typical owner-entrepreneur of large rural private enterprises to be a male, below the age of 40, to have better than average education, to have considerable experience outside of agriculture, and to have a network of contacts.[69] Even though these enterprises were located in villages, only 37 per cent of the owners reported farming as their last occupation before they went into private business. It is significant that almost 30 per cent of the owners belonged either to the Party or to the Communist Youth League, and many also had relatives working in the township government. Such affiliations are clearly useful in opening doors and gaining new contacts.

Although the average initial investment was relatively small (only RMB 32,000), 63 per cent of the initial capital was borrowed.[70] Of the 97 enterprises surveyed, 77 borrowed some of their initial capital from private lenders, and 52 had loans from the Agricultural Bank of China or the local Credit Co-operative. Since in China it is nearly impossible to borrow from the bank or from private lenders without personal contacts, the importance of *guanxi* is again apparent.[71] Subsequent investments were mostly self-financed, and all enterprises reinvested heavily in the early years of their existence. The lack of capital was the most serious problem facing these enterprises when they were first established, and it was still the most serious problem at the time of the survey.[72]

Why Do Private Enterprises Remain Small?

Notwithstanding the fact that between 1984 and 1988 the average number of hired workers employed by private enterprises increased (e.g. for non-agricultural associations it rose from 2.4 to 3.8 workers), there is little evidence to suggest that the number of rural private enterprises with more than 16 hired workers has increased in Jiangsu.[73] It is safe to say that the number of rural private enterprises with 50 or more workers continues to be negligibly small. In the three counties where in-depth fieldwork was conducted during the late 1980s, the largest specialized household employed 30 hired workers and apprentices,[74] the largest association employed 77,[75] and there were at most only one or two large private enterprises in each township. Successful private enterprises find it extremely difficult to expand beyond a certain size and in fact are under pressure, perhaps self-imposed, to remain relatively small.

In China, as in other socialist countries, there is a hierarchy of enterprises. At the top are the state enterprises, followed by large urban collectives and then rural enterprises. Among rural enterprises, there is also a hierarchy: township enterprises are at the top, then come village enterprises, and private enterprises are at the absolute bottom. The higher an enterprise is in the hierarchy the easier it is to obtain scarce materials and human resources but the more tightly its operation is controlled by the state (or by local governments). Another way of describing this is to say that the lower an enterprise is in the hierarchy the more it is market orientated—i.e. the more it is responsible for finding its own raw materials and marketing its own products. Thus rural private enterprises, at the very bottom of the totem pole, are totally market orientated. They operate with relatively little day-to-day interference from the state or from local governments but face enormous difficulties in obtaining important raw materials, land, capital, technology, and in gaining access to regional and national markets. These difficulties have made it hard for private enterprises to grow large.

Private enterprises are given the lowest priority for scarce resources. Since they operate outside the planned economy, private enterprises cannot easily buy goods that are centrally controlled, and when they have the opportunity to buy, the cost is often excessive and the quality of the goods inferior. Power and fuel, always in short supply, are also difficult to obtain, and private enterprises have more than their share of black-outs and brown-outs. Although private enterprises are eligible for bank loans, they rarely get any without the support and approval of local governments, and in periods of tight credit they are the first to be abandoned. In fact, it is even difficult for private enterprises to get bank accounts.[76] Because land is collectively owned, it is also difficult for private enterprises to gain access to additional land beyond the private plots allotted to rural households. Using their market power, state enterprises and large collective enterprises regularly force the smaller and financially weaker private enterprises to sell products to, or process goods for, them on credit and then refuse to pay on time, if at all. Because of the fear that private enterprises cannot handle large orders and are more likely to default on delivery, many enterprises in China simply refuse to buy from them, so marketing is also a problem. To overcome this problem, some private enterprises have tried to obscure their ownership by calling themselves TVEs.

While these operational difficulties have made production more costly and more difficult for private enterprises, they are not insurmountable. Indeed, many private enterprises, by responding rapidly to the needs of consumers and by being more efficient than their competitors in the collective and the state sectors,[77] have overcome them and become successful. A far more serious obstacle to the growth of private enterprises, however, is the government's ambiguous policy towards private ownership.

Despite assurances from the government that private enterprises have a role in a socialist economy, private ownership is still viewed with suspicion if not hostility. The government's position is that, because they have a gap-filling role, small private enterprises may be tolerated, but large ones should not be because, if left unchecked, capitalism will return and labour exploitation will intensify in the countryside. But at what size does a private enterprise become ideologically too large? And should size be measured by employment, by output or sales, by the value of capital employed, by profit, or by a combination of the above?

In the early 1980s the government defined too large as 'in excess of seven hired workers',[78] but did not enforce the regulation. Nevertheless, its existence has put strong pressures on rural private enterprises to refrain from becoming too large. Given the government's continued commitment to Marxist principles and in view of China's recent history, it is hardly surprising that rural entrepreneurs feel insecure and uncertain of their future, particularly if they have become successful and relatively large.[79]

One manifestation of the insecurity is the practice, apparently wide-spread among successful private enterprises, of entrepreneurs donating funds to charities (e.g. to help support five guarantees families) or to local governments for community development.[80] Of course such donations may be simply acts of generosity motivated by a desire to share one's good fortune, but one suspects they are also made in part to retain the protection and the good will of local governments. The need for protection also makes it easy for local governments to impose special fees on private enterprises, and in some parts of China these fees have become a serious burden.[81]

Another manifestation of insecurity is the observed reluctance of some private enterprises to expand much beyond the size ceiling stipulated by the government even though it is not enforced. It is not uncommon to find successful private enterprises, after reinvesting heavily in their early years, suddenly stop growing once they have reached the size beyond which it may be politically unsafe—i.e. the fear that 'large trees catch the wind'. In 1987, one survey found sharp declines in the growth of the older private enterprises (those in operation for five or more years).[82] In the course of our fieldwork, the question of expansion was discussed with several successful private entrepreneurs, all of whom had reinvested heavily when their business first started, and they all expressed little interest in expanding beyond their present size of about a dozen employees.

An alternative to the no growth option is to grow but with one foot in the collective sector. To make their large size or contemplated expansion politically more acceptable, some private enterprises have offered part ownership to their village or township government.[83] (And, of course, once an enterprise is partly owned by the community, it is no longer counted in the statistics as a private enterprise.) This was in fact the route followed by Xiaojian Bamboo Crafts Factory, one of the private en-terprises surveyed, which started life as an association in 1984 but was reorganized as a joint collective-private enterprise (with Xiaojian Village as one of the shareholders) in 1986, just before the enterprise signed a five-year contract to supply bamboo brooms to a Japanese importer.[84] In addition to protection, the village also contributed land to the partnership.

Having a foot in the collective sector may also be economically mo-tivated. In rural China, it is very difficult to obtain land and bank loans without the support and assistance of local governments. Since 1986 the allocation of collective land to non-agricultural uses has required not only the consent of the village and township governments but also of higher levels of government. To get bank loans, many private enterprises must borrow in the name of the village or use village assets as collateral. It is also the case that large private enterprises often have difficulties getting sufficient raw materials to operate near full capacity. Thus better access to supplies may be another motivation to merge with the collective sector. In other words, given China's political and economic environment, to take on the local community as a partner may improve the enterprise's chances of success.

But merging with the collective sector is not a step that is taken lightly, since most private owners also recognize that if they become indebted to, or too dependent on, the local government (e.g. if the expansion is done largely with government assistance) there is the danger that they may also lose much of their autonomy.

In order to encourage private businesses to reinvest and expand, China formally legitimized the status of large private enterprises in April 1988 when the 1982 Constitution, which had recognized only individual enterprises (*geti qiye*, i.e. private enterprises with seven or fewer employees), was amended to provide for private enterprises (*siying qiye*), formally defined as 'a profit-seeking economic organization that employs eight or more persons and whose property is privately owned'.[85] Shortly thereafter the Provisional Regulations Concerning Private Enterprises was promulgated which assured owners of *siying qiye* of their rights to own and inherit private properties and of protection from 'unlawful exactions'.[86] The regulations also imposed restrictions on *siying qiye*, e.g. they are required to reinvest at least 50 per cent of after-tax profits and to limit the salary of enterprise managers to less than ten times the average earnings of staff and workers.

The amended constitution and the new regulations will not quickly eliminate the reluctance of private enterprises to reinvest and expand.[87] In the past government policy allowed *siying qiye* to operate without constitutional protection, and policy can just as easily make life difficult for *siying qiye* under the amended constitution. In fact, the new regulations provide the government with the legal basis to harass *siying qiye* on a wide range of matters, e.g. tax compliance, labour code compliance, record-keeping. In view of the many policy shifts in the past, it is understandable why many private entrepreneurs continue to fear that government policy might change. In fact, in 1989–90, a year after the constitution was amended, private enterprises were harassed and harshly criticized, and many disappeared under the pressure.[88]

In the politically more relaxed and flexible environment of the 1980s, private enterprises have made a modest comeback in rural Jiangsu and their reappearance has helped to invigorate the rural economy. Rural private enterprises have created jobs for a large number of underemployed farmers, provided rural households, particularly those in poorer areas, with much needed additional income, and enhanced competition in both the rural factor and rural output markets. They have also enriched rural life by providing consumers with greater varieties of goods and services and satisfying needs that had previously been ignored. The opportunity cost of developing private enterprises, in most cases, has also been relatively low as they rely primarily on resources that were previously unused or underutilized (surplus labour, untapped household savings, and local materials). Most private enterprises have remained small, but a few have become large suggesting that they have successfully penetrated the regional and national markets. Despite the fact that large private enterprises were given official recognition in 1988, political discrimination against private activities has not disappeared. Little wonder private entrepreneurs remain insecure. Unless ongoing reforms change China's political environment in ways unforeseen at the moment, it is not likely that the private sector will grow much larger or become more dynamic.

NOTES

1. For discussions of rural non-agricultural development in Wenzhou see Zhongguo shehui kexue yuan jingji yanjiusuo (ed.), *Zhongguo xiangzhen qiye de jingji fazhan yu jingji tizhi* (*The Economic Development and Economic System of Rural Enterprises in China*) (Beijing: Zhongguo jingji chubanshe, 1987), part 1, and Yu Binghui, 'Zhengzhi, jingji, shehui dui Wenzhou moshi de zai kaocha (A re-examination of the political, economic, and social aspects of the Wenzhou model)', *NJYS* 2 (1988), 9–18.

2. In China, private businesses with seven or fewer employees (excluding family members) are called 'individual enterprises', and only those with eight or more employees are considered 'private enterprises'. In this chapter, we do not make this distinction. All private businesses, regardless of size, are called private enterprises.

3. Zhang Renshou *et al.*, 'Wenzhou moshi de tese ji qi yiyi (The characteristics and the meaning of the Wenzhou model)', *NJW* 9 (1986), 4.

4. Luo Hanxian, 'Wenzhou moshi yu shichang jingji (The Wenzhou model and market economy)', *NJW* 9 (1986), 10.

5. However, some Wenzhou cadres have become private entrepreneurs, and many private enterprises are operated by family members of cadres. See Yia-Ling Liu, 'Reform From Below: The Private Economy and Local Politics in the Rural Industrialization of Wenzhou', *CQ* 130 (June 1992), 305.

6. This and the other statistics in this paragraph are calculated from *JJN* (1987), 3. 107–12.

7. The extent of privatization is suggested by the change in the ownership structure of fixed capital. In 1983, of the fixed capital in rural Jiangsu, one-quarter was owned by teams and slightly over 12 per cent by households. By 1985, production teams owned less than 9 per cent and households owned over 27 per cent of the fixed capital in rural Jiangsu. Of course, included in

these figures are both agricultural and non-agricultural fixed capital. See Zhang Yixin, 'Guanyu nongcun jiating jingying fazhan he zhuanhua de tantao (A probe into the development and transformation of family production in rural areas)', *NJW* 7 (1987), table 3.

8. Township-Village Interview Notes.

9. The 'fear that tall trees catch the wind (*pa shu da zhao feng*)'—i.e. a successful person or venture is more likely to be criticized and attacked.

10. Village Background Papers. For example, in the villages we visited in Ganyu County, people still remembered that in the 1970s when some farm households started to fish and to produce fishnets commercially, they were designated 'upstart households (*baofa hu*)', i.e. households that wanted to become rich quickly, and 'cut off' as a 'tail of capitalism'. Conditions in Wenzhou were apparently different. Yia-Ling Liu argues that the reason why private economy developed first in Wenzhou was because it 'had a body of local cadres sympathetic to the private economy who were able to marshal a collective resistance to the state's encroachment' on private activities during 'the difficult years between 1949 and 1978' (see his 'Reform From Below', 308–9). In other words, Wenzhou's private sector developed rapidly in the 1980s partly because it was never eliminated and partly because since 1978 local cadres have continued to shelter private activities from government interference.

11. Mo Yuanren (ed.), *Jiangsu xiangzhen gongye fazhan shi* (*History of the Development of Rural Industry in Jiangsu*) (Nanjing: Nanjing gongxueyuan chubanshe, 1987), 250–1.

12. Ibid. 254–5.

13. The most important of these are in State Council 1984 Documents No. 24, 25, 26, and 27. See *XQZFX* 375–83 and 678–85.

14. Ibid. 681–2.

15. Ibid. 683.

16. Chinese planners classified consumer goods and producer goods into three categories according to their importance. Category 3 agricultural and sideline products and category 3 industrial goods included a large number of minor and miscellaneous commodities that were formally under local control and outside the state plan.

17. e.g. more than 3,000 private enterprises in Liaoning wore 'red hats' in 1988 (see Cheng Xiangqing *et al.*, 'Siying qiye fazhan mianlin de zhuyao wenti (Main problems confronting the development of private enterprises)', *NJW* 2 (1989), 24). Apparently, in the late 1980s, many of the collective enterprises in Wenzhou were also private enterprises in disguise (see Yia-Lin Liu, 'Reform From Below', 295). The problem may be less serious in Jiangsu, since its local governments removed many bogus TVEs from the collective sector in the mid-1980s. See Luo Xiaopeng, 'Ownership and Status Stratification', in William A. Byrd and Lin Qingsong (eds.), *China's Rural Industry* (Oxford: Oxford University Press, 1990), 150.

18. *JXQTZ* (1987), 2–3 and 126–65.

19. In 1988 rural Jiangsu had over 657,000 walking tractors, nearly 19,000 motor vehicles, and over 116,000 motor barges (see *JSN* 134).

20. In contrast, three-quarters of all rural non-agricultural employment in 1986 was concentrated in the south (32.9 per cent in SDR, 29.0 per cent in NDR, and 12.0 per cent in SMR). For more details, see Table 4.13.
21. Township-Village Interview Notes. In the late 1980s, more than 2,000 metric tons of fishnet (with a value of about RMB 10 million) were processed each year, mostly in rural households.
22. For a discussion of how private enterprises are organized in Wenzhou, see Yu Binghui, 'Zhengzhi, jingji, shehui dui Wenzhou moshi de zai kaocha', 9–18.
23. In the late 1980s, about one-half of the buttons used in China were produced in one township (Qiaotou) in Wenzhou. See ibid. 10.
24. In the early 1980s, specialized households were frequently called 'key point households (*zhongdianhu*)'. For an early survey of specialized households, see Zhonggong Foshan diwei bangongshi, 'Foshan diqu zhuanyehu zhong-dianhu bai hu diaocha (An investigation of 100 *zhuanyehu* and *zhongdianhu* in the Foshan area)', *JD* 1 (1983), 64–73.
25. More specifically, the new economic associations are a component of joint household enterprises (*lianhu qiye*) and the non-farm specialized households are a component of individual enterprises (*geti qiye*).
26. *JNZXTZ* (1986), 1.
27. In 1986, Jiangsu reported 200,722 specialized households, of which 153,461 were engaged in non-agricultural activity. In the same year, 595,000 individual enterprises were registered in rural Jiangsu.
28. *Zenyang ban hao nong gong shang lianhe qiye (How to Successfully Operate Agricultural–Industrial–Commercial Integrated Enterprises)* (Zhongguo shedui qiye bao, 1982), 13 and 30–2. For a discussion specific to Jiangsu, see Wu Rong and Wu Defu, 'Suzhou shi fazhan nongcun chanqian chanhou fuwu de zuofa (Methods used in Suzhou *shi* to develop activities that provide services to agriculture)', *JD* 4 (1985), 54–61.
29. For example, in Yiyang *xian* in Hunan Province, more than one-third of the households participating in associations were formerly specialized households. See Yang Mingguang, 'Yiyang diqu nongcun xin jingji lianheti de qingkuang (The situation of new economic associations in rural Yiyang)', *JD* 3 (1984), 10.
30. Nearly three-quarters of 510 associations surveyed in 1987 involved the pooling of capital. See Nongcun diaocha bangongshi, 'Noncun gaige yu fazhan zhong de ruogan xin qingkuang (Several new situations in the midst of rural reform and development)', *NJW* 4 (1989), 40.
31. A survey of 272 representative villages in 1985 uncovered 59,486 associations, of which 1,900 (3.2 per cent) involved rural households and collective units and 68 (0.1 per cent) involved rural households and state enterprises. In other words, 96.7 per cent of the associations involved only rural households. See Nongcun diaocha lingdao xiaozu, 'Quanguo nongcun shehui jingji dianxing diaocha qingkuang zonghe baogao (Summary report of the survey of typical social and economic areas in rural China)', *NJW* 6 (1986), 11.
32. Apparently some associations do not own any fixed capital or have a regular place of operation. For example, about 15 per cent of the associations in Yiyang *xian*, Hunan Province, had no fixed capital or regular place of

operation. See Yang Mingguang, 'Yiyang diqu nongcun xin jingji lianheti de qingkuang', 12.

33. In some parts of China, e.g. Sichuan and Hunan, collective enterprises contracted to groups of individuals are also considered new economic associations (for example, see ibid. and Li Hongru and Zhang Ning, 'Jiangyou xian nongcun xin jingji lianheti diaocha (An investigation of new economic associations in the countryside of Jiangyou County)', *JD* 3 (1984), 5–9). This may have been the case in Jiangsu as well in the early 1980s, but since the mid-1980s TVEs contracted to individuals or a group of individuals are not included as *xin jingji lianheti* in the statistics. See *JNZXTZ* 1.

34. In all, 16,325 new economic associations were recorded in rural Jiangsu in 1986, of which 14,375 were engaged in non-agricultural activity. In the same year, some 45,000 joint household enterprises were registered in Jiangsu.

35. This association is one of 362 registered in Haitou Township in 1987. Of the 362, 334 were fishing associations. All information in this and the following two paragraphs are from Enterprise Interview Notes.

36. Members of these households had previously worked as temporary labour for fishing companies operated by the county.

37. This distribution is very similar to the findings of the national survey of representative villages. See Nongcun diaocha lingdao xiaozu, 'Quanguo nongcun shehui jingji dianxing diaocha qingkuang zonghe baogao'.

38. In 1986, 48.2 per cent of the rural associations in China were in industry and 14.7 per cent in transportation.

39. Detailed data for 1984 are not available. But in 1985, for example, the average specialized industrial household employed 3.4 workers, of which 1.56 were hired workers. By contrast, in the same year, the average industrial association employed 9.5 workers, of which 3.7 were hired. (See Tables 6.4 and 6.5 for the source of this data.)

40. Specialized households employed 63,213 hired workers and new economic associations employed 23,014.

41. The numbers of other types of specialized agricultural household also declined, except for specialized fishing (including fish culture) households which increased by about 1,000.

42. The number of associations in farming declined from 674 in 1984 to fewer than 50 in 1988.

43. Of all private activities, the per capita income of those specializing in cultivation was among the lowest (see Tables 6.4 and 6.5).

44. The common practice in Jiangsu was to divide contract land into two categories: grain ration or subsistence land (*kouliang tian*) and responsibility land (*zeren tian*). (Apparently this practice is also common in other parts of China; e.g. see 'Nongcun gongfuye dui nongye de buchang ji qi dui zeren zhi de yingxiang (The use of rural sideline industry to subsidize agriculture and its effects on the responsibility system)', *JD* 2 (1983), 55.) All rural households contracted some subsistence land (the amount determined by household size, i.e. population) and some responsibility land (the amount determined by the number of total workers or the number of non-agricultural workers in the household). With rural non-agricultural activities more profitable than

farming, many rural households wanted to reduce the amount of land they had contracted. For example, of the 1,266 rural households surveyed in one Jiangsu county (Jintan), 475 said that they wanted to reduce the amount of land they had contracted while only 81 households wanted to contract more land (see Jean C. Oi, 'Commercializing China's Rural Cadres', *Problems of Communism* (Sept.–Oct. 1986), 4). But it is also the case that while many wanted to return some of their responsibility land they had contracted, few if any wanted to give up the subsistence land. And, in fact, very little land was actually returned for redistribution. One indication that little land changed hands in Jiangsu is that land transfer companies (*tudi zhuanbao gongsi*) did not spring up in Jiangsu as they apparently did in some parts of China. For a description of how one such company operated, see ibid. The evidence of land transactions in other parts of China suggests that participation in the land market has been relatively limited (see Justin Yifu Lin, 'Rural Factor Markets in China After the Household Responsibility System Reform', in Bruce L. Reynolds, *Chinese Economic Policy Economic Reform at Midstream* (New York: Paragon House 1989), 179.

45. That the Jiangsu government expected farms to become substantially larger, with perhaps many in the 50+ *mu* range, is suggested by the fact that its Bureau of Agricultural Machinery projected a 39 per cent increase in the province's stock of medium/large tractors between 1983 and 1990 but only a 16 per cent rise in the number of walking tractors. (See Samuel P. S. Ho, *The Asian Experience in Rural Nonagricultural Development and its Relevance for China* (Washington, DC: World Bank, 1986), 63–4.)

46. This decline occurred in other parts of China as well. For example, see Lu Gaoxin, 'Nongcun geti gong shang hu weihe zai zheli da fudu xiajiang? (Why the large decline in individual industrial and commercial enterprises?)', *NJW* 5 (1987), 58–9.

47. In many parts of Jiangsu, the walking tractor is used more for transportation than for farming. Most walking tractors are owned by individuals. (For example, of the 17 walking tractors in Xiaojian, one of the six villages we surveyed, 13 were owned by specialized households.) Between 1984 and 1988, the number of walking tractors in rural Jiangsu increased from 445,000 to 657,000. The cost of a walking tractor in the late 1980s was about RMB 3,000–4,000.

48. The average number of hired workers used by specialized households in non-agricultural activities increased from 0.33 in 1984 to 0.79 in 1988 and the number used by non-agricultural associations increased from 2.44 to 3.81.

49. e.g. one specialized household we surveyed in Haitou Township employed fifteen people (five family members and ten hired workers) in 1987 but provided work through a putting-out system to over 15,000 farm households (some worked full-time but most on a part-time basis) within a 35 km. radius.

50. In 1987, a survey of ninety-seven private rural enterprises in eleven provinces found that on average the ratio between fixed capital and working capital was 1:0.8. For rural collective enterprises, the ratio was 1:1. See Nongcun diaocha bangongshi, 'Dui bai jia nongcun siying qiye diaocha de chubu fenxi (A preliminary analysis of a survey of one hundred private rural enterprises)',

NJW 2 (1989), 20.

51. The Chinese term is 'specialized credit households (*xinyong zhuanyehu*)'. The traditional credit association (*hui*) has also re-emerged, if it ever totally disappeared from the countryside. See Yu Binghui, 'Zhengzhi, jingji, shehui dui Wenzhou moshi de zai kaocha', 10–11.

52. The regulations governing loans to private enterprises are very restrictive. For example, before they would consider lending to a private enterprise, banks usually require acceptable collateral or the guarantee of either the community, a state enterprise, or a collective enterprise. There is evidence that, in many parts of China, the amount of loans to private enterprise has declined over the years. See Cheng Xiangqing *et al.*, 'Siying qiye fazhan mianlin de zhuyao wenti', 25.

53. These figures are from the 1985 survey of 3,400 rural households. See *JJN* (1986), 3. 65.

54. This figure is from the 1985 survey of 1,535 households of urban workers. See ibid. 3. 64.

55. In 1985, the average rural household had 3.92 persons and 1.98 workers, i.e. the average rural worker supported slightly less than two people. In the same year, the average urban worker in Jiangsu supported 1.67 persons.

56. On the assumptions that the return to capital is 10 per cent of invested capital and the return for taking risk is 2 per cent of total capital used in the enterprise, Chinese government economists estimated the wage earned by the average owner of a large private enterprise to be slightly more than three times that of the average hired labour. See Nongcun diaocha bangongshi, 'Dui bai jia nongcun siying qiye diaocha de chubu fenxi', 21.

57. The annual survey of rural households found the average gross income of rural households in Jiangsu to be RMB 2,876 in 1984, RMB 2,984 in 1985, and RMB 4,679 in 1988. The average net income of rural households was RMB 2,086 in 1984, RMB 2,156 in 1985, and RMB 3,396 in 1988. See *JJN* (1986), 3. 65, and *JJN* (1989), 3. 100–1.

58. The annual survey of rural conditions showed that in 1987 the production expenses of 510 associations increased by 52.1 per cent, largely because of rapid rises in the prices of raw materials. See Nongcun diaocha bangongshi, 'Nongcun gaige yu fazhan zhong de ruogan xin qingkuang', 41.

59. The number that left the specialized household category was larger, since 78,000 represents the net decline.

60. Zhongguo shehui kexue yuan jingji yanjiusuo (ed.), *Zhongguo xiangzhen qiye de jingji fazhan yu jingji tizhi*, 60–1.

61. Hebei, Henan, Shandong, and Guangdong are other provinces where private enterprises were particularly well developed.

62. In Wenzhou, private enterprises that used hired workers employed, on average, 3.2 workers, even fewer than in Jiangsu.

63. More precisely, the 1984 State Council Document No. 26 states that private enterprises, with permission of the local government, may hire 'one or two helpers', and those with special skills may also accept 'two or three but at most five apprentices' (or a total of seven hired workers). The document was aimed at individually operated rural industry and commerce (*nongcun geti*

gong shang ye), which presumably includes both specialized households and associations. See *XQZFX* 683.

64. In 1987, of the 4,573 private enterprises that employed eight or more non-family workers, 4,280 were located in rural areas (*JJN* (1989), 8. 39).

65. In 1984, the Agricultural Development Research Centre of the State Council and the Agricultural Policy Research Office of the CCP Central Committee jointly selected nearly 300 representative villages in twenty-eight provinces to use as permanent sites for their annual survey of rural conditions. Of these sites, 120 are located in the following eleven provinces: Guangdong, Fujian, Zhejiang, Jiangxi, Hunan, Hubei, Hebei, Shanxi, Liaoning, Shaanxi, and Yunnan. All large (as defined in the text) private enterprises in these 120 villages were surveyed. All survey findings reported in this and the following paragraphs are from Nongcun diaocha bangongshi, 'Dui bai jia nongcun siying qiye diaocha de chubu fenxi'.

66. The distribution by the number of hired workers was: below 7, 7.2 per cent; between 8 and 20, 63.9 per cent; between 21 and 50, 22.7 per cent; between 51 and 99, 4.1 per cent; and 100 and above, 2.1 per cent. The largest employer used 208 hired workers. The distribution by total capital was: below RMB 50,000, 10.3 per cent; RMB 50–100,000, 9.3 per cent; RMB 100,000–200,000, 52.6 per cent; RMB 200,000–500,000, 18.6 per cent; RMB 500,000–1 million, 6.2 per cent; RMB 1 million and above, 3.1 per cent. The enterprise with the most capital had RMB 4.30 million.

67. The literature on entrepreneurship is a large one, and there is no agreement on a definition. Here, we closely follow Joseph Schumpeter's view of entrepreneurship. See his *The Theory of Economic Development* (Cambridge, Mass.: Harvard University Press, 1934).

68. These traits are repeatedly described in the Chinese literature on successful private enterprises. For example, see Yu Binghui, 'Zhengzhi, jingji, shehui dui Wenzhou moshi de zai kaocha', and Cheng Xiangqing *et al.*, 'Siying qiye fazhan mianlin de zhuyao wenti'.

69. Two-thirds of the owner-entrepreneurs were below the age of 40, only 9.3 per cent were functionally illiterate (as compared to 21.6 per cent of the workers in the same villages), and two-thirds had work experience that would have given them the opportunity to establish a network of contacts (e.g. in state enterprises or as state or local cadres, buyers/salesmen, or teachers).

70. Twenty enterprises started entirely on borrowed funds, and they probably accounted for the bulk of the borrowed capital.

71. Of the fifty-two enterprises that borrowed from the Agricultural Bank of China or the local Credit Co-operative (basically a branch of the Bank), seventeen borrowed in the name of their village and used village assets as collateral. This would be impossible without considerable *guanxi*.

72. The second most serious problem, initially and at the time of the survey, was the lack of fuel and raw materials.

73. Rural enterprises that employed more than eight non-family workers numbered 4,280 in 1987, 544 more than in 1985 (*JJN* (1989), 8. 39).

74. There were two, a brick-making enterprise in Yixing and a glass fibre enterprise in Zhangjiagang.

75. A metal solvent enterprise in Zhangjiagang.
76. Several TVEs we surveyed allowed private enterprises to use their bank account numbers for a fee.
77. That private enterprises are often found to be more efficient than TVEs and SEs in the same industry has been pointed out by many observers. See, for example, Zhongguo shehui kexue yuan jingji yanjiusuo (ed.), *Zhongguo xiangzhen qiye de jingji fazhan yu jingji tizhi*, 68, and Nongcun diaocha bangongshi, 'Dui bai jia nongcun siying qiye diaocha de chubu fenxi', 21. However, at the aggregated national level, private enterprises have lower labour productivity and lower gross profit (profit + taxes) per worker. But comparison at the aggregated level is not meaningful since private enterprises are much more labour intensive and are concentrated in different industries than SEs.
78. The ceiling was set at seven hired workers probably because this number was used by Marx to separate petty proprietors from capitalists. See ibid. 73.
79. That peasants continue to feel insecure about holding visible property was confirmed by all the cadres we interviewed. Indeed one senior cadre commented that some rural residents were even afraid to deposit their savings in banks.
80. We were told by one particularly successful private entrepreneur that in the past few years he contributed RMB 100 to help repair the Great Wall, RMB 1,900 to help relocate the local elementary school, RMB 200 in support of the poor families (*wubao hu*) in his village, RMB 5,000 to help the village construct an agricultural trade centre, and purchased RMB 2,500 of government bonds. In addition, whenever the local tax department asked him to contribute a little extra so as to achieve its annual collection quota, he always complied.
81. Nongcun diaocha bangongshi, 'Nongcun gaige yu fazhan zhong de ruogan xin qingkuang', 43. Also see Yia-Ling Liu, 'Reform From Below', 304, and Jean-Louis Rocca, 'Corruption and its Shadow: An Anthropological View of Corruption in China', *CQ* 130 (June 1992), 410.
82. Nongcun diaocha bangongshi, 'Nongcun gaige yu fazhan zhong de ruogan xin qingkuang', 41.
83. The arrangement is not always so subtle. Apparently, in Wenzhou, 'to reduce the anxiety and fend off bureaucratic harassment, many large private partnership enterprises have developed the "power share". By means of this device, owners of private enterprises give free shares to powerful cadres in exchange for political protection and favours.' See Yia-Ling Liu, 'Reform From Below', 305.
84. Enterprise Interview Notes.
85. Alison E. W. Conner, 'China's Provisional Regulations Governing Private Enterprises', *East Asian Executive Reports* (Oct. 1988), 9–11.
86. See ibid. for a discussion of these regulations.
87. For example, despite the change in the constitution, most private enterprises in Wenzhou still liked to be known as 'local collective enterprises'. Yia-Ling Liu quotes one Chinese official saying in 1990 that the promulgation of regulations for private enterprises in 1988 did not appear to have had much

effect 'in increasing the entrepreneurs' confidence in state policy or in reducing political discrimination against the private sector'. See Yia-Ling Liu, 'Reform From Below', 302.

88. Alison E. W. Conner, 'Private Sector Shrinking Under Intense Criticism and Increasing Controls', *East Asian Executive Reports* (Dec. 1989), 10–13.

7

Township-Village Governments, Township-Village Enterprises, and Rural Development

THE purpose of this chapter is to explore the links between township-village governments (TVGs) and township-village enterprises (TVEs) and those between TVEs and rural development.[1] To provide the necessary institutional background, the chapter opens with a description of how rural China was reorganized after the commune system was dismantled in the early 1980s. It then examines the priorities and objectives of TVGs as revealed by the formal criteria used to measure success when evaluating rural cadres. The financial links between TVEs and TVGs and the extent to which TVGs are dependent on TVEs for funding are discussed in the third section, and the chapter concludes with a discussion of the links between TVEs and rural development, specifically the financial support TVEs have provided to agriculture, to collective welfare, and to rural industry.

Government Administration at the Township and Village Levels[2]

Two major institutional changes that altered the organization of activities at the grass-roots level were introduced in the early 1980s. One was the widespread adoption of *baogan daohu* or *da baogan*, the most radical version of the household responsibility system, under which farming was contracted to individual households.[3] The other change was the dismantling of the communes in 1983, which led to the separation of the Party, the government, and the economy (*dang*, *zheng*, and *jing*). The adoption of the household responsibility system has produced far-reaching changes in agricultural organization and management as well as in other aspects of rural life. For example, it has transferred most of the decision-making power in agriculture from cadres to individual farm households.[4] This and the growth of private non-agricultural activities have combined to loosen the tight control that rural cadres once had over the rural economy. However, the separation of *dang*, *zheng*, and *jing* has not produced the anticipated results.

The separation of party affairs, government administration and economic management was intended 'to change speedily the situation in which the Party does not handle party affairs, governments do not handle government affairs, and government administration is fused with and inseparable from enterprise and commune management'.[5] The township (*xiang*) government was re-established in 1983, and the commune was replaced by the township party committee, the township people's government, and the township economic commission (*jingji lianhe weiyuanhui*).[6] At the same time, the production brigade (*dadui*) was replaced by the village branch of the township party committee, the village people's committee (*cunmin weiyuanhui*), and the village economic co-operative (*cun jingji hezuoshe*).[7] In theory, the village people's committee is supervised by the township government and the village economic co-operative by the township economic commission. And, in turn, the township organizations report to and are supervised by their respective counterparts at the county level. In other words, since the mid-1980s, three leading bodies (*santao banzi*) have existed in the countryside: the Party, the government, and the economic commission (co-operative).

Fig. 7.1 presents the structure of the three organizations, abstracting from local details and variations, and it suggests a clear division of labour among the three. In principle, the Party is concerned with party discipline and organization, mass organizations, and the militia. Civil administration, the development of social and economic infrastructure, education, health, and family planning are the primary concerns of the government. The economy, particularly the TVEs and the other collectively owned business activities, is managed by the economic commission through its agricultural, industrial, and sideline corporations. This division of labour is also supposed to exist at the village level.

However, the evidence from Jiangsu suggests that the division of labour among the three organizational channels exists only in theory.[8] In practice, the local Party secretary is not only in charge of Party work but manages both the local government and the local economic commission. Take Sigang Township as an example. In 1988, the Party secretary of Sigang had overall responsibility for the township, made all appointments,[9] and also took personal charge of industry because of its paramount importance in the township. The Party secretary was assisted by two deputies. One deputy Party secretary was concurrently head of the township government and in charge of agriculture, sidelines, small town construction, and family planning. The other deputy was concurrently director of the economic commission and in charge of economic affairs but with special emphasis on industry. The administrative arrangements in the other townships we surveyed were more or less the same as that in Sigang. At the village level, the division of labour among the three

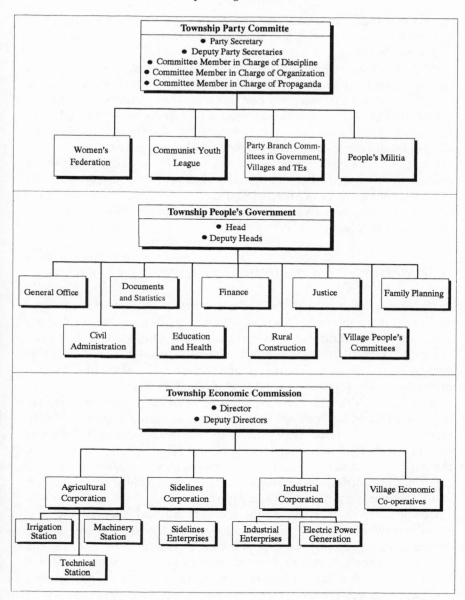

FIG. 7.1. Organization of Township Party Committee, Township People's Government, and Township Economic Commission

organizational channels was even less clear. The village Party secretary was usually also the head of the village committee or the head of the village co-operative.

When asked to comment on the separation of *dang*, *zheng*, and *jing*, one township Party secretary responded by saying 'in theory, the division cannot be clearly explained; in practice they [Party affairs, government administration, and economic management] cannot be clearly divided; accordingly they cannot be separated cleanly at the lower [township and village] levels (*lilun shang jiang bu qing, shiji shang fen bu qing, long de xiamian gao bu qing*)'.[10] This view that *dang*, *zheng*, and *jing* cannot in practice be separated is shared by many. Because *dang*, *zheng*, and *jing* are not separated, they are treated in this chapter as a single entity. Thus, when we use the term TVG, we mean not just the government *per se* but all three leading bodies.

Objectives of Township-Village Governments

Party cadres, particularly local Party secretaries, are the principal decision-makers in rural China. In addition to running the local Party branches and government bureaucracies, they also manage the rural economy, frequently taking direct responsibility for the collectively owned non-agricultural sector. At one time, rural cadres were judged primarily by their commitment to political ideology and their administrative skills. Although these qualities are still taken into account, entrepreneurial and business abilities are now much more important. In the more market-orientated environment of the 1980s, a resourceful Party secretary with good entrepreneurial skills can have a profound impact on the pace of rural development and on the economic welfare of those in his or her jurisdiction. Increasingly the performance of the local economy is the criterion used to assess rural cadres.

By the late 1980s, most counties in Jiangsu had adopted the post responsibility system (*gangwei zeren zhi*). Under this system, annual targets are assigned to townships and villages and then used at year-end to assess the performance of Party secretaries and other senior cadres at the local level. It is safe to assume that the targets with the most weight are the objectives considered most important by local governments. In other words, the priorities of local governments may be deduced from the content and structure of the post responsibility system. Since the year-end assessment determines the income and benefits of local cadres as well as their job security and political status, it is also safe to assume that objectives with the greatest weights are also the ones that cadres will work the hardest to achieve.[11]

The assessment method most widely used in Jiangsu is the 100-point assessment system (*baifen kaohe zhi*). Under this system, targets are specified for each post, and points are awarded for meeting these targets. Points are allocated to targets according to their importance. The method is called the 100-point assessment because the points allocated to the targets normally add up to 100.[12] When targets are under- or overfulfilled, points are deducted or added in the same proportion. A cadre's income is determined by the total number of points awarded to him or her and the value of each point (which is usually set at a level so that the base salary is equivalent to 100 points).[13] In some parts of Jiangsu, the value of each point (or the base salary) is predetermined, and in others determined at year-end. But in either case the value of each point (or the base salary) is linked closely to the average income level of that locality.[14] Some townships also have regulations that restrict the size of income differentials among cadres and between cadres and the masses.

Targets assigned to the principal cadres at the township and village levels are numerous and quite specific,[15] but they fall roughly into two main categories: economic targets and non-economic targets. Economic targets focus on the township's (or village's) performance in agriculture, sidelines, and industry. The two most important targets in agriculture and sidelines are gross output and contract sales to the state. Of the many industrial targets (e.g. gross output, output–sale ratio, revenue collected, profit), profit or a composite profit indicator is normally the most important. For example, in assessing village party secretaries in Sigang Township in 1988, the composite profit target accounted for 70 per cent of the weight allotted to industrial performance.[16]

Of the non-economic targets, the most important are those associated with birth control. Points are awarded when the number of births does not exceed the assigned quota and when there is no second-order birth. In many places, when birth control targets are not achieved, points are not only withheld but are also deducted.[17] Another non-economic target that is important in some areas is land use, i.e. making sure that land is used according to township regulations so that farm land is not diverted illegally to non-agricultural uses (particularly housing).

Table 7.1 presents the points allotted to the different targets, arranged by categories, used to assess rural cadres at the township and village levels in two Jiangsu townships. Abstracting from local variations, the data suggest that in the 1980s rural cadres were judged primarily by their success in economic management and in promoting local development. In assessing village Party secretaries, 80 per cent of the weight was assigned to economic targets in Sigang Township and 65 per cent in Hufu Township. Furthermore, these percentages do not fully reflect the importance of economic work since Table 7.1 does not include bonus points, most of

TABLE 7.1. *Criteria Used to Assess Township and Village Cadres in Two Jiangsu Townships, 1988*

	Points awarded for meeting targets			
	Sigang			Hufu
	Township Party secretary[a]	Village		Village Party secretary
		Party secretary	Head	
Economic targets	95	80	—	65
Agriculture	17.5	15	—	} 45[d]
Sidelines	17.5	15	—	
Agricultural construction fund	5	—	—	—
Industry (TVEs)	50	50	—	20
Per capita income	5	—	—	—
Non-economic targets	5	20	100	35
Birth control	5	20	40	10
Land use	—	—	30	2
Education, health, welfare	—	—	5	8
Social and civil order	—	—	10	6
Militia work	—	—	5	2
Miscellaneous[b]	—	—	10	7
Total[c]	100	100	100	100

[a] 200 points were distributed in the original document. To facilitate comparison, the number is converted to 100.

[b] Includes such activities as Party work and work in support of the Communist Youth League.

[c] Excludes bonus points.

[d] Divided in each village between agriculture and sidelines (including forestry) in proportion to their 1987 GOV.

Sources: Zhangjiagang Party Committee General Office, 'Guanyu 1988 nian xiangzhen jiguan ganbu zeren zhi kaohe yijian (Some suggestions about the assessment of cadres in township [party and government] organizations under the responsibility system)' (14 Mar. 1988); Sigang Township, 'Sigang zhen cun ganbu zeren zhi hetong (Contract under the responsibility system for village cadres in Sigang Township)' (1 Jan. 1988); and People's Government of Hufu Township, 'Hufu zhen 1988 niandu baifen kaohe xize (Detailed standards, 100-point assessment, Hufu Township, 1988)'.

which were to award exceptional economic achievements. The weight given to economic targets in the assessment of township Party secretaries may be even greater.[18]

The distribution of points among economic sectors reflects by and large the structure of the local economy. Thus, for example, industrial targets related to TVEs are assigned greater weight in regions where rural industry is important than in regions where it is not. With the exception of birth control, non-economic targets (including Party work) receive substantially lower priorities, and this is reflected not only in the fewer points allotted to these targets but also in the fact that those held primarily responsible for meeting these targets are frequently not the Party secretary.

If the economic targets are set at levels that are not more difficult to achieve than the non-economic targets,[19] then the distribution of points shown in Table 7.1 suggests that self-interest would direct local Party secretaries to devote most of their energy and time to economic work, i.e. support agriculture so that farm output and quota (contract) sales to the state are fully met, seek out new economic opportunities so as to improve rural income, and improve the performance and profitability of community-owned enterprises.

The priority of local governments, as revealed by the content and structure of the post responsibility system adopted to assess cadres, indicate that in the mid- and late 1980s the most important objective of local governments in Jiangsu was local economic development. The importance given to economic development, particularly the development of the rural non-agricultural sector and TVEs, is confirmed both by what rural cadres told us in interviews and by the emphasis given to economic work in official documents issued during this period.[20]

Local Public Finance, Township-Village Enterprises, and Rural Development

It is easy to understand why TVGs attach such importance to rural non-agricultural development and to TVEs. Because TVGs have no independent tax power, are not adequately funded by senior governments, and are not permitted to operate with a deficit, they can play a constructive role in rural development only if they have a source of extrabudgetary revenue.[21] With the adoption of the household responsibility system and the consequent removal of agricultural assets from government control, the collectively owned non-agricultural assets (the TVEs) have become the TVGs' most reliable and productive source of extrabudgetary revenues. The growth of TVEs is also favoured by governments at the county and provincial levels since it expands their revenue base. Understandably, the increased importance of TVEs as a source of local revenue and economic power has significantly enhanced local governments' interest in their management and development.

This section discusses why local governments value TVEs and why they give TVE development such high priority. We do this by examining China's public finance at the local level and the relative importance of TVEs as a source of local revenue. The section is organized as follows. To place our discussion in perspective, a brief description of the Chinese fiscal system and recent fiscal reforms are provided. The redistributive impact of the national budget is then examined, and this is followed by a discussion of how fiscal reforms have affected central–local fiscal relations and why provincial and country governments are under fiscal pressure to expand their local revenue base and to tap extrabudgetary funds, of which the most important source is the income controlled by locally owned enterprises. The section concludes by examining the structure of revenue and expenditure at the township and village levels, and the data collected from three township governments (TGs) and five village governments (VGs) in Jiangsu show that the fiscal pressure to expand the local revenue base and to tap extrabudgetary funds that exist at the provincial and county levels are as intense, if not more so, at the township and village levels.

The National Budget and Fiscal Reforms

Because the Chinese government wanted the national budget to be a powerful allocation and redistribution instrument, it developed a highly centralized fiscal system in the 1950s under which the revenues and expenditures of provincial and county governments were made part of a consolidated national budget, so that in effect the only independent budgetary administration was at the central level.[22]

In brief, the situation in the early 1970s was as follows. Taxes collected by all levels of government as well as the profits of state enterprises and large urban collectives were all included in the national budget. Budgetary allocations to local governments were earmarked for specific purposes and had to be used accordingly. However, several sources of funds remained outside the national budget, the most important being funds controlled by enterprises, surtaxes and various fees collected by local governments, and funds controlled by rural communes. These extrabudgetary funds were at first very limited, but the size increased over the years as changes were made in the fiscal system. Nevertheless, in the late 1970s they were still relatively small, amounting to about 35 per cent of budgetary revenue.[23]

In the 1980s, the fiscal system was changed in substantial ways. Reform measures adopted in the early and mid-1980s (e.g. profit remittance contracts (*baogan*) and the conversion of profit remittance to tax) allowed extrabudgetary funds to expand rapidly, which of course reduced the relative importance of the consolidated national budget and the government's control of resources.[24] Between 1978 and 1989, the share of budgetary revenue in national (material) income dropped from 37 per cent to 22 per cent,[25] and the share of investment by state-owned units financed directly through the budget declined from 83 per cent to only 13 per cent.[26]

The revenue-sharing relations between the different levels of government also changed in the 1980s. In 1980, China introduced a system called revenue-sharing by sources. Although local governments continued to collect virtually all taxes for the national budget, the revenue from some taxes was retained totally at the local level, some was shared between the provinces and the central government, and some went entirely to the central government.[27] In some provinces revenue-sharing was applied to all revenues. Profit remittances were changed to direct taxes after 1983–4 but the revenues generated by these taxes continued to accrue to different levels of government according to enterprise ownership. A revenue-sharing arrangement was also introduced between the province and subprovincial administrative units down to the county level.[28] Then in 1988 Jiangsu and some of the other more developed provinces adopted the progressive revenue-sharing responsibility system (*shouru dizhen baogan tizhi*). Under this system, budgetary revenue collected in the province in the current year up to an amount equal to 105 per cent of the actual budgetary revenue collected in the previous year (R_o) is divided between the province and the central government according to a predetermined sharing ratio (Jiangsu's share in the late 1980s was 41 per cent) and the remainder (i.e. revenue above R_o) is kept entirely by the province.[29]

The Budget and the Redistribution of Income

Despite revenue-sharing and the decline in the relative importance of the national budget, the budget continued to be an important redistributive instrument in the 1980s. This was because all revenue-sharing schemes, those between the central government and the provinces as well as those between the provinces and the subprovincial administrative units, were constructed to reflect the division of revenues in the base year.[30] In other words, under the various sharing schemes, prosperous areas that produced budgetary surpluses in the base year continued to produce surpluses after the adoption of revenue-sharing, and most poorer areas that had budgetary deficits (i.e. received a net inflow of resources from the budget) in the base year continued to receive a net inflow of resources. In the 1980s, the budget redistributed income both between provinces and between regions within provinces.

Because Jiangsu was a relatively rich and industrialized province, it consistently generated far more revenue for the national budget than its budgetary expenditure. In 1988, for example, Jiangsu's budgetary revenue was RMB 11.5 billion and its budgetary expenditure was RMB 7.9 billion, or 68 per cent of revenue.[31] In other words, in that year, the national budget extracted a net flow of RMB 3.6 billion from Jiangsu. Since not all parts of Jiangsu were equally prosperous the budget also redistributed income between localities within the province.

Table 7.2 presents the 1988 per capita budgetary revenue and per capita budgetary expenditure for Jiangsu's eleven regions (*diji shi*). Regions with high per capita national (material) income (e.g. Wuxi, Suzhou, Changzhou, and Nanjing) generated far more in budgetary revenue than they spent in budgetary expenditure. By contrast, the poor regions in north Jiangsu (e.g. Huaiyin, Xuzhou, Yancheng, and Lianyungang) collected only slightly more in revenue than they spent in budgetary expenditure. In the 'have' regions, on a per capita basis, budgetary revenue exceeded budgetary expenditures by RMB 351 in Wuxi, RMB 239 in Suzhou, RMB 244 in Changzhou, and RMB 219 in Nanjing. In the 'have not' regions, the per capita surplus was RMB 5 in Lianyungang, RMB 2 in Yancheng, RMB 32 in Xuzhou, and RMB 7 in Huaiyin. In other words, the huge surplus transferred out of Jiangsu by the budget came almost entirely from the rich southern third of the province.

The eleven regions in Jiangsu administered 64 counties (*xian* or *xianji shi*) and 41 urban districts (*shixia qu*) organized into 11 municipalities.[32] In 1988, of the 64 counties, 24 were in deficit in the sense that the revenues they generated for the budget were less than what they spent in budgetary expenditure, i.e. the national budget injected a net inflow of resources to these counties. Except for one, all 24 counties were in north Jiangsu (defined in Chapter 3 as the north mountainous (NMR) and the north coastal (NCR) regions), where per capita national (material) income was substantially below the provincial average (RMB 1,543 in 1988).[33] Since the budget did not inject a net flow of resources into any region in north Jiangsu, the redistribution among counties occurred primarily within each region, with resources flowing from its more prosperous urban districts (*shixia qu*) to its poorer counties. For example, in

TABLE 7.2. *Per Capita Budgetary Revenue and Expenditure[a] by Administrative Region, Jiangsu, 1988*

Region	Population (million)	Per capita national (material) income (RMB)[b]	Per capita budgetary		Per capita surplus (RMB)	Number of counties in region	Number of[c] deficit counties in region
			Revenue (RMB)	Expenditure (RMB)			
Wuxi	4.07	2,988	479	128	351	3	0
Suzhou	5.51	2,759	350	111	239	6	0
Changzhou	3.17	2,289	365	121	244	3	0
Nanjing	4.88	2,223	374	155	219	5	1
Zhenjiang	2.54	1,912	204	105	99	4	0
Yangzhou	9.08	1,349	112	69	43	10	1
Nantong	7.72	1,323	142	70	72	6	0
Lianyungang	3.18	1,212	98	93	5	3	3
Yancheng	7.50	1,017	69	67	2	7	4
Xuzhou	7.44	1,004	111	79	32	6	5
Huaiyin	9.40	893	77	70	7	11	9

[a] Budgetary revenue collected and budgetary expenditures made by county and township governments.
[b] National (material) income is the value added of the material products sectors.
[c] A deficit county is one where budgetary expenditures > budgetary revenues.

Source: Calculated from data in *JJN* (1989), 3. 103–5, 3. 109–11, and 3. 130–2.

Huaiyin, its urban districts generated a surplus of RMB 173 million in 1988, and nearly two-thirds of this was used to subsidize its rural counties.[34] At least in a 'have' province like Jiangsu, intra-provincial redistribution through the national budget occurred primarily within each region, and there was little or no transfer between regions.

Since the urban districts in north Jiangsu were neither large in size nor highly industrialized, the extent of redistribution within each region was limited and far from sufficient to offset the underlying differences in resource endowments and levels of development. In 1988, the per capita expenditure allocated through the budget in the 'have not' regions was substantially lower than that in the 'have' regions. However, because of redistribution within each region, the variation in per capita expenditure was substantially smaller than that in per capita revenue.[35]

Fiscal Pressure on Local Governments

Fiscal reforms in the 1980s increased substantially the fiscal pressure on local governments in both rich and poor regions. While fiscal reforms changed the revenue-sharing arrangements between the different levels of government, they did not change significantly the division of expenditure responsibilities.[36] The central government continued to be responsible primarily for debt servicing, national defence, and capital construction, and local governments continued to be responsible primarily for agriculture, social services (education and health), and administration. Because of this division of responsibilities, the fiscal decline in the 1980s affected the finances of local governments more adversely than that of the central government.

The central government was able to adjust to the fiscal decline by reducing the budgetary outlays for capital construction, a cutback that was a part of the adopted reforms which called for the shift of resources to enterprises. However, local governments found it impossible to adjust to the fiscal decline by reducing their budgetary expenditures since they were mostly current obligations, including price subsidies on urban food grains. To make matters worse, policies introduced by the central government in the 1980s (e.g. higher national standards on health and education) further increased the financial burden on local governments, and, on several occasions, the central government also transferred new current expenditure obligations to local governments. For example, when Jiangsu was put on the progressive revenue-sharing responsibility system in 1988, the province was also told to take over the losses sustained by local FTCs and by enterprises involved in the distribution and processing of urban food grains.[37]

Because their expenditure burden has increased while their share of resources has declined, many local governments at the provincial and county levels are chronically short of budgetary funds. Indeed, after meeting the expenditure obligations imposed on them by central government policies, most local governments have little budgetary funds left for discretionary use. Thus, despite fiscal decentralization, local autonomy in the allocation of budgetary expenditures has remained limited.

Not surprisingly, the increased fiscal pressure has forced local governments to seek ways to increase their revenues. Promoting local economic development, particularly the growth of locally owned enterprises, is the most direct way to expand the local revenue base since the income taxes paid by these enterprises belong to local governments. A second popular method is to use extrabudgetary funds to finance government expenditures. Since the most important source of extrabudgetary funds is the income controlled by locally owned enterprises, this method also requires the development of local industries. In other words, local governments at both the provincial and county level have strong financial reasons for promoting the development of locally owned enterprises, including TVEs. This is undoubtedly one, and perhaps the main, reason why TVGs have been under pressure from provincial and county governments to develop TVEs and why they reward rural cadres so generously when TVEs under their management develop and prosper (see Table 7.1).

The evidence presented above suggests that despite fiscal decline and the transfer of fiscal control to local government in the 1980s, the national budget continues to be an important redistributive instrument. Each year, huge amounts of resources are transferred by the national budget from Jiangsu, primarily from its most prosperous southern counties, to other parts of China. The budget is also used to redistribute income within the province. In Jiangsu, for example, resources are transferred from the more prosperous urban districts in north Jiangsu to their surrounding rural counties. Reforms in the 1980s have put local governments under intense fiscal pressure forcing them to take a strong interest in the development and the protection of locally owned industries, the source of much of their budgetary and extrabudgetary revenues. In consequence, provincial and county governments have become more involved than ever in the economy.

Total Receipts and Outlays of Township and Village Governments

This section examines the financial resources available to TVGs, and the evidence suggests that the pressure to develop locally owned enterprises also exists at the township and village levels and is one reason why TVGs have become so involved in the development and management of TVEs.

Township Finance

The separate accounts for budgetary and extrabudgetary receipts and expenditures for three township governments are reproduced in Table 7.3, and they show three major revenue source: (i) allocations from the national budget, (ii) education surcharges authorized by the provincial government, and (iii) locally generated and controlled revenue (the township fund). The last two sources are of course extrabudgetary revenue, and Table 7.3 shows that they are far more important in the rich and more developed townships than in the poor ones. Comparing the three townships, Sigang is by far the most prosperous, Hufu is ranked next, and Haitou is the least developed.[38] In 1987, of the total financial resources available to each township government, extrabudgetary revenue accounted for 93 per cent in Sigang, 75 per cent in Hufu, and 25 per cent in Haitou. The data also show that the township fund is by far the most important extrabudgetary revenue source.

The township government collects taxes for the state from economic units (the most important being TVEs) under its jurisdiction,[39] and it also makes expenditures using funds allocated from the national budget. However, the size of the budget allocation is not determined by the amount of tax revenue collected by the township government. Hufu, a relatively prosperous township, collected more than RMB 5.2 million in taxes for the state in 1987 but received in the same year only RMB 0.41 million in budget allocation.[40] In contrast, Haitou, a relatively poor township, received slightly more in budget allocation (RMB 0.979 million) than it collected for the state in taxes (RMB 0.907 million). Thus, from the township's viewpoint, its finances are determined not by how much taxes it collects but by what it receives in budget allocation, i.e. item II.A in Table 7.3 (receipts from budget allocation). In 1987, Sigang, the most prosperous of the three townships, received the smallest allocation (RMB 190,000), Hufu received the next smallest (RMB 409,000), and Haitou, the least developed, received the most (RMB 979,000). On a per capita basis, the more appropriate comparison, Sigang received RMB 6.67 per person in state allocation, Hufu RMB 16.6, and Haitou RMB 28.28. In other words, the amount of budget allocation received by each locality was related inversely to its level of development.

Allocations from the national budget must be used by each township according to principles and rules established by the province. In Jiangsu, townships use state funds primarily for administration (e.g. salaries for state cadres), social services (pension and relief), education, and agricul-

TABLE 7.3. *Receipts and Expenditure of Three Township Governments,*[a] *Jiangsu, 1986 and 1987 (RMB 000)*

	Sigang		Hufu	Haitou	
	1986	1987	1987	1986	1987
I. Total (II + III + IV)					
A. Receipts	1,685	2,903	1,635	963	1,305
B. Expenditure	1,665	2,953	1,620	944	1,182
C. Surplus (deficit)	20	(50)	15	19	123
II. Budget allocations					
A. Receipts	192	190	409	595	979
B. Expenditure	192	190	409	590	894
Administration	50	52	39	53	76
Agriculture and irrigation	40	61	36	130	275
Education and health	5	6	257	318	456
Pension and relief	62	68 }		{ 79	76
Others	36[c]	2 }	78	{ 9	12
C. Surplus (deficit)	—	—	—	5	85
III. Special education surcharge					
A. Receipts	409	538	350	98	53
B. Expenditure	405	539	338	89	68
Salaries	186	225	137	50	50
Others	218	314	201	39	18
C. Surplus (deficit)	4	(1)	12	9	(15)
IV. Township fund					
A. Receipts	1,084	2,175	876	270	273
Remittances from	838	895	863	228	155
TEs	813	874	855	220	n.a.
Others	25	21	8	8	n.a.
Contributions from TEs	160	1,215	—	—	—
Others[b]	87	65	13	42	118
B. Expenditure	1,068	2,224	873	265	220
Administration	171	201	225	158[d]	71
Agriculture and irrigation	267 }		{ 82	—	—
Public construction	411 }	1,796	{ 173	—	—
Road	143	n.a.	20	—	—
Small town development	268	846	153	—	—
Sidelines	57	45	—	—	—
TE development	—	—	353	61	27
Education and health	129	104	3	—[d]	95
Others	33	78	36	46	27
C. Surplus (deficit)	16	(49)	3	5	53

[a] Sigang is in Zhangjiagang (1 of 6 counties in Suzhou *shi*), Hufu is in Yixing (1 of 3 counties in Wuxi *shi*), and Haitou is in Ganyu (1 of 3 counties in Lianyuangang *shi*). In 1987, total population was 28,493 in Sigang, 24,607 in Hufu, and 34,623 in Haitou.
[b] Township government's share of collected tax revenue, fines, etc.
[c] Of this, RMB 30,000 was used for road construction.
[d] In this year, expenditure for education and health is probably included in administration.

Sources: Compiled from Township-Village Interview Notes and Township-Village Background Data.

ture. However, it is important to note that the national budget funds agriculture and education much more generously in poor than in rich localities. In fact, in some of the more developed townships, budget allocations are sufficient only to support the government bureaucracy and to pay for social services. In fact, in the more prosperous townships the bonuses and allowances paid to state cadres are also partly or totally financed out of extrabudgetary revenue.[41] Finally, it should be noted that, in both poor and rich localities, very little if any of the budget allocation is used to promote rural non-agricultural development.

To understand why paying the bonuses and allowances of state cadres from extrabudgetary revenues is important, a brief explanation of China's cadre system may be necessary. China has two categories of cadres: state cadres (*guojia ganbu*) and local cadres (*difang ganbu*). Cadres at the township level are state cadres while those at the village and team levels are local cadres. State cadres are paid according to a uniform national pay scale and their salaries are included in the national budget. However, local cadres (*difang ganbu*) are paid from extrabudgetary funds, so their incomes vary across localities in accordance with local income levels, which have become very high in the more industrialized regions. Using extrabudgetary funds to finance the bonuses and allowances of state cadres in the more developed localities allows them to earn incomes more or less in line with those of local cadres without at the same time increasing the burden on the state budget. This of course means that state cadres in rich townships earn substantially higher incomes than those in less developed townships, and also higher incomes than cadres at more senior levels whose salaries are funded entirely by the state.[42] Linking state cadres' incomes to the local income level also has the effect of increasing the incentives of state cadres to promote economic development at the local level.

To foster rural education, Jiangsu has authorized its townships to collect an education surcharge (*jiaoyu fujiafei*) to help pay for the salaries of local teachers (*minban jiaoshi*), the repair of school buildings, and the purchase of school equipment. In townships where rural industry is developed and profitable, the surcharge has been paid entirely by TVEs (usually calculated as a percentage of annual revenue).[43] In less developed townships where there are few profitable TVEs, the amount that could be collected from enterprises has been insufficient, so households have also been required to pay a surcharge which has been collected by the villages and remitted to the townships.[44] Table 7.3 shows that rich townships such as Sigang have relied primarily on the education surcharge to finance rural education, whereas in poor townships like Haitou where little revenue could be raised locally for education, the state has been obliged to shoulder a much higher share of the cost of educating the rural young.

The township fund consists of the following revenues: various remittances from township enterprises (TEs) and other income-producing ventures (e.g. cinemas), contributions from TEs, the township's share of the collected tax revenues, and miscellaneous fines. Table 7.3 shows TEs are by far the most important revenue source.

The fiscal link between TEs and the township government is strong both in prosperous townships that have many profitable TEs like Sigang and in relatively poor townships that have few profitable TEs like Haitou. In the late 1980s, the common practice in Jiangsu was for TEs to remit to the township government a fixed quota profit and a percentage of the above-quota profit.[45] In 1987, TEs in Jiangsu remitted to TGs on average about one-third of their after-tax profits.[46] Township enterprises also pay a management fee (*guanlifei*) (essentially a tax since TEs receive little or no management service in return) to the township industrial corporation, of which it keeps a portion and transfers the remainder to the county rural enterprise bureau (*xian xiangzhen qiye guanliju*). When necessary, the township government may also call on its TEs to make contributions. For example, in Haitou TVEs paid a monthly fee of RMB 10 per worker into a farm construction fund. When Sigang needed funds to construct a major road in 1987, it turned to its enterprises and raised RMB 1.2 million (56 per cent of all township funds raised that year) in contributions (*xiangzhen qiye jijin*). Raising funds this way also has a tax advantage since enterprises may deduct their contributions as expenses and thus reduce their income tax.

There are reasons to believe that TVEs provide more funds to TVGs than suggested by the officially reported remittances and contributions. For example, several TVEs we surveyed reported larger amounts of funds transferred to the TVG than what the local government reported. It is common practice for TVGs to disguise, whenever possible, some of their legitimate and not-so-legitimate operating costs (e.g. banquets, liquor and cigarettes, transportation) as enterprise business expenses (paid usually by one of their more profitable TVEs). By doing this, of course, the TVG not only covers some of its expenses but also reduces the amount of income tax its profitable TVEs have to pay.

Compared to TEs, all other revenue sources are of secondary importance. In 1986, the public finance system was extended to the township level in the more developed parts of Jiangsu. Under the new scheme, township governments were permitted to retain a share of the tax revenue they collected for the state. The most common practice was for the township government to retain a percentage of any revenue collected above a predetermined quota.[47] The quota and the retention share varied by township. However, as yet revenue-sharing has not developed into an important revenue source, and it is easy to understand why. While most township governments are conscientious about meeting their tax quotas, there is little incentive for them to collect large amounts of above-quota taxes from community-owned businesses.[48] As one senior township cadre remarked, 'why should the township collect more, since what is not collected belongs entirely to the township?'.[49] And, furthermore, exceeding the quota by large amounts may also lead to higher tax targets in future years. Fines provide TGs with another source of revenue, but in absolute magnitude they are not very important.[50]

The township fund has special importance for township governments because it is the only fund over which they have discretionary control. In the more prosperous townships, it is frequently the only fund available to promote local development, and not surprisingly it is used largely for this purpose. In Sigang and Hufu, for example, more than two-thirds of the expenditures from the township fund were used for economic and small town development. In less developed townships, however, after paying the basic administrative and social expenses, there is usually little left in the township fund for development purposes. For example, Haitou was able to allocate less than 13 per cent of its township fund, about RMB 27,000, to TVE development in 1987.

Village Finance

Because villages do not receive any allocation from the national budget, their outlays are funded entirely by extrabudgetary revenues. Thus, community-owned businesses are even more important to local governments at the village than at the township level. The strong fiscal link between village enterprises and VGs is clearly evident in Table 7.4 where the receipts and expenditures of five Jiangsu villages are presented. To put the data in Table 7.4 in context, it should be noted that three of the five villages (Zhashang, Yangquan, and Xiaojian) are located in relatively industrialized townships in the south, and the other two (Haiqian and Haiqi) are in Haitou, a far less developed township in the north.[51] In terms of per capita household income, Zhashang with an income of RMB 1,313 per person was highest in 1987, followed in descending order by Yangquan (RMB 1,210), Haiqian (RMB 876), Xiaojian (RMB 618), and Haiqi (RMB 569).

The fiscal consequence of *da baogan* was that it removed agricultural assets from collective control and thereby made it more difficult to extract revenue from agriculture to finance village operation. However, in the 1980s, it was still possible to raise funds by imposing levies on households (e.g. 'the two funds and one fee'[52]), but in Jiangsu only villages desperately short of revenue did so. Of the five villages in Table 7.4, only one (Haiqian) collected fees directly from households. Collecting fees from households was not only time consuming and difficult but the amount that could be collected was usually quite small. Private businesses paid a management fee, but much of it went to the industrial corporation at the township level. Sometimes private businesses were also asked to make contributions to help finance special village projects, but this was not a reliable source of revenue. Besides, senior governments disapproved of the proliferation of this type of levy.

The lack of revenue alternatives left village enterprises the only reliable source able to provide sizeable revenues for village governments. The data in Table 7.4 certainly confirm this. Profit remittances and management fees from VEs accounted for nearly all the revenue (excluding loans) raised by VGs in Zhashang, Yangquan, and Xiaojian. Haiqi had no village enterprise but obtained all its revenue from its most important community-owned asset (prawn ponds), and, in Haiqian, remittances from its few small VEs accounted for over 40 per cent of the revenue collected by its VG.

Many VGs in south Jiangsu, because they have a large number of profitable VEs, have access to discretionary funds that are significantly larger than those available to TGs in the less developed parts of Jiangsu. For example, in 1987 the village governments in Zhashang and Yangquan collected respectively RMB 0.76 million and RMB 0.59 million from their VEs, while the government in Haitou, a less developed township but certainly not among the poorest in Jiangsu, was able to raise only RMB 0.33 million from all its extrabudgetary revenue sources.

A part of the revenue collected in each village is the township's share of the management fee and this of course is turned over to the township. The rest is used for cadres' salaries, social services, and village development (investment in VEs and in village infrastructure). Because village

TABLE 7.4. *Receipts and Expenditure of Five Village[a] Governments in Jiangsu, 1987 (RMB 000)*

	Sigang Township	Hufu Township		Haitou Township	
	Zhashang	Yangquan	Xiaojian[h]	Haiqian	Haiqi
Receipts	1,504.3	593.0	85.4	101.3	64.3
VE remittances	764.5	593.0	85.4	42.8	—
Liangjin yifei[b]	—	—	—	37.0[j]	—
Education surcharge[c]	—	—	—	18.5[k]	—
Other revenues	189.8	—	—	3.0	64.3[i]
Loans	550.0	—	—	—	—

TABLE 7.4. *Continued*

	Sigang Township	Hufu Township		Haitou Township	
	Zhashang	Yangquan	Xiaojian[h]	Haiqian	Haiqi
Expenditure	1,504.3	593.0	85.6	101.3	64.2
Cadres' wages and subsidies	148.0	⎫143.0⎧	12.0	18.0	11.7
Management fee	119.1	⎭ ⎩	—[l]	2.0	4.9
Support of agriculture	42.4	58.0	28.1	14.0	2.0
Family planning[d]	39.2	22.0	9.3	—	—
Public welfare[e]	6.1	6.0	10.5	9.0	7.5
Pension payments	102.3	—	—	—	—
Education surcharge remitted	—	—	—	17.5	9.6
VE development[f]	735.0	250.0	—	—	—
Village construction[g]	111.2	62.0	17.3	40.8	25.0
Others	201.0[m]	52.0	8.5	—	3.5
Surplus (deficit)	—	—	(0.2)	—	0.1

[a] There are 16 villages in Sigang, 22 villages in Hufu, and 18 villages in Haitou. In 1987 there were 2,566 people in Zhashang, 1,162 in Yangquan, 939 in Xiaojian, 3,720 in Haiqian, and 2,510 in Haiqi.

[b] 'Two funds and one fee'. However, when villages in Jiangsu have sufficient income from other sources, they usually do not collect these fees.

[c] In Haitou, a portion of the special education surcharge was contracted to its villages in the form of remittance quotas. Apparently some villages did not collect the surcharge directly from households because they had sufficient income from other sources.

[d] Included here are the bonuses given to one-child families.

[e] Subsidies given to the *wubao hu* (households that enjoy the 5 guarantees) and families of martyrs.

[f] Funds provided to VEs (usually in the form of interest-free loans) for capital construction and as working capital.

[g] Includes road improvements, village beautification, house renovation subsidies, construction of public buildings, etc.

[h] The figures in this column are for 1986.

[i] Of this, RMB 63,900 was in fees paid to the village by those who had contracted village-owned prawn ponds.

[j] Each household was assessed RMB 5 for every *mu* of responsibility land plus the monetary value of its social labour obligations (calculated at RMB 30 for each male worker and RMB 24 for each female worker in the household).

[k] Each household was assessed RMB 5 per person.

[l] Included with 'others'. In 1987, Xiaojian paid RMB 3,516 in management fees.

[m] Of this, RMB 117,400 was used to purchase state bonds.

Sources: Township-Village Interview Notes and Township-Village Background Data.

cadres are paid salaries that are closely linked to local income levels, much more is spent on cadres' salaries and benefits in rich villages (e.g. Zhashang) than in poor ones (e.g. Haiqian and Haiqi).[53] The rich villages also spent considerably more on social services (e.g. pensions) and on village development. It is also interesting to note that because Zhashang was rich and owned several profitable enterprises, it was able to borrow substantial sums in 1987, providing the village with even more resources to develop its economy.

Extrabudgetary funds are the primary source of revenue for governments at the township and village levels. Because the most important source of extrabudgetary revenue is locally owned enterprises, extrabudgetary funds are far greater in the better-endowed and the more industrialized localities than in the poorer and less developed areas. It is easy to understand why many TVGs believe that local development and prosperity can be achieved only through the development of TVEs. Townships and villages receive little allocations from the national budget, so rural communities must rely largely on their own resources to finance local development, and, since the implementation of the household responsibility system, the most reliable source of such revenue has been the TVE. Not surprisingly most TVGs consider the development of a profitable TVE as a necessary first step to local development.

Financial Links Between Township-Village Enterprises, Township-Village Governments, and Rural Development

How important were the financial links between TVEs and the rest of the economy? The available provincial data are less complete than the township and village data presented above but they nevertheless provide a sense of the magnitude of the financial flows generated by TVEs. Table 7.5 presents the principal financial flows from TVEs to the rest of the economy for Jiangsu and the country as a whole.

TVEs are linked financially to the rest of the economy by the income they contribute to rural households, by the taxes they pay to the state, by their remittances to TVGs, and by development expenditures financed from their retained earnings. These links are discussed briefly below.

The impact TVE development has had on rural income will be discussed in detail in Chapter 8. Suffice it to say for now that the level of TVE development is a principal if not the primary determinant of per capita household income in the countryside. In 1987, TVEs in Jiangsu injected RMB 6.68 billion into the rural economy as wages and salaries.

TABLE 7.5. *Principal Financial Flows Involving TVEs, Jiangsu and China, 1987*

	China		Jiangsu	
	RMB (million)	% of net profit	RMB (million)	% of net profit
Salaries and wages paid to rural households	42,758	—	6,681	—
Direct and indirect taxes paid to the state	13,073	—	2,326	—
Net profit	18,775	100.0	1,710	100.0
1. Remitted to TVGs	7,206	38.4	585	34.2
Used for:				
Supporting agriculture	846	4.5	46	2.7
TVE development[a]	2,418	12.9	150	8.8
Collective welfare	509	2.7	34	2.0
Small town development	406	2.2	43	2.5
Education	472	2.5	45	2.6
Unidentified	2,555	13.6	267	15.6
2. Retained by enterprise	11,013	58.6	1,125	65.8
Used for:				
TVE development[a]	7,579	40.4	751	43.9
Collective welfare	1,294	6.9	184	10.8
Education	446	2.4	56	3.3
Unidentified	1,694	9.0	134	7.8
3. Unidentified	556	3.0	0	0

[a] Allocated to TVEs for use as working and fixed capital.
Source: *ZNN* (1988), 317–19.

This works out to RMB 128.4 per rural person, or about 20 per cent of Jiangsu's 1987 rural per capita income of RMB 626.5.[54] TVEs were of course much more developed in Jiangsu than in most other parts of China. Thus, for the country as a whole, the amount of salaries and wages TVEs paid per rural person was much smaller, only RMB 74.1.

Jiangsu's TVEs earned slightly more than RMB 4 billion in gross profit (net profit plus all taxes) in 1987. Of this, 58 per cent (or RMB 2.33 billion) went to the state as taxes, 17 percentage points higher than the national average. That TVEs in Jiangsu shouldered a heavier tax burden was a reflection of China's steeply progressive income tax scale and of the fact that TVEs in Jiangsu earned significantly larger profits than TVEs in most other parts of China. The taxes collected from TVEs accounted for a substantial share of all revenues mobilized through the budget in

Jiangsu. Taxes paid by TVEs were nearly 22 per cent of the budgetary revenue collected in Jiangsu.[55] By comparison, agricultural tax contributed only 3.2 per cent to Jiangsu's budgetary revenue. Some of the taxes collected of course remained in the countryside and were used to finance local budgetary expenditures. But judging from data presented earlier in the chapter, except in poor counties located primarily in the north, a significant share of the taxes collected from TVEs was removed from rural Jiangsu.

The use of after-tax profit (i.e. net profit) is decided locally, based largely on the needs and priorities of TVGs. In 1987, of the net profit generated by TVEs in Jiangsu, about one-third was remitted to TVGs and two-thirds were retained by enterprises. The retained profit was used primarily further to develop the enterprises. The only other major expenditure was collective welfare, which accounted for 16 per cent of the retained profit. Of the RMB 585 million remitted to TVGs, 55 per cent was used for the following purposes: support of agriculture (7.9 per cent), collective welfare (5.8 per cent), small town development (7 per cent), education (7.6 per cent), and TVE development (26.7 per cent). A large portion of the unidentified expenditures went to cover salaries and administrative and overhead expenses. At the national level, the net profit of TVEs was used in a roughly similar manner.

A longer perspective of how the net profits of TVEs were used is presented in Table 7.6. Between 1978 and 1988, in Jiangsu as well as in China as a whole, at least two-thirds of the net profits were used to promote rural development in the following three ways: support of agriculture, support of TVE development, and support of collective welfare.

Support of Agriculture

A substantial share of the net profits earned by TVEs was used to subsidize (*bu*) and to build (*jian*) agriculture. In China as a whole, the total support given by TVEs to agriculture between 1979 and 1989 was equivalent to about one-third of the total state investment in water conservancy, forestry, and meteorology during the same period.[56] Included in 'support of agriculture' (Table 7.6) were funds used (i) to purchase farm equipment and machineries, (ii) to finance capital construction (e.g. the repair of farm fields and the construction of irrigation, roads, and bridges), and (iii) to support poor brigades (villages).

In Jiangsu, the share allocated to the support of agriculture was about one-fourth of net profit in the late 1970s, and, of the funds allocated, 30 per cent was used to purchase agricultural equipment, 50 per cent for farm capital construction, and 20 per cent to support poor brigades.[57] However, the amount allocated to support agriculture declined sharply in

TABLE 7.6. *TVE Net Profit by Use, Jiangsu and China, 1978–1988 (RMB million)*

	1978	1980	1982	1983–5	1986	1987	1988
China							
Net profit[a]	8,810	11,840	11,550	41,788	16,103	18,775	25,918
Distributed by use:							
TVE development[b]	3,090	4,700	4,800	19,094	8,023	9,997	13,819
Support of agriculture	2,640	2,270	1,400	2,901	692	846	1,160
Collective welfare	400	680	900	4,688	1,446	1,804	2,641
Distributed to commune members	n.a.	n.a.	n.a.	2,956	n.a.	n.a.	n.a.
Education	n.a.	n.a.	n.a.	n.a.	n.a.	917	n.a.
Small town development	n.a.	n.a.	n.a.	n.a.	n.a.	406	n.a.
Unidentified	2,680	4,190	4,450	12,149	5,942	4,805	8,298
Jiangsu							
Net profit[a]	966	1,808	1,441	5,484	1,474	1,710	2,491
Distributed by use:							
TVE development[b]	383	860	685	3,015	850	901	1,389
Support of agriculture	219	299	132	301	45	46	121
Collective welfare	39	89	117	655	167	217	308
Distributed to commune members	n.a.	198	183	260	n.a.	n.a.	n.a.
Education	n.a.	n.a.	n.a.	n.a.	n.a.	100	n.a.
Small town development	n.a.	n.a.	n.a.	n.a.	n.a.	43	n.a.
Unidentified	325	362	324	1,253	412	403	673

[a] Gross profit minus all indirect and direct taxes.
[b] Allocated to TVEs for use as working and fixed capital.

Sources: Compiled from data collected in the field and in *ZNTN* (1985), 190; *ZNN* (1984), 123–44; *ZNN* (1985), 181–2; *ZNN* (1986), 228–30; *ZNN* (1988), 318–19; *ZTN* (1985), 205, and *ZTN* (1989), 242–3.

the early 1980s, and between 1982 and 1988 averaged only slightly more than 5 per cent of net profit. The decline was also observed at the national level.

Support of agriculture declined partly because rapid economic growth in rural Jiangsu reduced the number of villages that needed financial help, and partly because, given China's macro environment, it was not attractive to invest more heavily in agriculture. However, the decline in the level of support provided by TVEs to agriculture was probably less dramatic than suggested by the figures in Table 7.6. Since the mid-1980s, some TVGs, particularly those in south Jiangsu, have used TVEs to subsidize agriculture in ways that may not be reflected in the figures presented in Table 7.6.

In those parts of Jiangsu where rural non-agricultural activities were developed and the opportunity cost of farming was high, peasants were often reluctant to work their contract land intensively. This problem became quite serious in the mid-1980s, when the terms of trade turned against agriculture and farm profits began to decline sharply.[58] To ensure that cultivation, particularly grain production, was not neglected, some of the more developed villages began to experiment with ways to use TVEs and the income they generated to make agriculture economically more attractive.[59] The most direct way was to provide farm households with an income subsidy. For example, some villages paid farm households a fixed subsidy for every *mu* of land cultivated (and the size increased substantially in the 1980s);[60] some paid farm households a subsidy for every kg. of grain they sold to the state; and some did both.

A more indirect way was to merge farming with the more lucrative collective industrial sector (*nonggong yitihua*). For example, some TVEs in Jiangsu have taken over the responsibility of agricultural production by hiring and organizing agricultural households into workshops (*chejian*) that specialize in cultivation. In one township, 286 specialized grain-producing households with 946 *mu* of land were merged with various TVEs.[61] The participating households were given production norms, i.e. a minimum yield (an output quota), input norms, and a base wage per unit of land cultivated. To receive the base wage, the minimum yield had to be achieved. In addition, households were responsible for all above-norm production expenses. Earnings from the quota output went to the enterprise but above-quota production belonged to the household. TVEs also paid the agricultural tax and the collective accumulation and public welfare funds on behalf of the participating households. In effect this arrangement allowed specialized grain-producing households to earn a higher income than they could otherwise, given the cost of inputs and the official purchase price of grain.

Another popular way of subsidizing agricultural production (particu-

larly in south Jiangsu) and of ensuring that farming is undertaken successfully is to use agricultural service workshops (*nongye fuwu chejian*) to
assist part-time farmers to work their contract land. Workers in these
workshops are paid a fixed salary either by the village or by VEs to do
most of the farm work (e.g. land preparation, ploughing, irrigation, and
pest control) under village supervision.[62] Thus, in villages where these
service workshops exist, individual households are responsible only for
the planting, the occasional weeding, and the harvesting of their contract
land. Since the amount of land contracted to each household in many
parts of south Jiangsu is very small, household members can easily
complete these chores by working, on average, a few hours a week. By
relying on agricultural service workshops, the vast majority of villagers in
parts of south Jiangsu have been able to work full-time in non-agricultural
jobs while still farm as a sideline.

More recently, some counties in Jiangsu have experimented with
more direct measures to support agriculture. For example, in 1987,
Zhangjiagang *shi* established an agricultural development company, and
all enterprises under its jurisdiction (i.e. locally owned state enterprises,
large collective enterprises, and TVEs) must contribute a percentage of
their annual sales to the company.[63] The collected funds are then used by
the company to invest in agriculture and to provide credits to agriculture.

Support of Collective Welfare

Between 1983 and 1988, expenditure on collective welfare accounted for
about 12 per cent of the net profit earned by TVEs. In poor townships,
very little is spent on collective welfare other than welfare payments to
poor households (e.g. supporting the five-guarantees households (*wubao
hu*)).[64] However, in the more prosperous villages, some of the funds
provided by VEs were used to pay in part or in full the medical expenses
of village members. Moreover, starting in the mid-1980s, villages in south
Jiangsu began to use funds from village and team enterprises to provide
pensions to retired cadres and enterprise workers.

Pensions were paid either by the enterprise directly, by the village
using funds contributed by the enterprise for the purpose, or by the
village from its operating fund (which of course was funded largely if not
completely by the enterprises in the village).[65] In 1987 some counties in
south Jiangsu began to experiment with pension funds. In Zhangjiagang,
for example, TVEs made monthly contributions of RMB 10 per worker
to the fund, and retired workers received a pension from the fund. The
size of the pension depended on the worker's age at retirement and the
length of his or her service with the enterprise. Similar schemes also
existed in the Wuxi region, where in 1987 approximately 39 per cent of

rural workers were covered by the pension system. The size of the pension varied by village according to its level of development.[66] In other words, the level of TVE development in a locality determined the extent and the quality of its collective welfare.

TVE Development

The bulk of the net profit earned by TVEs was used for TVE development. Between 1982 and 1988, over 54 per cent of the net profit earned by TVEs in Jiangsu (over RMB 6.8 billion) was ploughed back into the TVE sector as additional working or fixed capital. At the national level, the percentage was 49 per cent. Most of this was from profits retained by TVEs, but TVGs also reinvested a portion of the profits they received from TVEs.

It is easy to understand why such a high share of the net profit is reinvested further to develop the TVE sector. TVEs are community owned, so they are the most reliable source of extrabudgetary revenue for TVGs, and this fact cannot be overemphasized. TVEs ease the tight budget constraints facing all TVGs and thus give them not only greater operational flexibility and therefore a better chance of successfully completing the tasks assigned to them by senior governments, but also a degree of independence. For these reasons alone TVGs would regard TVEs as their most important community-owned assets. But developing TVEs is viewed not only as a way of easing budget constraints. It is also considered as a way out of poverty. In fact, with a fixed amount of land and low agricultural prices, many TVGs see developing TVEs as the best way to promote local development and to raise local income. Finally, since their income, status, and professional success are all linked to the local income level and to the overall economic performance of the collective unit, rural cadres also have a vested interest in developing TVEs.

TVGs are directly involved in the planning and the management of rural development, with township and village cadres the principal decision-makers and entrepreneurs. Because of fiscal decentralization, TVGs are under considerable pressure from senior governments to develop the local economy, particularly the rural non-agricultural sector, and to do so without much support from the state. Thus, in the 1980s, rural communities relied primarily on their own (extrabudgetary) resources to develop the rural non-agricultural sector.

But self-reliance, when combined with unequal initial conditions (e.g. differences in resource base and in levels of development) and relatively low-factor mobility, has produced a highly uneven pattern of rural non-agricultural development. As was shown in Chapter 3, the south delta region, with about 17 per cent of Jiangsu's rural population, accounted for nearly 60 per cent of the rural industrial growth in the 1980s. The better-endowed areas are able to develop more rapidly because they can allocate more resources to rural development in general and TVE development in particular.

Because of the strong fiscal links between TVEs and TVGs, differences in the discretionary financial resources available to TVGs far exceed differences in per capita income or levels of development. A comparison of Sigang and Haitou gives an indication of the magnitude of the differences at the township level. In 1987, the ratios of per capita income, per capita gross value of agricultural and industrial output, and per capita discretionary revenue in Sigang to those in Haitou were respectively 1.5:1, 4:1, and 9.4:1.[67] Redistribution of financial resources through the national budget reduced the difference in spending capacity between Sigang and Haitou, but not sufficiently to eliminate it. In 1987, the ratio between the total financial resources available to Sigang's government and those available to Haitou's government was 2.6:1, still substantially larger than the difference in their per capita incomes.[68]

The inequities in fiscal capacity and in the level of economic development are likely to perpetuate and widen. Rich townships like Sigang and rich villages like Zhashang and Yangquan, because they already have a sizeable number of community-owned enterprises, can collect substantially more revenue than is needed to support the rural bureaucracy and to provide minimum social services. Consequently they have more resources for development purposes, particularly for TVE development and for community development. In addition, in these localities the community-owned enterprises also use the bulk of their retained profits for TVE development.

In contrast, villages and townships in poor areas must use the bulk of their allocated funds and locally mobilized extrabudgetary revenue to support the rural bureaucracy and to provide a minimum level of local services. Indeed, in poor areas fiscal pressures may force TVGs to squeeze TVEs, leaving them with little or no resources for expansion and development.[69] In any case, poor regions have few profitable TVEs to squeeze. In other words, in poor regions rural non-agricultural development cannot generate the momentum it needs to become self-perpetuating. To escape the vicious circle of a small rural non-agricultural sector, low income, low investment, and little or no rural non-agricultural growth, either more external resources must be provided or the population must be permitted to migrate to the more prosperous and developed regions.

NOTES

1. These links have also been examined by others, in particular see William A. Byrd and Lin Qingsong (eds.), *China's Rural Industry: Structure, Development, and Reform* (Oxford: Oxford University Press, 1990), chs. 16 and 17.
2. The focus here is the township and the village. For discussions of economic management at the county level, see Y. Y. Kueh, *Economic Planning and Local Mobilization in Post-Mao China*, Contemporary China Institute Research Notes and Studies No. 7 (London: School of Oriental and African Studies, University of London, 1985), and Y. Y. Kueh, 'Economic Reform in China at the *"xian"* Level', *CQ* 96 (Dec. 1983), 665–88.
3. The contract is between the production team (the landowner) and its member households. When the system was first implemented, the amount of land contracted to each household was determined either by population (household size), the number of workers in the household, or a combination of the two. (In some Jiangsu villages where farm land was first contracted to households according to population, so many rural workers subsequently left farming for employment in TVEs that the burden of farming for some of the larger households became so great that land had to be redistributed again in the mid-1980s, this time according to labour.) Collectively owned farm assets other than land, e.g. draught animals and agricultural tools and equipment, were usually divided and distributed to member households—i.e. transferred to private ownership. In effect, under *da baogan*, collectively owned land is now cultivated and managed on a long-term basis by individual households. After meeting its sales quota, taxes, and contributions to the team for collectively provided services, the household is free to dispose of its agricultural output as it wishes. In effect, China's agriculture has been decollectivized. Much has been written about the *baogan daohu* system. For more details, see e.g. Elizabeth J. Perry and Christine Wong (eds.), *The Political Economy of Reform in Post-Mao China* (Cambridge, Mass.: Council on East Asian Studies, Harvard University, 1985), chs. 1 and 2.
4. e.g. village cadres no longer assign farm tasks or allocate work points. One immediate consequence of this was that fewer cadres were needed to manage collective activities, thus making possible a reduction in full-time staff.
5. CCPCC and SC, 'Guanyu shixing zheng she fenkai jianli xiang zhengfu de tongzhi (Notice concerning the separation of government administration from commune management and the establishment of township governments)' (12 Oct. 1983), in *ZJN* (1984), 9. 9.
6. The discussion in this and the following paragraphs is based on fieldwork conducted in Jiangsu. However, townships and villages in other parts of China are organized more or less similarly.
7. In some villages in Jiangsu, the village co-operative is called 'village economic commission (*cun jingji lianhe weiyuanhui*)' or 'agricultural, industrial, and commercial united company (*nong gong shang lianhe gongsi*)'.
8. Many have come to the same conclusion. See e.g. Zou Fengling, 'Xiangzhen qiye chengbao hou de jige wenti (Several problems after the contracting out

of rural enterprises)', *JQ* 3 (1986), 63; and Wang Jianhua, 'Xiangzhen qiye de suoyouzhi wenti (On the problem of ownership of rural enterprises)', *NJW* 8 (1985), 15.

9. In principle, village heads and Party secretaries are elected respectively by villagers and by Party members. But in fact they must also have the approval of the township Party committee, so these positions are *de facto* Party appointments.

10. Township-Village Interview Notes.

11. In our interviews with county, township, and village cadres, we were told repeatedly that the income of village Party secretaries was determined by the three leading bodies at the township level (*xiang santao banzi*) and that of village cadres other than Party secretaries was determined by the three leading bodies at the township level in consultation with the relevant village Party secretary. Township cadres were assessed by the leading bodies at the county level in a similar manner. We were also told that the method used to determine income was the post responsibility system (*gangwei zeren zhi*). Undoubtedly its was one important determinant of a cadre's income, but it would be naïve to believe that income was determined solely by a cadre's performance as measured by the post responsibility system and that other factors (political and personal) were not taken into consideration.

12. In some areas they add up to 200 points, and the system is called the double 100-point assessment.

13. In some villages (e.g. Zhashang in Sigang Township), the base salary was paid when cadres were awarded 80 points, and in some the 100-point assessment was used only to determine the size of the bonus. These minor differences do not alter the fact that all the different methods linked income to performance, and that performance was judged according to established criteria.

14. Linking cadres' income to their work performance and to the general economic performance of the collective unit appears to have worked fairly well in Jiangsu, where rapid development of TVEs has kept the rural collective sector strong, but in parts of China where the collective sector is weak it has posed problems. For a discussion of some of the problems see Richard J. Latham, 'The Implications of Rural Reforms for Grass-roots Cadres', in Perry and Wong (eds.), *Political Economy of Reform*, 168–71.

15. The assessment scheme was first adopted in south Jiangsu and then gradually implemented in north Jiangsu. As a rule, the assessment systems used in south Jiangsu are more complex than those in north Jiangsu.

16. In Sigang, the composite profit indicator is the sum of after-tax or book (*zhangmian*) profit and profits outside the book (*zhangwai lirun*), consisting mainly of contributions to local governments other than remitted profits.

17. Thus, where birth control is an important target, it is conceivable that cadres may end the year with a very small income. For example, in mid-1988, the 1987 assessment of village cadres in Haitou Township was still not finalized, in part because the 1987 birth control targets were not achieved and the consequence of this for the cadres' incomes was so severe that the township was reluctant to implement the assessment.

18. In Sigang, it was 95 per cent, but it is unclear how representative this is. However, interviews in other townships suggest that leading Party cadres were also judged largely by their success in managing the local economy and promoting its development.

19. The major factors considered when setting targets for lower units were (i) targets assigned by the immediate senior unit, (ii) the performance of lower units in the previous year, and (iii) local economic conditions. None of the rural cadres we interviewed complained about unrealistic economic targets. Several of the non-economic targets (e.g. party work and militia work) appear relatively easy to achieve but several others (e.g. birth control) are known to be difficult to achieve. The one child per couple policy and the restrictions placed on house construction are among some of the most difficult policies to implement in rural China. If some of the non-economic targets are perceived to be significantly more difficult to achieve than economic targets, then cadres may find it profitable to strive even harder to exceed the economic targets (where a large number of bonus points can be earned) in the hope that overachievement in the economic areas will compensate possible failures in the non-economic areas.

20. Many of these documents were discussed in Ch. 2, and a large number of them may be found in *XQZFX*.

21. Strictly speaking TVGs cannot borrow to finance a deficit, but they have found ways to avoid these restrictive rules, e.g. by borrowing through TVEs.

22. A good discussion of the development of the Chinese budgetary system is in Nicholas R. Lardy, *Economic Growth and Distribution in China* (London: Cambridge University Press, 1978), ch. 2. For a description of the budgetary system as it existed in the late 1980s, see World Bank, *China, Revenue Mobilization and Tax Policy* (Washington, DC: World Bank, 1990).

23. Song Lina and He Du, 'The Role of Township Governments in Rural Industrialization', in Byrd and Lin Qingsong (eds.), *China's Rural Industry*, 343.

24. For details on some of the reforms and the problems created by the rapid expansion of extrabudgetary funds see Barry Naughton, 'False Starts and Second Wind: Financial Reforms in China's Industrial System', in Perry and Wong (eds.), *Political Economy of Reform*, 223–52, and Zhao Yujiang, 'Management of Extrabudgetary Funds', in Bruce L. Reynolds (ed.), *Reform in China: Challenges and Choices* (Armonk, NY: M. E. Sharpe, 1987), 130–41.

25. *ZTN* (1990), 34 and 230.

26. *ZTN* (1985), 416, and *ZTN* (1990), 154.

27. To encourage tax collection, the sharing rate applied to the revenue above the amount collected in the previous year is usually more favourable to the province. For details of China's revenue system, see World Bank, *China, Revenue Mobilization*, 85–90 and Annex 2, and Christine P. W. Wong, 'Central–Local Relations in an Era of Fiscal Decline: The Paradox of Fiscal Decentralization in Post-Mao China', *CQ* 128 (Dec. 1991), 699–701. For a description of the revenue-sharing arrangement between Jiangsu and the central government, see Penelope B. Prime, 'The Impact of Self-Sufficiency

on Regional Industrial Growth and Productivity in Post-1949 China: The Case of Jiangsu Province', Ph.D. diss., University of Michigan, 1987, app. A.

28. Jiangsu was the first province to decentralize the fiscal system to subprovincial regions, and over the years more and more revenue and expenditure responsibilities were decentralized to localities. Of the total financial resources (*caili*) controlled by Jiangsu, the share controlled directly by the provincial government declined from 46.4 per cent in 1978 to only 22.9 per cent in 1989 (*JJN* (1990), 6. 7).

29. *JJN* (1990), 6. 5. The original announcement was that the new system would be used for three years, from 1988 to 1990. Its use has since been extended and the system is still in place.

30. e.g. the developed and prosperous localities normally have lower revenue-sharing ratios than the less developed localities.

31. *JJN* (1989), 3. 84.

32. See the Appendix for an explanation of Jiangsu's administrative structure.

33. The exception was Jiangpu, one of several poor counties in the Nanjing region with a per capita (material) income of around RMB 1,050 in 1988.

34. Calculated from *JJN* (1989), 3. 131.

35. To be specific, the coefficent of variation of per capita revenue in 1988 by county was about 50 per cent greater than that of per capita expenditure (1.35 as compared to 0.91).

36. This and the following paragraphs draw heavily on Wong, 'Central–Local Relations', 701–6.

37. *JJN* (1990), 6. 7. These additional current expenditure obligations apparently exceeded the additional revenue that Jiangsu received under the new sharing arrangement.

38. Selected economic indicators for the three townships are presented in Table A.1.

39. State enterprises located in the townships we surveyed paid their taxes directly to the county or to the region (*diji shi*).

40. Of the revenue collected for the national budget, the most important were the industrial and commercial tax and the income tax, and of course they were collected primarily from TVEs. For example, in 1987, the industrial and commercial tax accounted for 89 per cent of the taxes collected by the Haitou township government. These and the other data in this paragraph are from Township-Village Interview Notes.

41. In other words, budget allocations are used only to cover the base salary. Township-Village Interview Notes.

42. That state cadres in the more developed regions can earn higher incomes at the township level goes a long way towards explaining why they sometimes resist promotion that moves them from the township to the county level.

43. e.g. in Hufu TVEs paid 0.8 per cent of their revenue to the township as an education surcharge.

44. In Haitou, what each household paid was determined by the per capita income of the village. The most that households paid was about RMB 5 and the least was RMB 4. However, a small number of villages that had excess

revenue from other sources (e.g. Haiqi) paid the surcharge from this revenue and did not collect from households.

45. See Ch. 4. In some cases, TEs remitted a percentage of their total sales to the township government.

46. In 1987 profit remittances accounted for 34 per cent of the net (after-tax) profits earned by TVEs in Jiangsu. For China as a whole, TVEs remitted on average about 38 per cent of their net profits. See *ZNN* (1988), 318.

47. In some parts of Jiangsu, the township government is also permitted to retain a small percentage of the tax revenue collected within the quota.

48. e.g. the amount of above-quota tax revenue Sigang retained in 1987 was only RMB 30,000, or about 0.3 per cent of what it turned over to the state in taxes.

49. Township-Village Interview Notes. In fact, in counties where TVEs have developed rapidly, a consequence of the revenue contract (*baogan*) is to reduce the amount of income tax TVEs actually pay to the state. Although local tax bureaux collect income tax from TVEs in accordance to the rates stipulated by the income tax law, county governments would refund some of the tax they collected if they did not need the revenue to meet their tax quotas.

50. However, they may be important in poor townships where revenue sources are few. For example, in 1987, the excessive birth fee (*chaosheng zinu fei*), basically a fine paid by villages that did not meet their birth control targets, collected by Haitou amounted to nearly RMB 82,000, or about 30 per cent of its township fund.

51. See Appendix for a more detailed description of these villages.

52. *Liangjin yifei*. Villagers in China are obliged to contribute to a collective accumulation fund (*gongjijin*) and a collective welfare fund (*gongyijin*) and to pay a management fee (*guanlifei*).

53. Because of differences in endowment and levels of development, village Party secretaries are likely to receive very different incomes even within the same township. For example, in Sigang the income received by the highest-paid village Party secretary (RMB 8,840) in 1987 was three times that of the lowest-paid (RMB 2,920). Yet the lowest-paid village Party secretary in Sigang was still paid much more than the highest-paid village Party secretary in Haitou (RMB 1,600). In poor townships, there is much less variation in the income received by village Party secretaries. This is because cadres' incomes do not fall below a floor, which in 1987 was RMB 1,000 in Haitou.

54. Jiangsu had a rural population (*xiangcun renkou*) of 52.05 million in 1987 (*JJN* (1988), 3. 15 and 3. 27).

55. For the country as a whole, TVEs were less important as a revenue source. In 1987, the budgetary revenue was RMB 234.66 billion (*ZTN* (1988), 748), of which RMB 13.07 billion (see Table 7.5) or 5.6 per cent came from TVEs.

56. *BR* (27 Aug.–2 Sept. 1990), 18. For Chinese views of the policy of using industry to subsidize agriculture (*yi gong bu nong*), see Chen Liangbiao, 'Xiangzhen qiye yi gong bu nong de lilun jichu ji qi duice yanjiu (A study of the theoretical foundation for and the policy of using industry to subsidize

agriculture)', *NJW* 3 (1987), 42–6, and Pan Shui *et al.*, '"Yi gong bu nong" bixu jianchi ("To use industry to subsidize agriculture" must be upheld)', *NJW* 8 (1985), 17–19.

57. In China as a whole, 40 per cent was used to purchase equipment, 35 per cent for farm capital construction, and 15 per cent in support of poor brigades (see *ZNTN* (1985), 190).

58. See the discussion in Ch. 3.

59. Although the evidence suggests that the more developed and industrialized villages tend to be more involved in agricultural management, partly because they are financially more able to intervene actively and partly because they need to intervene more actively to ensure that crop cultivation is not neglected for other more lucrative activities, village cadres in the less developed regions are still involved to some extent in agricultural management. For example, in most Jiangsu villages several cadres are in charge of agricultural production (assisted usually by several agricultural extension workers), and their primary responsibility is to organize the preparation of seedlings, irrigation, and pest control in a unified manner (*tongyi zuzhi*), i.e. either to co-ordinate the work or to arrange a small group of workers to do the work under village supervision but at the expense of individual households. Where ploughing has been mechanized, it is sometimes also done in a unified manner. For example, in several of the villages we surveyed, collectively owned tractors were contracted to individuals who operated them on the condition that they must serve agriculture during the peak agricultural months, when the village organized both the collectively and the privately owned tractors for ploughing. However, individual households paid for the tractor service. Finally, village cadres with primary responsibility for agriculture also have the responsibility to make sure that households pay their agricultural taxes and fulfil their contract sales to the state.

60. e.g. the amount of subsidy given to households for every *mu* of land cultivated in Nandu Village in central Jiangsu increased from RMB 10 in the early 1980s to RMB 55 in the late 1980s (*JJN* (1990), 7. 36).

61. This paragraph draws heavily from Jean C. Oi, 'Commercializing China's Rural Cadres', *Problems of Communism* (Sept.–Oct. 1986), 5–6. The example here involves grain production, but it is not unusual for TVEs in Jiangsu to employ farmers as workers to specialize in vegetable farming and animal husbandry (e.g. see Ch. 5).

62. The agricultural service workshop exists sometimes as a separate unit in the village but it is usually incorporated within one or several village enterprises. Among the villages we surveyed, both organizational forms were found. For example, Yangquan Village in Hufu Township operated an agricultural service workshop and paid its eight members RMB 1,100 each in 1988 for about seven months of work. Because of the ill effects of agricultural chemicals on health, the composition of the agricultural service workshop was changed periodically. In Sigang Township, agricultural workshops were part of TVEs and workers were paid directly by the enterprises.

63. Township-Village Interview Notes. In the late 1980s, Zhangjiagang mobilized about RMB 4 million annually in this fashion. However, we were told that

locally owned state and collective enterprises were extremely unhappy about the arrangement, and many of them resisted paying. In fact, to collect, the agricultural development company had to threaten some of the enterprises with court actions.

64. Childless and infirm old persons who are guaranteed food, clothing, medical care, housing, and burial expenses by the village.
65. *ZNN* (1988), 41–2.
66. Township-Village Interview Notes. The size of pension may even vary by team. For example, in Zhashang, where team enterprises were important, retired workers and cadres in team 10 and team 12 received pensions of RMB 600 per person in 1986. All other retired workers and cadres in the village received pensions of only RMB 200 per person.
67. Calculated from data in Tables A.1, A.3, 7.3, and 7.4.
68. Since villages receive no budget allocation, differences between their spending capacities remain very large. For example, the ratio between per capita village funds (all village receipts except loans) in Zhashang and those in Haiqi was 14.5:1 in 1987, much larger than the difference between their levels of development.
69. Whether fiscal predation is a serious problem is difficult to say since there is little hard evidence. However, the temptation is certainly there for cadres to squeeze TVEs for short-term benefits to the government at the cost of long-term growth to the enterprise.

8

Rural Non-agricultural Development, Rural Income, and Income Distribution

THIS chapter examines how rural non-agricultural development in the 1980s affected income and income distribution in rural Jiangsu. It begins with a discussion of the growth and the changing composition of rural household income and consumption since 1978. It then examines rural income distribution. The size distribution of income in six Jiangsu villages at different stages of development are derived from survey data and compared. Drawing on these village data and on the available time series data on per capita income, we speculate on the likely impact rural non-agricultural development may have had on rural income distribution.

Income of Peasant Households

In the two decades prior to 1978, because of low procurement prices, rising agricultural production costs, restrictions on rural non-agricultural activities, and increases in rural population, per capita peasant[1] income in Jiangsu improved slowly despite substantial growth in agricultural output.[2] Peasants derived their income from three main sources: the collective, household sideline activities, and other sources. In Jiangsu, between 1957 and 1978, per capita peasant income in current prices increased from RMB 84 to RMB 155, and of the 71 *yuan* increase, 64 *yuan* came from increases in income received from the collective and 7 *yuan* from increases in sideline income.[3] Income from other sources did not change. Since the income from TVEs accounted for less than 4 per cent of the income distributed by the collective in 1978, the source of nearly all the increase in peasant incomes in the two decades prior to 1978 was collective agriculture.[4]

The quality of rural growth improved significantly after 1978. The reform measures adopted in the late 1970s and the early 1980s to improve work incentives and to increase employment opportunities in the countryside (e.g. increased procurement prices, reduced direct collective control of resources, the vigorous development of non-agricultural activities, and greater production and marketing freedom for rural households) together brought about a dramatic rise in per capita rural income.

The best available estimates of per capita rural income and per capita rural consumption are provided by government surveys of record-keeping peasant households.[5] These surveys define household income to include distributed collective income (e.g. wages and distributed profits from TVEs, collective welfare payments, bonuses distributed by the collective, and income from activities under unified accounting[6]), income from household (private) activities, and other income except loans. Included in other income are remittances from within China and from abroad, income earned by rural residents working outside the commune or the rural economy (e.g. as temporary workers in county or state enterprises) and government subsidies. Survey findings have been published annually since the early 1980s, and the data for selected years for both Jiangsu and China are presented in Table 8.1. Until 1989–90, when the economy went into a severe recession, per capita peasant income and per capita peasant consumption increased rapidly in nominal and in real terms. Between 1978 and 1988, there was a fourfold increase in nominal per capita peasant income and in nominal per capita peasant consumption at the national level. In Jiangsu, the increase was even more dramatic, with both nominal per capita income and nominal per capita consumption rising more than fivefold.[7] The reform measures that brought about such extraordinary increases in income also produced strong inflationary pressures, particularly after 1985. When these price increases are taken into account, the rise in per capita income and in per capita consumption in the countryside, though reduced, was still significant. At the national level, real per capita peasant income increased by 14.5 per cent annually between 1981 and 1984 and by 2.3 per cent annually between 1985 and 1988. In rural Jiangsu, the average growth rate of real per capita income was 17.4 per cent during the years 1981–4 and 6.3 per cent during the period 1985–8, and in these same two periods the average growth rate of real per capita consumption was an incredible 14.5 per cent and 10.4 per cent respectively. We shall show below that the main reason why per capita income and per capita consumption grew so much faster in Jiangsu than in the country as a whole was that Jiangsu had a significantly larger and stronger rural non-agricultural sector.

The available evidence also indicates that since the implementation of rural reforms the economic position of nearly everyone in the countryside has improved in an absolute sense. Table 8.2 presents, among other distributions, the percentage of surveyed peasant households in different per capita income groups in selected years between 1980 and 1989. In 1980 about 87 per cent of all surveyed peasant households in China had a per capita income of less than 300 *yuan*. Nine years later, in 1989, only 16 per cent of the surveyed peasant households had a per capita income of less than 300 *yuan*. Even after adjusting for the 80 per cent rise in rural

TABLE 8.1. *Per Capita Income of Surveyed Peasant Households, China and Jiangsu, 1978–1990*

	1978	1980	1984	1986	1988	1990	Average annual growth rate (%)		
							1981–4	1985–8	1989–90
China									
Sample size[a]	6,095	15,914	31,375	66,836	67,186	66,960			
Average household size	5.74	5.54	5.37	5.07	4.94	4.80			
Average number of workers	2.27	2.45	2.87	2.95	2.95	2.92			
Per capita income (RMB)									
Current prices	134	191	355	424	545	630	16.8	11.4	12.2
1980 prices	142	191	328	349	360	340	14.5	2.3	1.8
Per capita consumption (RMB)									
Current prices	116	162	274	357	477	538	13.9	14.9	5.9
1980 prices	124	162	253	294	315	290	11.8	5.7	−4.1
Jiangsu									
Sample size[a]	576	985	1,544	3,400	3,400	3,400			
Average household size	5.20	4.93	4.66	4.32	4.26	4.14			
Average number of workers	2.48	2.47			2.84	2.78			
Per capita income (RMB)									
Current prices	155	218	448	561	797	884	19.8	15.7	4.5
1980 prices	165	218	414	462	527	477	17.4	6.3	−4.8
Per capita consumption (RMB)									
Current prices	140	195	360	499	747	787	16.7	20.1	2.8
1980 prices	149	195	333	411	494	425	14.5	10.4	−7.2

[a] See n. 5 in text for a discussion of how survey samples were selected.

Sources: ZTN (1984), 472; (1985), 572; (1987), 697–8; (1989), 747; (1990), 313 and 315–16; *Zhongguo tongji zhaiyao (Statistical Survey of China)* (1991), 48–9; *JJN* (1987), 3. 101–2; *JTN* (1990), 387–8, and *JTN* (1991), 325–6. Per capita household income in 1980 prices was obtained by deflating per capita household income in current prices by the index of rural retail prices.

TABLE 8.2. *Distribution of Peasant Households, Production Teams, and Villages by Per Capita Income, China and Jiangsu, Selected Years*

A. Distribution of surveyed peasant households (% of total)

Yuan	China			Jiangsu
	1980	1986	1989	1986
500+	1.6	28.7	53.2	53.3
400–500	2.9	16.5	15.6	17.9
300–400	8.6	21.7	15.6	15.6
200–300	25.3	21.8	10.9	10.6
150–200	27.1	7.0	2.8	2.1
100–150	24.7	3.2	1.3	0.5
0–100	9.8	1.1	0.6	0.1
Total	100.0	100.0	100.0	100.0

B. Distribution of production teams or villages in Jiangsu

1978			1986		
Yuan[a]	Number of production teams	% of total	Yuan	Number of villages	% of total
151+	21,715	6.92	1000+	820	2.28
101–150	87,493	27.89	800–1,000	3,063	8.50
51–100	151,048	48.16	500–800	16,180	44.89
0–50	53,413	17.03	400–500	8,117	22.52
Total	313,669	100.00	300–400	6,050	16.79
			200–300	1,706	4.73
			100–200	102	0.28
			0–100	2	0.01
			Total	36,040	100.00

[a] Includes distributed collective income only.

Sources: ZTN (1987), 697; (1988) 822; (1990), 312; *JNJSFTZ* 13; *1978 nian Jiangsu sheng nongye tongji ziliao* (*Agricultural Statistical Material, Jiangsu Province, 1978*), 273, and *Jiangsu sheng 1986 nian nongcun zhuhu diaocha ziliao huibian* (*Compilation of Material on Survey of Rural Households, Jiangsu Province, 1986*), 12.

retail prices between 1980 and 1989, the evidence still suggests that the bulk of the rural population experienced an absolute improvement in their real income in the 1980s. In other words, the number of poor rural households declined. If we define a poor household as one with a real per capita income below RMB 150 (in 1980 prices), then poor households accounted for about 35 per cent of the surveyed peasant households in China in 1980 but less than 16 per cent in 1989.[8] In 1978, when collective activities were still the main source of income in rural Jiangsu, 93 per cent of Jiangsu's 313,669 production teams reported an average per capita distributed collective income of less than RMB 150 (or RMB 160 in 1980 prices). By contrast, in 1986 less than 3 per cent of the surveyed peasant households in Jiangsu had a per capita income of less than RMB 165 in 1980 prices, and of the 36,040 villages in the province only 104 had a per capita income below RMB 165 in 1980 prices.[9]

Table 8.3 shows that rapidly rising income has produced the expected changes in consumption patterns in both rural Jiangsu and rural China. Although per capita consumption of basic food increased steadily through-out the 1980s, the expenditure on food as a percentage of total consumption, in accordance with Engle's law, declined sharply.[10] In one decade (1978–88), food expenditure as a share of total consumption declined by more than 14 percentage points (from 67.7 to 53.4 per cent) in rural China and by nearly 17 percentage points (from 62.3 to 45.7 per cent) in rural Jiangsu.[11] With the exception of expenditures on clothing and fuel, expenditures on other non-food items as shares of total outlay all displayed a rising trend.

The consumption of semi-luxury goods (e.g. liquor) and consumer durables (e.g. television sets) increased dramatically. In rural Jiangsu, the annual per capita consumption of liquor increased from 1.5 kg. in 1978 to 8.3 kg. in 1988, and the number of television sets per 100 households rose from 5.4 in 1984 to 35.2 in 1988.[12] The most significant and certainly the most noticeable change in household behaviour is the increased share of total outlays allocated to housing.[13] Decades of austerity had created a huge pent-up demand for better houses and furnishings, and since 1978 peasants have used their new freedom and increased income to improve their living conditions. Per capita living space in rural Jiangsu increased from 12 square metres in 1980 to nearly 24 square metres at the end of the decade.[14] Expenditures by Jiangsu's peasant households on housing as a share of total outlay increased from below 8 per cent in 1978 to over 27 per cent in 1988.

The rapidly rising per capita household consumption, the declining share of expenditure on food, and the rising shares of expenditure on semi-luxury items and consumer durables all suggest that the standard of living and the quality of life have improved considerably in the post-

TABLE 8.3. *Consumption Pattern and Ownership of Consumer Durables, Rural Households, China and Jiangsu, Selected Years*

	1978	1982	1984	1986	1988	1990
China						
Per capita consumption (RMB in current prices)	116	260	274	357	477	538
% of consumption						
Total	100.0	100.0	100.0	100.0	100.0	100.0
Food	67.7	60.5	59.0	56.3	53.4	54.9
Clothing	12.7	11.2	10.4	9.5	8.6	8.4
Fuel	7.1	5.6	5.5	5.2	4.6	4.5
Housing	3.2	10.3	11.7	14.4	14.9	12.9
Other goods	6.6	10.2	11.0	11.5	12.8	11.9
Cultural and other services	2.7	2.2	2.4	3.1	5.7	7.5
Bicycles per 100 households	30.7	51.5	74.5	90.3	107.5	118.3
TVs per 100 households	—	1.7	7.2	17.3	31.4	44.4
Jiangsu						
Per capita consumption (RMB in current prices)	140	261	360	499	747	787
% of consumption						
Total	100.0	100.0	100.0	100.0	100.0	100.0
Food	62.3	55.5	53.0	49.5	45.7	51.6
Clothing	12.5	11.1	9.7	8.7	6.8	7.3
Fuel	5.9	4.5	5.6	4.3	3.8	3.1
Housing	7.8	16.3	19.0	23.5	27.5	20.7
Other goods	8.1	10.4	10.7	11.3	11.9	11.5
Cultural and other services	3.4	2.2	2.0	2.7	4.2	5.8
Bicycles per 100 households	19.8	n.a.	89.8	117.9	145.6	159.1
TVs per 100 households	—	n.a.	5.4	18.9	35.2	54.4

Sources: ZTN (1984), 471–4; (1988), 822–5; (1989), 742–3; (1990), 312, 316, 324; *Zhongguo tongji zhaiyao* (*Statistical Survey of China*) (1991), 49; *JSN* 181; *JTN* (1989), 363, 367; *JTN* (1991), 327, 329; *JJN* (1989), 4. 102.

reform period. But the living conditions of individuals are not a function of private consumption alone; they are also affected by government expenditures, or in the case of rural China by social collective consumption.[15] For example, expenditure on health and education paid for by collective funds is not included in household consumption but is ultimately consumed by households and therefore should be taken into account when evaluating changes in the standard of living of individuals.[16] With the introduction of the household responsibility system and the weakening of the finances of collectives, social collective consumption began to erode, particularly in the poorer regions.[17] However, the rise in private rural per capita consumption during the 1980s was so large that, even with some erosion in social collective consumption, there is little doubt that the average peasant in Jiangsu (and in China) lived significantly better at the end of the 1980s than at the beginning of the decade.

The contribution of rural non-agricultural development to the rise in rural per capita income has been substantial in both Jiangsu and the country as a whole. Table 8.4 compares the estimated distribution of per capita income of peasant households by income source for selected years, and it shows that in the 1980s the share of agricultural income in total peasant income declined sharply. Between 1980 and 1989, income derived from agriculture (defined broadly) as a share of total per capita household income, dropped from 78 to 60 per cent in China and from 75 to just 50 per cent in Jiangsu. In other words, income from non-agricultural sources has become increasingly important in rural areas, and its growth is a main reason for the rapid rise in per capita peasant income.[18] Between 1980 and 1989, non-agricultural sources accounted for 49 per cent of the increase in per capita peasant income in China and nearly 58 per cent of the increase in Jiangsu. The significantly larger contribution from non-agricultural sources in Jiangsu is the main reason why per capita rural income (and per capita rural consumption) growth was significantly higher in Jiangsu than in the country as a whole prior to 1989 and also why it fell more sharply in Jiangsu in the 1989–90 recession when rural non-agricultural activities went into a severe decline (see Table 8.1).

Increases in peasant income came from increased agricultural output, higher agricultural prices, and better and more employment opportunities in collective as well as private rural non-agricultural activities. There is no doubt that rising agricultural prices and the widespread adoption of the household responsibility system in the late 1970s and the early 1980s had a dramatic positive impact on agricultural output and peasant income. Between 1980 and 1983, 54 per cent of the increase in peasant income in China and 45 per cent of that in Jiangsu may be attributed directly to rising agricultural prices and production. After 1983, when farm costs began to rise more rapidly than output prices, growth of agricultural production and income slowed considerably. In terms of their contribution to income, there was also a noticeable shift from farming to other agricultural activities. Farming accounted for 70 per cent of the increase in per capita agricultural income between 1980 and 1983 but only 52 per cent between 1983 and 1989. The declining importance of farming as a source of income is also clearly evident in Jiangsu. In part the decline reflects the cap placed by the government on the prices of some crops (especially grain) but it is also a reflection of the fact that, with less direct collective control over agricultural resources, peasants shifted their resources from

TABLE 8.4. *Distribution of Per Capita Income of Rural Households by Source, China and Jiangsu, Selected Years*

	China 1980	China 1983	China 1989	China Increases 1980–3	China Increases 1983–9	Jiangsu 1980	Jiangsu 1983	Jiangsu 1989	Jiangsu Increases 1980–3	Jiangsu Increases 1983–9
Per capita income (RMB)										
1980 prices	191	295	335	104	40	218	340	487	122	147
Current prices	191	310	601	118	292	218	356	876	138	520
% from[a]										
Agricultural activities	78.1	68.8	60.0	53.7	50.7	74.9	63.5	50.4	45.3	41.3
Farming[b]	61.6	52.5	39.9	37.8	26.6		56.1	35.1		20.6
Other activities	16.5	16.3	20.1	16.0	24.1		7.3	15.4		21.0
Non-agricultural activities	21.9	31.2	40.0	46.3	49.3	25.1	36.5	49.6	54.7	58.7
Household activities	4.8	10.2	22.2	19.0	34.9	3.8	10.7	19.2	21.6	25.0
Handicrafts and industry	1.5	1.8	3.2	2.4	4.7	1.8		3.5		
Construction and transport	} 3.3	2.2	4.6	} 16.6	7.1 }	2.0		5.4		
Other activities		6.2	14.4		23.1 }			10.3		
Collective income[b]	6.5	11.6	9.4	20.0	7.1	11.6	16.8	23.5	25.2	28.1
TVEs	3.1 }	11.6 }	6.4 }	20.0 }	7.1 }	8.4		21.5		
Miscellaneous[c]	3.4 }		3.0 }			3.2		2.0		
Associations	0.0	0.3	0.6	0.7	0.9	0.0	0.0	0.1	0.0	0.2
Other sources[d]	10.7	9.1	7.8	6.6	6.5	9.7	9.0	6.8	7.9	5.3

[a] Calculated from current price data.

[b] The estimated distributed income from collective cultivation in 1980 (before the adoption of the household responsibility system) is included under farming.

[c] Includes the distributed income earned in other collective non-farm activities, collective welfare payments, and bonuses awarded by collectives.

[d] All other receipts except loans. The principal items included are remittances from within China and from abroad, government subsidies, and wage earnings of household members from outside the rural sector (before 1983, the commune).

Sources: Calculated from data in *ZTN* (1986), 673–4; (1988), 827; (1990), 313, 315; *JTN* (1990), 387–8. The 1983 distribution for Jiangsu is probably less reliable than the others since only gross income figures were available for that year. The net income figures for 1983 were estimated from the gross figures by using the ratios between net income and gross income from other years.

cultivation where the profit margin was low and getting lower to other more lucrative agricultural activities (e.g. fishery and forestry).

Peasants derive rural non-agricultural income from two main sources: (1) distribution from the collective sector (the most important being the wages and bonuses distributed by TVEs) and (2) private activities. In the post-reform environment, both increased rapidly. Because of its larger rural non-agricultural sector, the contribution of rural non-agricultural development to rural income has been significantly greater in Jiangsu than in China as a whole. But what really sets Jiangsu apart from the rest of China is the size and the growing strength (particularly after 1983) of its TVE sector. Nearly 60 per cent of the increase in Jiangsu's rural per capita income between 1983 and 1989 was accounted for by increases in rural non-agricultural income, of which the most important component was the distributed income from the rural collective sector. And, between 1980 and 1989, it is estimated that nearly 95 per cent of the increase in the distributed income from the rural collective sector was accounted for by increases in the wage and bonus earnings distributed by TVEs. Table 8.4 shows that before 1983 the pace of rural income growth in Jiangsu was roughly similar to that in other parts of China. From 1983 to 1989, the growth of real rural income was considerably stronger in Jiangsu than in most other parts of China, and the primary reason for this difference was Jiangsu's larger and stronger TVE sector.

Economic Reform, Rural Non-agricultural Development, and Rural Income Distribution

While there is little doubt that rural reform and increased participation in rural non-agricultural activities have contributed immensely to the rapid rise in rural income, their effects on income distribution are more uncertain.[19] In many less developed countries, agricultural income is distributed unequally primarily because farm land is distributed unequally. However, when land ownership is not extremely concentrated, the unequal distribution of agricultural income may be offset by non-agricultural income to produce a relatively equal distribution of total income. In other words, those with little land may be able to compensate for their lack of agricultural income with income earned from non-agricultural activities. This apparently was the case in Japan, South Korea, and Taiwan, and is one reason why some development economists believe that rural non-agricultural development is likely to reduce both poverty and income inequality.[20]

However, in mainland China, where land is collectively owned and distributed in an extremely egalitarian manner, where rural reforms have changed income differentials between agriculture and non-agricultural activities and those within each sector substantially, and where rural non-agricultural activities have developed unevenly in the countryside, rapid rural non-agricultural development may not have been income equalizing. In the absence of direct evidence of changes in rural income distribution, we shall explore the impact of rural non-agricultural development on income distribution as follows. The size distribution of income from various sources is first investigated at the village level. Building on the insights gained from the village data, we then examine the available time series data on interregional income distribution and hypothesize on how rural income distribution might have changed in the past decade in Jiangsu. Because household income data that can be used to derive size distribution of income are not always available and because the data that are available are flawed, the conclusions reached in this section should be treated as hypotheses to be tested when more and better data become available.

Income Distribution at the Village Level

Selected households in six Jiangsu villages (Haiqi, Haiqian, Xiaojian, Yangquan, Zhashang, and Jingxiang) participated in a year-long record-keeping survey, and the results of the survey were used to compute the average and inequality measures presented in Table 8.5. Before discussing the findings, a few words about the survey are necessary.[21] Because the surveyed households, though randomly selected, included only those that had a literate member who was able and willing to keep a journal of daily receipts and outlays, poor households were underrepresented. Consequently we believe the survey findings to be biased; specifically, we believe per capita household income is biased upwards and income inequality measures are biased downwards. Furthermore, we believe the bias to be greater the poorer the village. In assessing the evidence below, readers should keep the degree and the direction of these biases in mind.

To put the survey findings in context, a few words about the surveyed villages are also in order. Although all six villages are in regions that experienced rapid economic changes in the 1980s, they differ significantly in size, in prosperity, in endowment, in economic structure, and in their level of rural non-agricultural development.

Haiqian and Haiqi, the two coastal villages in Haitou Township in north Jiangsu, are the least industrialized and the most dependent on agriculture (broadly defined) for income. Haiqi is also by far the poorest village of the six we surveyed. Because of their coastal location, a substantial number of the workers in these two villages were employed full-time in coastal fishing, so they are not typical north Jiangsu villages.[22]

TABLE 8.5. *Average Per Capita Household Income, Inequality Measures of Per Capita Household Income, and Components of Per Capita Household Income, Survey of Six Jiangsu Villages, 1987*

	Haitou		Hufu		Sigang	
	Haiqi	Haiqian	Xiaojian	Yangquan	Zhashang	Jingxiang
Per capita cultivated area						
Average (*mu*/person)	0.46	0.40	0.55	0.60	0.64	1.10
Coefficient of variation	0.69	0.92	0.23	0.36	0.44	0.24
% of family labour (in standard days) allocated to agriculture	56.4	22.7	22.0	17.8	4.9	18.5
Average per capita household income (*yuan*)	1,135	1,307	1,302	1,915	1,612	1,894
Income inequality measures						
Share of lowest quintile (%)	8.59	9.59	12.51	9.80	12.49	12.07
Share of highest quintile (%)	43.88	37.62	27.58	33.26	33.23	32.59
Coefficient of variation	0.780	0.549	0.298	0.489	0.414	0.380
Gini index	0.367	0.294	0.164	0.249	0.214	0.226
Standard deviation of log of income	0.252	0.215	0.126	0.183	0.157	0.172
Income components (% of total)						
Agricultural income	61.30	32.83	26.17	26.35	16.84	23.32
From cultivation	13.48	17.10	10.43	8.29	15.14	19.41
From other agricultural activities	47.82	15.73	15.74	18.06	1.70	3.91
Non-agricultural income	38.71	67.17	73.83	73.65	83.16	76.68
From household activities	26.86	58.02	13.99	30.63	12.71	18.33
From collective activities[a]	8.06	6.37	52.88	38.52	67.43	54.36
From transfers and remittances	3.79	2.78	6.97	4.50	3.03	4.00
Total income	100.00	100.00	100.00	100.00	100.00	100.00

[a] Mostly wages, salaries, and other distributions from TVEs.

Source: Household Record-Keeping Survey.

The importance of fishing in the two villages also explains their relatively low per capita cultivated area (and the wider variance) even though they are in north Jiangsu where cultivated land is relatively abundant.

Sigang Township, where Jingxiang Village and Zhashang Village are located, is in a prosperous agricultural region in south Jiangsu and has one of the most highly developed rural non-agricultural sectors in the province. In 1987, most residents in these two villages were engaged in some form of non-agricultural activity, and farming was only a sideline for many households. Nearly every household in these two villages had a member who worked in a TVE. Some also had members who worked in county enterprises. In addition, many residents worked in enterprises that were operated by their production teams.[23] In other words, the rural non-agricultural sector in Jingxiang and Zhashang was highly developed and was comprised largely of collective activities.

Xiaojian and Yangquan are in Hufu Township, a grain deficit mountainous region adjacent to Taihu Lake and near the border between Jiangsu and Zhejiang Province. In 1987, Hufu was less developed than Sigang but more developed than Haitou. Individual non-agricultural activities were more important in Hufu than in Sigang. In other words, the rural non-agricultural sector was less dominated by collective activities in Hufu's villages than in Sigang's. Reflecting its mountainous environment, rural non-agricultural activities in Hufu were also more resource based than in Sigang.

Four commonly used measures of income inequality are presented in Table 8.5.[24] Although the four do not always give an unambiguous ranking of income distributions in the six villages, they nevertheless indicate that, in 1987, income was distributed significantly more unequally in Haiqi and Haiqian than in the other villages. Taking sample biases into account would further strengthen this conclusion since Haiqi and Haiqian are also the poorest of the six villages.[25] After Haiqi and Haiqian, Yangquan had the most unequal income distribution. Xiaojian Village, in Hufu Township, had the most equal income distribution.

Table 8.5 compares the principal sources of income in the six villages, and it reveals that the three villages (Haiqi, Haiqian, and Yangquan) with the most unequal distribution of income also derived the largest share of their household income from household activities (i.e. agricultural income and non-agricultural income from self-employment). For example, Haiqian derived one-third of its household income from agriculture (17 per cent from cultivation and 16 per cent from other agricultural activities) and 58 per cent from non-agricultural household activities. By contrast, Zhashang derived less than 30 per cent of its household income from household activities (17 per cent from agriculture and 13 per cent from non-agricultural activities) and over two-thirds from collective activities (mostly wages, salaries, and distribution from TVEs). It would appear that a major influence on income distribution at the village level in Jiangsu is the importance of collective activities. Since TVEs are the most important rural collective activity, their prominence in and near a village is likely to have an immense impact on its income distribution. Specifically, income distribution is likely to be more equal in villages where TVEs are major employers. This is because TVEs draw their workers primarily from the local region, because jobs in TVEs are distributed fairly equitably among local households, and because wages paid by TVEs in the same area generally fall within a relatively narrow range.[26] The practice of second distribution (*erci fenpei*), under which rural governments distribute part of the profits remitted by collective enterprises to teams for redistribution to agricultural households (i.e. those with contract land), also contributes to a more equal distribution of income.[27]

Table 8.6 presents inequality measures for each of the principal components of per capita household income by village, and it shows that in 1987 both agricultural and non-agricultural income was distributed significantly more unequally in Haiqi and Haiqian than in the other villages. Agricultural income was distributed more unequally in Haiqi and Haiqian than the other villages for two reasons: (1) cultivated land was distributed more unequally (e.g. in Haiqi many households did not farm and therefore had no land) so income from cultivation was distributed more unequally, and (2) a very large share of the agricultural income came from off-farm activities where income was more variable.[28] Non-agricultural income was distributed more unequally in Haiqi and Haiqian because (1) household activities, where earnings were more variable, made up the bulk of the non-agricultural income and (2) wage income from collective activities was distributed very unequally, reflecting the fact that, because there were few TVEs in or near the two villages, only a few households received any wage income.

In contrast to Haiqi and Haiqian, the four southern villages were more developed and much less dependent on agriculture. Agricultural income was not only less important, it was also distributed relatively equally. The more equal distribution reflects the fact that in these villages, as in most parts of rural Jiangsu, land (crop land as well as forest land) is distributed fairly equally. A more important source of income inequality is the composition of non-agricultural income, particularly the relative importance of income from non-agricultural household activities, which is more variable and dispersed. Indeed, it appears that the primary reason why income was distributed more unequally in Yangquan than in the other three southern villages was the greater importance of its household non-agricultural activities and the greater variability of income from this source.[29]

TABLE 8.6. *Inequality Measures of Components of Per Capita Income, Six Jiangsu Villages, 1987*

	Total	Agricultural income	Non-agricultural income			
			Total	Household activities	Collective activities	Transfers and remittances
Haiqi						
Share of lowest quintile (%)	8.59	4.73	5.05	2.46	0.00	2.56
Share of highest quintile (%)	43.88	58.36	48.48	52.69	95.68	73.35
Coefficient of variation	0.780	1.234	0.852	1.009	2.600	1.785
Gini index	0.367	0.543	0.436	0.514	0.918	0.708
Haiqian						
Share of lowest quintile (%)	9.59	−1.64[a]	3.23	1.81	0.00	0.00
Share of highest quintile (%)	37.62	56.44	45.96	51.94	89.68	91.39
Coefficient of variation	0.549	1.105	0.807	0.979	2.114	2.449
Gini index	0.294	0.584	0.435	0.524	0.851	0.863
Xiaojian						
Share of lowest quintile (%)	12.51	12.30	11.25	9.78	6.61	1.89
Share of highest quintile (%)	27.58	26.61	30.10	44.24	34.99	62.61
Coefficient of variation	0.298	0.257	0.378	1.044	0.566	1.429
Gini index	0.164	0.151	0.203	0.352	0.299	0.632
Yangquan						
Share of lowest quintile (%)	9.80	7.26	6.94	2.84	3.64	1.40
Share of highest quintile (%)	33.26	32.21	38.13	73.59	33.83	53.46
Coefficient of variation	0.489	0.468	0.613	1.550	0.547	1.055
Gini index	0.249	0.270	0.321	0.662	0.321	0.547
Zhashang						
Share of lowest quintile (%)	12.49	11.75	10.74	8.52	9.56	0.27
Share of highest quintile (%)	33.23	34.37	35.11	43.99	37.62	76.31
Coefficient of variation	0.414	0.424	0.483	0.650	0.587	1.872
Gini index	0.214	0.227	0.249	0.336	0.289	0.726
Jingxiang						
Share of lowest quintile (%)	12.07	14.14	10.24	5.24	8.55	1.06
Share of highest quintile (%)	32.59	31.94	34.82	62.63	35.24	86.40
Coefficient of variation	0.380	0.323	0.445	1.485	0.460	2.248
Gini index	0.226	0.182	0.265	0.576	0.276	0.818

[a] Negative because in 1987 several households in the sample lost a considerable sum in prawn culture, a high-risk agricultural activity.

Table 8.7 presents the income share of different income components earned by the lowest and the highest quintile (ranked by per capita household income) of households. For example, in Haiqi the top quintile earned 23 per cent of all income from cultivation, 64 per cent of all income from agricultural activities other than cultivation, 15 per cent of all non-agricultural income from household activities, and 58 per cent of all non-agricultural income from collective activities.

Table 8.7 suggests that in rural regions where households are heavily dependent on agriculture (broadly defined), e.g. Haiqi and Haiqian, households are rich (i.e. those in the highest quintile) because they earned not only a disproportionate share of the agricultural income but also a disproportionate share of certain types of non-agricultural income. In other words a household is rich not only because it earns above-average agricultural income but also because it has access to one of the few non-agricultural wage jobs in the area or because it has the entre-preneurial ability or the market or technical expertise to be involved in one of the more lucrative household non-agricultural activities (e.g. crafts, trade, small-scale manufacturing or processing). In a rural region that is poor in terms of agricultural resources but with a modestly developed non-agricultural sector (e.g. Xiaojian and Yangquan), households in the top quintile do not earn a significantly disproportionate share of the income from agriculture. Rather, they are rich because they earn a disproportionate share of the income from non-agricultural sources (i.e. they have jobs in TVEs or operate their own businesses). In rural regions with both a prosperous agricultural sector and a highly developed non-agricultural sector, e.g. Zhashang and Jingxiang, households in the top quintile earn a disproportionate share of all types of income (other than transfers and remittances). In other words they are rich because they are involved in several activities and do better than average in all of them. Households in these areas can be involved in many activities because, having a diverse and developed local economy, there are more employ-ment opportunities.

In summary, the village data suggest that agricultural (broadly defined) income is generally more equally distributed than non-agricultural income. The exception is when agricultural income is composed largely of income from off-farm agricultural activities (e.g. prawn cultivation, forestry, and fishing) which tends to be much more variable than income from culti-vation. Within non-agricultural income, wage income is distributed more equally than self-employment income except where wage jobs are very scarce.[30] These findings suggest that both agricultural and non-agricultural income have equitable and inequitable components. Thus the impact of non-agricultural development on total income distribution depends on the type of non-agricultural activities that develop.

TABLE 8.7. *Income Shares of Various Income Components of Lowest and Highest Quintiles by Per Capita Household Income, Six Jiangsu Villages, 1987 (%)*

	Total	Agricultural income			Non-agricultural income			
		Subtotal	Cultivation income	Other income	Subtotal	Household activities	Collective activities	Transfers and remittances
Haiqi								
Share of lowest quintile	8.59	5.23	17.61	1.74	13.91	17.25	3.89	11.59
Share of highest quintile	43.88	54.64	22.95	63.58	26.84	15.27	57.50	43.56
Haiqian								
Share of lowest quintile	9.59	10.07	7.68	12.67	9.36	9.68	7.91	5.97
Share of highest quintile	37.62	33.29	32.48	34.17	39.74	44.36	0.00	34.23
Xiaojian								
Share of lowest quintile	12.51	14.43	14.55	14.35	11.82	19.45	9.49	14.25
Share of highest quintile	27.58	23.30	24.58	22.45	29.10	12.74	34.68	19.57
Yangquan								
Share of lowest quintile	9.80	14.71	14.76	14.69	8.05	4.86	10.56	8.29
Share of highest quintile	33.26	19.64	17.79	20.49	38.13	73.59	10.66	31.96
Zhashang								
Share of lowest quintile	12.49	21.12	20.40	27.58	10.74	15.32	10.22	3.14
Share of highest quintile	33.23	26.16	25.46	32.42	34.66	28.23	37.09	7.52
Jingxiang								
Share of lowest quintile	12.07	17.55	19.49	7.93	10.41	14.53	9.60	2.47
Share of highest quintile	32.59	25.25	23.87	32.11	34.82	50.80	29.64	31.96

Source: Household Record-Keeping Survey.

Changes in Income Distribution at the Village Level

How has reform and rural non-agricultural development affected rural income distribution? Rural reform has allowed the TVE sector to create more wage employment in the countryside, has given individuals greater control over resource allocation, has promoted the distribution of income according to work performance and productivity, and has encouraged individuals and households to explore new opportunities in the private sector. Economic theory tells us that these reform measures are likely to increase income inequality since individuals with the most entrepreneurial and technical skills and the better-endowed regions (in terms of natural resources, savings, and access to markets) would benefit the most from such reforms. However, in the case of rural development in China, there are other factors to consider.

One factor is that in the mid-1970s, because of regulations prohibiting or restricting certain types of economic activity in the countryside, rural resources (labour and entrepreneurship) may have been seriously under-employed so that the immediate impact of rural reforms was to allow the rural economy to return to its production possibility frontier. During this period the income of all rural residents is likely to have improved, and income distribution may also have improved (if the poor were also those with the most underemployed resources and therefore those who would receive a larger share of the gains from the move to the production possibility frontier), although, a priori, the result is indeterminable.

Another factor to keep in mind is the insight provided by the village data, which show that rural income comes from three main sources or sectors (agriculture, household non-agricultural activities, and collective non-agricultural activities), so that income distribution is a weighted average of within- and between-sector inequalities.[31] Village data suggest that, in Jiangsu, income variability differs significantly by income sources, e.g. in south Jiangsu income from household non-agricultural activities is distributed much more unequally than either agricultural income or income from collective non-agricultural activities. Thus the effects of reforms on income distribution at the village level would depend in part on the strength of the resulting rural non-agricultural development (i.e. how rapidly resources moved from agriculture to non-agricultural activities), in part on changes in income differentials within and between sectors, and in part on the relative strengths of the collective and the household non-agricultural sectors. Two plausible scenarios (A and B) are discussed below.

Scenario A assumes that the rural non-agricultural sector that emerged after reform was composed largely of household or private activities. In this case income inequality within the village is likely to increase over time because both between-sector and within-sector inequalities are likely to increase with development. Between-sector inequality is likely to widen because agricultural productivity is likely to lag behind non-agricultural productivity and, for the reasons mentioned earlier, reforms are bound to increase inequality both within agriculture and within the rural household non-agricultural sector. One suspects this was the likely scenario followed in most regions other than where TVEs were highly developed (e.g. the heart of *sunan*).

In scenario B, the assumption is that rural non-agricultural activities developed primarily within a collective sector where income distribution is more equal than in the decollectivized agricultural sector. In this case, income distribution within the village is likely to widen at first and then narrow. When labour first begins to shift from agriculture to the collective non-agricultural sector, income inequality increases because the higher non-agricultural income is received by only a small number of the workers. But as more and more workers move to the collective non-agricultural sector, inequality will eventually narrow because it is also the sector with the smaller internal inequality. Villages in the heart of *sunan* probably followed scenario B.

Unfortunately, household income data at the village level either do not exist or are not available in sufficient detail to derive the inequality measures for the past decade or longer that would be needed to document the changes in rural income distribution. What is available for analysis are official per capita income measures for selected households in five of the six villages where we conducted fieldwork (Haiqi and Haiqian in Haitou, Yangquan and Xiaojian in Yixing, and Zhashang in Sigang).[32] The selected households were those that participated in the 1987 record-keeping survey. Data were collected for selected years both before and after the implementation of the household responsibility system (*da baogan*).[33] Before discussing the findings in Table 8.8, a few words of caution about the data may be appropriate.

The income measure available before *da baogan* (i.e. the implementation of the household responsibility system) is distributed collective income, and that after is household income.[34] Since these are different income measures, conclusions based on a comparison of the data should be treated with considerable caution. For the purpose of inter-temporal comparison, the most serious problem with the data in Table 8.8 is that the income figures before *da baogan* exclude the income from sideline activities. How important was sideline income before the implementation of *da baogan*? In Hufu and Sigang, we were told that before *da baogan*, particularly before 1978 when private initiatives were still officially discouraged, sideline income was negligible. Apparently, the underground economy was much less developed in Hufu and Sigang than in some other

TABLE 8.8. *Measures of Per Capita Income and of Income Inequality Identical Households in Five Jiangsu Villages, Selected Years*

	Village estimates				1987 survey estimates	
	Distributed collective income before *da baogan*		Household income after *da baogan*		Household income	
	1975	1980	1985	1987	Including imputed rent	Excluding imputed rent
Haitou						
Haiqi						
Average (*yuan*)	69	99	728	1,078	1,135	1,085
Coefficient of variation	0.2620	0.2855	0.6114	0.7057	0.7801	0.8011
Gini index	0.1478	0.1616	0.3286	0.3706	0.3667	0.3759
Standard deviation of log of income	0.1111	0.1265	0.2342	0.2644	0.2519	0.2588
Haiqian						
Average (*yuan*)	88	373	831	1,183	1,307	1,246
Coefficient of variation	0.2060	0.5291	0.2553	0.3793	0.5493	0.5666
Gini index	0.1173	0.2957	0.1494	0.1911	0.2939	0.3047
Standard deviation of log of income	0.0855	0.3298	0.1194	0.1392	0.2154	0.2250
Hufu						
Yangquan						
Average (*yuan*)	206	450	935	1,662	1,915	1,810
Coefficient of variation	0.2711	0.2256	0.2994	0.5156	0.4894	0.5130
Gini index	0.1470	0.1319	0.1653	0.2673	0.2487	0.2590
Standard deviation of log of income	0.1096	0.0969	0.1214	0.2192	0.1826	0.1909
Xiaojian						
Average (*yuan*)		270		920	1,302	1,206
Coefficient of variation		0.3654		0.3187	0.2977	0.3232
Gini index		0.2106		0.1865	0.1640	0.1775
Standard deviation of log of income		0.1991		0.1481	0.1260	0.1369
Sigang						
Zhashang						
Average (*yuan*)		803[a]	1,266	1,467	1,612	1,442
Coefficient of variation		0.5016[a]	0.3868	0.3155	0.4143	0.4573
Gini index		0.3011[a]	0.2136	0.1734	0.2138	0.2321
Standard deviation of log of income		0.2505[a]	0.1560	0.1283	0.1567	0.1706

[a] 1982.

Sources: Household Record Keeping Survey and Village Background Data.

parts of China.[35] However, private income was substantially more important in the poorer county of Haitou in the mid-1970s, where rural households found it difficult to subsist on distributed collective income alone, and had to supplement it with private income earned largely from coastal fishing. We were told that before *da baogan* per capita private sideline income in Haitou may have been as high as one-half the level of per capita distributed collective income, or roughly one-third of per capita household income. At the national level, in 1978, per capita private income from sideline activities was about 40 per cent of per capita distributed collective income or 27 per cent of per capita household income.[36] Sideline income was probably less important in earlier years, but the 1978 data suggest that it certainly was not negligible. In other words, the per capita income figures for the pre-*da baogan* years in Table 8.8 are downward biased. Furthermore, if private income was distributed more unequally than collective income before *da baogan*, as we believe it was, then the inequality measures for the pre-*da baogan* years are also biased, i.e. income was distributed more unequally than the measures in Table 8.8 suggest.

The post-*da baogan* official per capita household income data were estimated by village accountants based in part on village records and in part on their knowledge of each household. Besides these estimates, we also have the household survey data for 1987. A comparison of the official village estimates with the survey results (Table 8.8) suggests that the income figures derived from the survey are generally higher. A part of the difference may be explained by the fact that, unlike the village estimates, the survey estimates include imputed rent of owner-occupied dwellings as income. But the difference also reflects the fact that village estimates of private income were often on the low side.[37]

The income data for individual rural households suggest that while nearly everyone in the surveyed villages benefited from the reform measures and from the increased opportunity for non-agricultural employment, they did not benefit equally. In fact, a comparison of the ranking of surveyed households by per capita income suggests that one probable consequence of rural reform was that the relative position of many households in the village changed dramatically.[38] In other words, some households that were relatively poor became relatively rich and vice versa. These changes are to be expected, since even though the surveyed households were from the same village and therefore faced a similar economic environment and similar changes, the ability to respond to the changes and the new opportunities they produced was bound to vary among households as they differ significantly in terms of age–sex structure, skills, and entrepreneurial ability.

How did rural reform and increased participation in rural non-agricultural activities affect income distribution within each village? As expected, Table 8.8 suggests that the impact varied substantially—in some villages income distribution apparently improved and in others it deteriorated—and that the direction of change depended to a significant degree on the importance of the rural non-agricultural sector and the relative strength of private (household) and collective non-agricultural activities. In the three villages (Haiqi, Haiqian, and Yangquan) where household activities (in agriculture and in the non-agricultural sector) predominated (scenario A), income distribution worsened as expected.[39] In Xiaojian and Zhashang where over one-half of the per capita household income in 1987 was composed of wages, salaries, and distributions from the collective TVE sector (53 per cent in Xiaojian and 67 per cent in Zhashang), income distribution improved. In these two villages nearly every household in 1987 had a member who was a wage-earner in a nearby TVE. It was noted earlier in this chapter that income earned by TVE workers does not vary greatly within a locality, certainly less than the income earned from household activities. Thus, during the 1980s, as a preponderant share of the village residents in Zhashang and Xiaojian moved from relatively low-paying household agricultural activities to better-paying jobs in TVEs, not only did average village income rise but income distribution within the villages also improved. In other words, Xiaojian and Zhashang were in the latter phase of scenario B, when income distribution becomes more equal.

In summary, depending on the organizational structure of the expanding non-agricultural sector, rural reform could make income distribution within a village either more or less equal. Where the reforms increased employment primarily in the private sector (i.e. household activities in either agriculture or the rural non-agricultural sector), income distribution is likely to become more unequal since, among the main components of income, self-employment income is usually distributed most unequally. Where the reforms increased employment primarily in a rapidly expanding collective TVE sector, income distribution is likely to become more equal (perhaps after an initial period of deterioration) largely because collective income is more equally distributed than self-employment income.

Changes in Income Distribution in Rural Jiangsu

Rural reforms have increased income-earning opportunities (in farming and in off-farm activities) in land-rich, prosperous regions as well as in land-poor areas. Thus the net effect of rural reforms on rural income distribution in Jiangsu depends not only on the income distribution of off-farm income and that of farm income and their relative rates of growth and importance in each region but also on the impact rural reforms have had on regional income differentials. Thus, depending on the spatial distribution of endowment and the pace and pattern of regional rural development, the results of the rural reforms may be either an increase or a decrease in rural income inequality. In the case of Jiangsu, we know that (1) income from non-agricultural activities has become increasingly important for peasant households in all regions, (2) the spatial distribution of rural non-agricultural activities is extremely unequal so that in some parts of the province rural income has increased dramatically while in others it has increased only modestly, and (3) the most prosperous agricultural regions also have the most developed rural non-agricultural sectors and have experienced the most rapid growth.[40] Thus there are strong reasons to believe that rapid development of the rural non-agricultural sector in Jiangsu during the 1980s was accompanied by an increase in rural income inequality. Even though the presence of large numbers of underemployed rural workers at the beginning of the decade and the growing importance of collective activities in Jiangsu's rural non-agricultural sector may have had equalizing effects on income distribution at the village or regional level in parts of Jiangsu, they were more than offset by increases in interregional income differentials so that income distribution in rural Jiangsu as a whole became more unequal.

Unfortunately, household income data collected by the annual province-wide survey of peasant households during the 1980s are not available to test the hypothesis that income inequality in rural Jiangsu has increased. What are available for analysis are various proxy measures of per capita rural household income by administrative division. To derive income inequity measures from this type of data, we would want the data to be as disaggregated as possible since the more disaggregated these data are, the closer they approximate the size distribution of household income. The most disaggregated data available are by county, which of course is still a considerable distance away from the household. However imperfect these data may be, they should nevertheless shed some light on the income distribution issue.

Table 8.9 presents the evidence on income distribution in Jiangsu. Five measures are provided, and of these only three are available for both 1978 and 1988. Of those in the latter group, the one closest to per capita household income is per capita national income (*guomin shouru*). Of the five measures available in 1978, the one closest to per capita household income is probably per capita distributed collective income. Table 8.9 shows that the closer the proxy measure is to per capita household income, the smaller is the variation, and that all available inequality measures indicate that income inequality increased between 1978 and 1988. It appears that rural reform and increased participation in rural

TABLE 8.9. *Inequality Measures of Various Per Capita Indicators[a] for 64 Counties in Jiangsu, 1978 and 1988*

	Per capita gross value of rural industrial output		Per capita gross value of rural agricultural and industrial output		Per capita national income		Per capita collective rural income		Per capita distributed collective rural income	
	1978	1988	1978	1988	1978	1988	1978	1988	1978	1988
Average (RMB)	120	1,904	317	2,886	282	1,314	103	n.a.	84	n.a.
Coefficient of variation	0.884	1.146	0.417	0.775	0.389	0.495	0.359	n.a.	0.338	n.a.
Standard deviation of log	0.404	0.451	0.184	0.278	0.150	0.194	0.152	n.a.	0.144	n.a.

[a] Except for national income (*guomin shouru*), these are per capita rural population figures. Rural population in each county was obtained by deducting its non-agricultural population (*fei nongye renkou*) from its total population. (The Chinese do not provide a precise definition of *fei nongye renkou* except to note that it includes primarily the non-agricultural population in cities and towns (*chengzhen fei nongye renkou*). See *JTN* (1991), 59.) In other words, rural population as it is used here includes those involved in non-agricultural activities at and below the township level. Per capita national income was derived by dividing the national income produced in each county by the total (rural and urban) population in the county. National income (*guomin shouru*) is the factor income generated in the material product sectors. Distributed collective income is collective income (in kind and in cash) distributed to commune members. It is less comprehensive than household income as it does not include income from sideline activities or income from outside the commune. Collective differs from distributed collective income in that it includes the income retained by the collective (e.g. *gongji jin* (collective accumulation fund), *shengchan fei jijin* (reserve for production), and other reserve funds).

Sources: Calculated from data in *JJN* (1989), 3. 103–20, and in *1978 nian Jiangsu sheng nongye tongji ziliao* (Agricultural statistical material of Jiangsu Province, 1978), 2–11, 34–9, 246–61.

non-agricultural activities have caused income inequality in Jiangsu to increase, at least regionally.

The evidence presented above suggests that income distribution in rural Jiangsu probably has become more unequal. Since the factors responsible for the increase in income inequality in Jiangsu, i.e. regional differences in resources and in the pace of rural non-agricultural development, are present to an even greater degree in China, income inequality probably has also increased at the national level. Certainly rural income differentials across provinces increased significantly in the 1980s. The coefficient of variation of per capita income of peasant households by province was 0.27 in 1980, 0.31 in 1985, and 0.38 in 1989.[41]

In the long term, income distribution in rural China is likely to become more unequal. Rural development in China is fuelled primarily by local resources so that the better endowed rural areas (those with better infrastructure, greater resources, more developed non-agricultural activities, and closer proximity to urban centres) will continue to grow more rapidly than the poorer areas. It is possible to moderate the trend towards increased inequality if more resources are allocated to help develop agriculture, sideline activities, and industry in the poorer regions and if peasants from the poorest regions are permitted to migrate and work in areas with higher productivity.

NOTES

1. In this chapter peasant and rural are used interchangeably. This is also the practice in China. For example, a rural resident who is engaged full-time in commerce or business is usually called a peasant-merchant or peasant-entrepreneur in China.

2. For China as a whole, per capita peasant income in 1977 prices rose from 102.8 *yuan* in 1957 (73 *yuan* in current prices) to 113 *yuan* in 1977, or about 0.5 per cent per year (see S. Lee Travers, 'Getting Rich Through Diligence: Peasant Income after the Reforms', in Elizabeth J. Perry and Christine Wong (eds.), *The Political Economy of Reform in Post-Mao China* (Cambridge, Mass.: The Council on East Asian Studies, Harvard University, 1985), 111). Real per capita peasant income probably increased more rapidly in Jiangsu than in China as a whole, but it was still considerably lower than the rate of increase in real GVAO, which averaged about 5 per cent a year in Jiangsu during this period (see *JSN* 19).

3. Derived from data in *JSN* 181, and Bu Ruizhi *et al.*, 'Jiangsu nongmin shouru ji zhichu de shuliang fenxi (A quantitative analysis of peasant income and expenditure in Jiangsu)', *NJYS* 4 (1988), 14.

4. In 1978, of the RMB 108 received from the collective, RMB 4 was distributed by TVEs. See *JSN* 181.

5. The peasant households included in the annual survey were selected as follows. Before 1984, the sample consisted of below average, average, and above average households selected from representative regions at different stages of development (i.e. below average, average, and above average). This procedure undoubtedly introduced unknown biases into the sample. After 1984, Jiangsu's sample of 3,400 peasant households was selected from 340 villages in 34 counties in the following manner. First, all 68 county-level administrative districts (64 counties and the suburban districts of 4 cities) in the province were ranked by per capita income (three-year averages), and 34 alternate counties (e.g. the first, third, fifth, . . . , 67th) were selected for inclusion in the survey. All villages in each chosen county were ranked by per capita income (three-year averages), and 10 villages, separated equally in their ranking, were selected for inclusion in the survey. Finally, all households in each selected village were ranked by per capita income, and 10 households, separated equally in their ranking, were selected for the survey. Sample households in other provinces were chosen in the same way. (The above information provided by JPASS.) Most studies of peasant income and consumption published in Chinese journals in recent years have relied on these surveys, e.g. see Tong Nong, 'Woguo nongmin shouru jiegou de xin bianhua (New changes in the income structure of Chinese peasants)', *JD* 4 (1985), 113–16; Zhang Pingquan, 'Xinjiang nongmin shouru fenpei chayi ji qi pingjia (An appraisal of income differentials among peasants in Xinjiang)', *NJW* 10 (1987), 39–43; and Zhao Hong and Chen Xun, 'Nonghu jingji xingwei fenxi (An analysis of the economic behaviour of peasant households)', *NJW* 11 (1987), 53–7.

6. The Chinese term is *tongyi hesuan*.

7. The rate of increase in nominal per capita rural household income was even greater in the villages we surveyed, about twice the provincial rate.

8. In 1989, 15.6 per cent of the surveyed peasant households in China had real income below RMB 167 (RMB 300 in 1989 prices deflated by the index of rural retail prices).

9. RMB 200 in 1986 prices deflated by the index of rural retail prices is equivalent to RMB 165 in 1980 prices.

10. e.g. per capita grain consumption in rural Jiangsu increased from 165 kg. in 1978 to 202 kg. in 1988. Per capita consumption of meat, fish, eggs, sugar, edible vegetable oil, etc., all increased significantly.

11. The decline was reversed during the severe recession of 1989–90.

12. *JTN* (1990), 389.

13. Analysis of provincial cross-section data shows that housing has the highest income elasticity (between 2.370 and 2.692, depending on the functional form used) among all categories of consumption expenditures. See Jacques van der Gaag, *Private Household Consumption in China*, World Bank Staff Working Papers No. 701 (Washington, DC: World Bank, 1984), 21.

14. Nationally per capita living space in the countryside increased from 9.4 square metres in 1980 to 17.2 square metres in 1989 (see *ZTN* (1990), 324).

15. They are also influenced by hidden subsidies. For example, in China both the government and the work unit provide households with goods and services

free of charge or at greatly reduced prices, and these subsidies are not reflected in household expenditure data. However, such subsidies are directed mainly at the urban population and do not affect rural residents significantly.

16. For information on rural education and health, see Dwight Perkins and Shahid Yusuf, *Rural Development in China* (Baltimore: Johns Hopkins University Press, 1984), chs. 7 and 9.

17. e.g. in 1975 85 per cent of all production brigades in China had co-operative health insurance that was financed in part or in total by the brigade's welfare fund. By 1981, the figure had declined to 58 per cent, and further declines followed in 1982 and 1983. See World Bank, *China, The Health Sector* (Washington, DC: World Bank, 1984), 70–1.

18. Much of the increase in per capita income in the villages we surveyed can also be traced to increased rural non-agricultural activities. In Sigang township, for example, the share of per capita rural household income from non-agricultural activities increased from below one-half in 1978 to over three-quarters in 1987 (Township-Village Background Data). Hufu and Haitou appear to have had similar experiences. This is supported as well by evidence from other areas. For example, see Gregory Veeck and Clifton W. Pannell, 'Rural Economic Restructuring and Farm Household Income in Jiangsu, People's Republic of China', *Annals of the Association of American Geographers*, 79/2 (1989), 275–92, and Nong mu yu ye bu shedui qiye ju erchu, '1982 nian renjun fenpei shouru sanbai yuan yishan de fu xian he fu dui qingkuang (Circumstances of wealthy counties and brigades with per capita income above 300 yuan in 1982)', in Jingji diaocha bianjizu, *Jingji Diaocha (Economic Investigation)*, 2 (1983), 77–80.

19. Despite the flood of data on the Chinese economy that has appeared since 1978, we still do not have direct evidence of size distribution of income at the national level or how it has changed over the years. For discussions of the available evidence, see Carl Riskin, *China's Political Economy* (Oxford: Oxford University Press, 1987), 223–56; Keith Griffin and Ashwani Saith, *Growth and Equality in Rural China* (Bangkok: ILO-ARTEP, 1981); Keith Griffin, 'Income Differentials in Rural China', *CQ* 92 (Dec. 1982), 706–13; E. B. Vermeer, 'Income Differentials in Rural China', *CQ* 89 (Mar. 1982), 1–33; and Irma Adelman and David Sunding, 'Economic Policy and Income Distribution in China', in Bruce L. Reynolds (ed.), *Chinese Economic Reform How Far, How Fast?* (Boston: Academic Press, 1988), 154–71. All the evidence suggests that the level of inequality in China is among the lowest in the world.

20. e.g. see Dennis Anderson and Mark Leiserson, 'Rural Nonfarm Employment in Developing Countries', *Economic Development and Cultural Change*, 28/2 (Jan. 1980), 227–48. For discussion of rural non-agricultural development and rural household income in Taiwan and Japan, see Samuel P. S. Ho, 'Off-Farm Employment and Farm Households in Taiwan', and Ryohei Kada, 'Off-Farm Employment and the Rural–Urban Interface in Japanese Economic Development', in R. T. Shand (ed.), *Off-Farm Employment in the Development of Rural Asia* (Canberra: National Centre for Development Studies, ANU, 1986).

21. See Appendix for a description of the surveyed villages and a discussion of the survey and some of its weaknesses and perceived biases.
22. e.g. of the sixty-three workers in the households we surveyed in Haiqi, thirty-two considered fishing their main occupation. In fact a considerable number of households in Haiqi (and to a lesser extent in Haiqian) did not cultivate any responsibility or subsistence land.
23. In the mid-1980s, each of Zhashang's twelve production teams owned and operated at least one enterprise. In 1987, the village merged some of the team enterprises and elevated them to village status.
24. Several measures are provided because alternative inequality measures often produce conflicting rankings of income distribution. For a summary discussion of this problem see Pan A. Yotopoutos and Jeffrey B. Nugent, *Economics of Development: Empirical Investigations* (New York: Harper & Row, 1976), 239–47. In all cases, the Gini index reported in this chapter was computed from individual observations using the following formula:

$$G = [2/(T - 1)]\sum_{t=1}^{T} ty_t - [(T + 1)/(T - 1)],$$

where y_t is the proportional share in income of the tth ranked household and T is the total number of households in the sample. A larger Gini index indicates greater inequality. All surveyed households reported positive total income, although a few did report negative income from agricultural (broadly defined) activities. Because those with negative income from agriculture were engaged in prawn culture, a high-risk activity, I believed the reported negative income from agriculture was real and not caused by the underreporting of income. Thus I included all surveyed households in the inequality calculations even though a few reported negative income from agriculture in 1987.
25. In 1987 the per capita household income in Haiqi and in Haiqian was substantially below that in either Jingxiang or Zhashang (see Table A.3).
26. See Ch. 5 for evidence on wage distribution in the surveyed regions.
27. The first distribution is wage (Mo Yuanren (ed.), *Jiangsu xiangzhen gongye fazhan shi* (*History of the Development of Rural Industry in Jiangsu*) (Nanjing: Nanjing gongxueyuan chubanshe, 1987), 414). Sometimes, more than two distributions are made. For example, one village enterprise we surveyed made three rounds of distribution. The first was wage, and since nearly every household in the village had a worker in the enterprise most households received some wage income. Each household then received a second round of distribution, the size of which was determined by the amount of land it cultivated. Since land in the village was distributed by population, the second round of income was in effect distributed by population. The third round was a distribution of profits, the size of which was determined at the end of the year. A household with a member in the enterprise received a full share of the distributed profit if that person was 20 years or older and 70 per cent of a full share if he or she was below 20 (Enterprise Interview Notes).
28. The coefficient of variation of income from agricultural activities other than cultivation was 1.5066 in Haiqi and 2.1974 in Haiqian.
29. Denise Hare, working with data from 249 farm households collected in five townships in a central Guangdong county, found that the high variability in

income from household non-agricultural activities is due largely to the variability in income from different types of household non-agricultural activities rather than in variability in earnings across people involved in the same type of activity (correspondence with the author). In other words, income variability is likely to be greater when there is a wider choice of household non-agricultural activity.

30. Similar findings have been reported by Denise Hare in 'Rural Non-agricultural Activities and their Impact on the Distribution on Income: Evidence from Farm Households in Southern China', unpublished paper (Nov. 1991).

31. See Gary S. Field, *Poverty, Inequality, and Development* (New York: Cambridge University Press, 1980), where it is shown that if the aggregate inequality index is statistically decomposable, then it is the weighted sum of within- and between-sector inequalities, with population shares as weights.

32. Jingxiang's record was incomplete (past estimates of per capita income were found for less than half of the surveyed households) and is not included in Table 8.8.

33. *Da baogan* was adopted in different years in different parts of Jiangsu. In the regions we surveyed, it was implemented either in 1982 or 1983.

34. Compared to rural household income, distributed collective income is less comprehensive as it does not include income from private sources (e.g. sideline income) nor income from outside the rural sector (e.g. remittances and transfers and earnings from work outside the rural sector).

35. See Anita Chan and Jonathan Unger, 'Grey and Black: The Hidden Economy of Rural China', *PA* 55/3 (1982), 452–71.

36. *ZTN* (1981), 431. By 1981, per capita private income had increased to 73 per cent of per capita distributed collective income or 38 per cent of per capita household income. Also see Vermeer, 'Income Differentials in Rural China', 16–17 for a useful discussion of the relative importance of private income.

37. This was the view of several of the village cadres we interviewed.

38. The flaws in the data make it impossible to draw definitive conclusions. When we compare the per capita income of households in a pre-*da baogan* year (when the income measure excluded private income) with that in a post-*da baogan* year (when the income measure included private income), the change in ranking (i.e. the relative position of each household) may be only reflecting the fact that different income measures were used to rank the households. But some of the changes in relative position were so dramatic that I find it difficult to believe that shifts in relative position did not occur. Since the government discouraged private activities in the 1970s, rural households generally limited their earnings from sideline activities so that, while most households earned some private income, few (if any) households earned a great deal of private income. Thus, before *da baogan*, a ranking of households by per capita household income is not likely to be dramatically different from a ranking by per capita distributed collective income.

39. Agricultural income and income from household activities in the non-agricultural sector accounted for 57 per cent of total per capita household income in Yangquan, 88 per cent in Haiqi, and 91 per cent in Haiqian (see Table 8.5).

40. The Chinese news media have reported numerous examples of villages,

townships, and counties that went from rags to riches in a matter of a few years. In our fieldwork in Jiangsu, local cadres also told stories of dramatic economic improvements that occurred in selected villages and townships in their regions. For example, out of 1,900 counties for which data were available, the per capita distributed income in 175 more than doubled between 1978 and 1981 (Nong mu yu bu shedui qiye ju erchu, 'Nongye shouru fanfan xian de tongji ziliao (Statistic materials on counties whose agricultural income doubled)', in Jingji diaocha bianjizu, *Jingji Diaocha* (*Economic Investigation*), 2 (1983), 72. See Ch. 3 for additional evidence from Jiangsu on the last two points.

41. Calculated from data in *ZTN* (1990), 314. The mean, the standard deviation, and the ratio of the highest average provincial per capita income to the lowest for selected years are presented in the table. Other measures also indicate an increase in rural regional inequality at the provincial level (e.g. see Chor-Pang Lo, 'The Geography of Rural Regional Inequality in Mainland China', *Transactions, Institute of British Geographers*, NS 15/4 (1990), 470–4).

	Per capita household income (RMB)		
	1980	1985	1989
Mean	205	413	641
Standard deviation	55.6	125.9	246.3
Maximum/minimum	2.79	3.16	3.77

9

Retrospect and Prospect

THE implementation of the household responsibility system in agriculture and the promotion of rural non-agricultural development were undoubtedly the two most successful reforms adopted in the 1980s. In one short decade they have rejuvenated and transformed rural China. Rural industry has introduced a degree of competition to the economy, and its vigorous growth has put constant pressure on the state sector to reform. Its rapid development and its ability to create employment have also made it easier for China to continue its economic reform.

Rural Non-agricultural Development in China

This study has shown that the policy of leave the land but not the countryside, by allowing private activities to reappear and by modifying or removing restrictions on rural non-agricultural activities, has unleashed an explosion of entrepreneurial activities by private individuals and by rural cadres acting on behalf of their communities. Between 1978 and 1989, the average growth rate of real RGVNAO was better than 20 per cent a year, and tens of millions of workers at or below the township level were absorbed by the rural non-agricultural sector.

Because the ratios of net to gross value of output are much smaller in non-agricultural activities than in agriculture and because much of the profits earned by TVEs are remitted to TVGs or reinvested in TVE development and therefore affect personal income only indirectly, the impact of rural non-agricultural activities (and of TVEs in particular) on personal income is much less significant than its role in production. None the less, in the late 1980s, an average rural household in China earned about 40 per cent of its income from rural non-agricultural activities.

The rapid growth of rural non-agricultural activities has produced dramatic changes in the rural economy. Agriculture, which had produced nearly two-thirds of the value of rural gross output at the beginning of the decade, accounted for only 45 per cent at the end. In one decade, the share of rural workers whose main occupation was outside of agriculture increased from 10 per cent to over 20 per cent. In other words, agriculture has become a sideline activity for a large number of rural workers. In addition, many rural workers have become involved in rural non-agricultural activities on a part-time basis.

Rural non-agricultural activities have not developed evenly in China. Development has been much more rapid in some provinces (e.g. Jiangsu, Guangdong, Zhejiang, Shandong) than in others, and within each province development has also occurred very unevenly. For example, in Jiangsu, where rural non-agricultural development has been extremely vigorous, most of the growth has concentrated in the populous and economically more developed southern third of the province. As in other countries, the uneven spatial distribution of rural non-agricultural activities in China can be explained largely by regional differences in population density, in agricultural development, and in proximity to urban industries. Furthermore, rural non-agricultural activities in the more developed and urbanized regions are significantly more productive, suggesting that they were either more efficient or were engaged to a greater extent in the more productive non-agricultural activities, or, most likely, a combination of the two possibilities.

Rural non-agricultural activities and agricultural production expanded simultaneously in the late 1970s and the early 1980s, and this happened despite a decline in the primary inputs (land and labour) allocated to agriculture. It was possible to shift resources from agriculture to non-agricultural activities without adversely affecting farm production during this period because rural workers were underemployed, because the household responsibility system motivated peasants to work harder and longer, and because of increased application of labour-saving and yield-raising inputs financed in part by earnings from rural non-agricultural activities. However, in the mid- and late 1980s, as farming became increasingly less profitable relative to other rural activities, agricultural growth slowed. In the long run, agricultural growth can be sustained only if more resources are invested in the sector and farming made more productive and profitable.

During the 1980s, the rural non-agricultural sector in most parts of China served primarily local and domestic needs. The exception was in south China. Because the government gave Guangdong and Fujian special authorities and privileges, rural enterprises in these two provinces were the first to benefit from the open policy, and their growth was largely stimulated by foreign investment and driven by export. When China formally adopted the so-called coastal development strategy in 1988, TVEs in Jiangsu and other coastal regions were given incentives to export and to utilize foreign investment similar to those given earlier to TVEs in south China. Since then, an increasing number of TVEs in the coastal regions have turned outward to the world market, and rural industry has become China's leading export sector. If the coastal development strategy is maintained, many rural enterprises in the coastal regions will become as internationalized as those in Guangdong.

In most parts of China, but particularly in the more developed regions, rural non-agricultural development has followed the so-called *sunan* (south Jiangsu) pattern, with the collectively owned industrial enterprises the primary source of growth, and with township and village cadres the entrepreneurs and enterprise managers. In the early and mid-1980s, to help manage their expanding economic activities, TVGs adopted the so-called contract responsibility system. The hope was that the new system would change the relationship between TVEs and TVGs from a hierarchical to a contractual one—i.e. replacing the direct power that TVGs have over TVEs by a set of incentives and regulations. In addition, it was hoped that the contract responsibility system, by linking reward to performance and by holding enterprise managers responsible for profits, would improve the economic performance of TVEs. The evidence from Jiangsu suggests that while the contract responsibility system may have improved enterprise performance it has not significantly altered the relationship between TVEs and TVGs. In most parts of rural Jiangsu, TVGs continue to be deeply involved in the management of their enterprises, and most TVEs enjoy only limited autonomy.[1]

TVGs have used TVEs to absorb surplus labour from agriculture. In the absence of a labour-market, rural cadres allocated TVE jobs in the first instance to their relatives and friends, and then, as more jobs became available, distributed them more or less equally among households in the community. Only after the supply of local workers was exhausted, did TVEs look outside the community for workers. Over time, as more outside workers were needed, a rural labour-market emerged. The evidence from Jiangsu suggests that TVEs often employ far more workers than they need. Apparently TVGs have used jobs in TVEs to transfer disguised unemployment from agriculture to the non-agricultural sector, and to reduce the income differentials between agriculture and rural non-agricultural activities. Intra-enterprise income differentials among Jiangsu TVEs are surprisingly narrow, suggesting that wages in many TVEs are not strongly linked to performance. What maintains work discipline in TVEs is the absence of job security.

In less developed and less urbanized regions, where rural communities have less collective resources to invest and where the environment is not hospitable to rural industry, successful rural non-agricultural development has come about frequently as the result of growth in the private sector. Even in Jiangsu, where rural non-agricultural development has occurred primarily through the development of collective activities, private entrepreneurship has made important contributions. While the vast majority of the private activities in rural China are tiny in size and produce goods mainly for local consumption, a small number of medium-size private enterprises have emerged by filling the cracks in the regional or national market not occupied by the larger state and collective enterprises. Since the mid-1980s, the level of private activities in rural Jiangsu and elsewhere in China has not increased significantly, and the number of private enterprises with sixteen or more hired workers in the late 1980s remains negligibly small. It is not difficult to understand why private activities, after an initial burst of growth, have developed slowly. Despite repeated assurances from the government that private activities have its full support, private entrepreneurs still face serious political discrimination and bureaucratic obstructions. Unless ongoing economic and political reforms significantly improve the environment for private entrepreneurs, it is difficult to see the private sector expanding rapidly in the future.

The rural non-agricultural sector has developed with little or no financial assistance from the state. The capital needed by private entrepreneurs has come primarily from their own savings and from loans from relatives and friends. The capital needed by TVEs came initially from the community surplus and from household savings, but, since the mid-1980s, reinvestment of profits and bank credits have become their primary sources of capital.[2] Technology and skills have come from the urban sector, either purchased outright by rural enterprises or obtained through linkage arrangements with urban units. Because rural communities in the more developed and urbanized regions had more collectively owned resources and better access to urban units, they were the first to develop TVEs and their TVEs have developed at a more rapid pace.

Rural enterprises are almost completely market orientated in the sense that they produce little according to the state plan and receive little of their material supplies from the state-run material-supply system. Near complete dependence on the market in a partially reformed planned economy has meant that rural enterprises have had to face severe marketing and supply difficulties. They have survived and prospered because they have been flexible in setting prices, have catered to the wishes of customers, have responded quickly and in innovative ways to market opportunities, have marketed their products aggressively, and have developed elaborate networks of connections and special arrangements to acquire energy and raw materials.

The prime motive force behind the rapid development of rural non-agricultural activities in China lies within the townships and villages themselves. It is the interests of the localities rather than central directives or encouragement that have driven rural non-agricultural development forward. Rural non-agricultural development gives the rural population greater choices, not only in terms of more, and a wider variety of, jobs but also in terms of variety and availability of goods and services. Rural cadres also attach great importance to rural non-agricultural development, particularly TVE development. This is because rural cadres consider the development of TVEs as the surest way to generate the revenue needed by TVGs and the best way to prosperity for their communities as well as themselves. Remittances to TVGs from community-owned enterprises have become very important, and in most regions they are the only important extrabudgetary revenue over which rural cadres have discretionary control. Much of the remittances has been ploughed back into rural enterprises as reinvestments, but substantial amounts have been used by TVGs to support agriculture, to provide education and social services, and to construct rural infrastructure (roads and the development of small rural towns). It is no exaggeration to say that, without rural enterprises, few TVGs in China could afford much in the way of a community development programme. Rural cadres also have a direct personal interest in developing TVEs, since their income, status, and professional success are all linked to the overall economic performance of their communities.

Rural non-agricultural development has brought to the surface two potential problems. Rural reforms and the rapid growth of rural non-agricultural activities have brought about a significant widening of income inequality. Income differentials have increased and are likely to continue to increase not because rural non-agricultural income is less equally distributed than agricultural income (in fact, because the rural non-agricultural sector in most regions is dominated by collective activities, it is likely to be more equally distributed) but because the pace and quality of non-agricultural development have varied and will continue to vary significantly by region. Thus the more developed regions are likely to benefit much more from rural non-agricultural development than the less developed regions. Finally, it should be noted that rural non-agricultural development has been achieved, at least in some parts of China, at considerable cost to the environment. A significant amount of prime farm land has been lost to rural industries, and industrial pollution has become a serious problem in some parts of rural China. The government has introduced measures to regulate land use and industrial pollution. But because governments at all levels are more concerned about jobs and growth than about the environment, the loss of land and the spread of industrial pollution are likely to continue unabated.

The successes of rural reforms and the optimism created by rapid rural non-agricultural growth in the early and mid-1980s were overshadowed in the late 1980s by inflation, slow progress in urban economic and price reforms, increased friction between SEs and TVEs, regional fragmentation, and rising political tension. Because they are competitors and because they operate under different rules, tensions have always existed between TVEs and SEs. During the 1980s, many in the state sector complained that TVEs expanded too rapidly, with little attention to efficiency or availability of markets and resources, and tried to restrain their growth.

In 1989, when the economy was in a deep recession, the hostility against TVEs again surfaced, and some government leaders openly attacked rural enterprises and advocated that they be suppressed in order to save state enterprises in difficulties (*bao quanmin qiye ya xiangzhen qiye*).[3] The argument went as follows. Under existing government policies, SEs were like caged tigers that were forced to compete for food (capital, raw materials, energy, markets) with freely roaming monkeys (TVEs), even though monkeys were not as productive as tigers.[4] Since it was not possible to uncage the tigers, the only solution was to cage the monkeys. For a time, rural enterprises were deliberately denied access to bank credit and to centrally controlled materials (e.g. coal, electric power, steel).[5] In Jiangsu, for example, because of shortages of centrally controlled materials, the price TVEs paid for coal more then doubled and that for steel tripled between 1988 and 1989.[6] The rise in energy prices alone reduced the profits of Jiangsu's TVEs in 1989 by RMB 2.5 billion, about 80 per cent of the actual profits earned that year.[7] As the decade came to a close, China's rural non-agricultural sector, particularly TVEs, faced an uncertain future.

To counter the criticisms, supporters of TVEs launched a spirited defence.[8] They argued that the development of TVEs represented a creative and economically sound response to China's unique conditions: a huge number of underemployed and unskilled rural workers, shortages of capital, a weak transport system, and congested urban centres. Specifically, they noted that while SEs may be caged tigers, they were also regularly fed raw materials that were collected for them by the government. By contrast, the monkeys did not even have easy access to many raw materials that were produced in the rural sector, but must scurry about the jungle looking for leftover scraps.[9] Furthermore, SEs paid the lower state prices for their raw materials while TVEs must pay the higher market or negotiated prices. They also wondered why there were such concerns about TVEs diverting capital and energy away from SEs when, in 1988, less than 4 per cent of the loans provided by the government-controlled banks went to TVEs and the TVE sector was a net supplier of energy to the country.[10]

While admitting to the wide gap in labour productivity between TVEs and SEs, supporters of TVEs argued that the comparison was not meaningful since SEs and TVEs operated under different conditions. In particular, SEs were capital intensive while TVEs were labour intensive. They were also quick to point out that while the state sector used five times more fixed capital per worker in 1987–8 than did the TVE sector, its labour productivity was only 1.7 times greater.[11] Furthermore, in the 1980s labour productivity rose more than three times faster in the TVE sector than in the state sector. In other words, if TVEs and SEs were to operate under similar conditions, TVEs would not necessarily be less efficient. Left unsaid was the implication that, since China was abundant in labour but scarce in capital, the labour-intensive path of industrialization followed by TVEs was more appropriate than the substantially more capital-intensive route taken by the state sector.

Towards the end of 1989, the government relented to the pressure and reversed its policy of suppressing rural enterprises. Rural industry had become too important to the social and economic stability of China to be long suppressed. In the late 1980s, rural enterprises of all levels of ownership (i.e. township, village, team, and individual) supported perhaps as many as 200 million people in rural China,[12] and were the primary source of revenue for governments at the grass-roots level.[13] More important from the perspective of the national government was that over one-half of the increase in budgetary revenue in the late 1980s was accounted for by taxes collected from rural enterprises.[14] Furthermore, the taxes paid by rural enterprises were net contributions since, unlike SEs, they had received little or no investment from the state. However, the most convincing reason for ending the suppression of rural industry was the dire consequences that would follow if the sector became moribund. To paraphrase one pro-TVE official:[15] if government policies fail to provide a stable livelihood for the 200 million[16] surplus workers in the countryside, then not only would the national goal of achieving a relatively comfortable standard of living (*xiaokang*) be just empty words, but tens of millions of rural workers would be driven by poverty to seek employment in the cities.

Political and economic conditions have improved considerably since 1989, and the future of the rural non-agricultural sector, particularly that of TVEs, now appears rather brighter. At the Fourth Session of the Seventh National People's Congress in March 1991, the government confirmed its intention to facilitate the sound development of TVEs.[17] In fact, the government now sees the development of TVEs as helping it to 'consolidate the economy of socialist public ownership in the countryside'[18] as part of its effort to develop gradually in rural China a dual management system, under which agriculture will be managed in part by households and in part collectively.[19]

What lies ahead for the rural non-agricultural sector? What are its problems and prospects?

Prospects

Chinese projections predict that, in the year 2000, China will have a rural labour force of about 450 million, of which 150 million will be engaged in cultivation, 100 million in other agricultural activities, and the remaining 200 million in non-agricultural activities.[20] These projections suggest a moderate movement of population from rural to urban areas and, at the same time, a dramatic decline in the share of rural workers engaged in agriculture, falling from about 80 per cent in 1987 to 56 per cent in 2000. In other words, these projections suggest that, by the end of this century, sufficient numbers of rural workers will have shifted from agriculture to other activities in the countryside for rural China to have an economic structure very much like rural Jiangsu's in the late 1980s.

No country has ever experienced such a large shift in one decade. Thus, these projections should not be viewed as realistic targets but as what Chinese policy-makers hope will happen in the not too distant future, perhaps in the first quarter of the next century. What is clear is that the Chinese government has decided to continue its policy of allowing peasants to leave the land but not the countryside and is hoping that in the next two or three decades it will shift a large number of rural workers from agriculture to the more productive rural non-agricultural activities. Is this a sustainable policy? Can rural non-agricultural development do for China what it has done for Jiangsu?

Because China's rural non-agricultural sector is dominated by TVEs, particularly the industrial enterprises, its viability and prospects are closely linked to those of TVEs. The record compiled by TVEs in the 1980s, remarkable as it is, is not unblemished. Even before the 1989–90 recession, many TVEs were unprofitable or barely profitable with little prospect for improvement, suggesting that many TVEs were inefficient and poorly managed. Scarce resources have been wastefully used by TVGs in attempts to replicate the successful experience of south Jiangsu in regions where conditions are dissimilar. Overinvestment has created duplications and excess capacity in some rural industries. Because they are widely dispersed and primarily serve small markets, many TVEs have little chance of growing larger and benefiting from scale economies. And finally, because of the uneven development of rural non-agricultural activities in general, and of TVEs in particular, vast regions of the country have experienced little or no rural non-agricultural development.

Advocates of the policy of leave the land but not the countryside admit to these problems, and recognize that effective solutions need to be found quickly if rural non-agricultural development is to continue to play a positive role in China's development and if rural enterprises are not to become an impediment to efficient economic growth as some SEs have.[21] What follows is a brief discussion of some of the factors that are likely to influence the effectiveness of the policy of leave the land but not the countryside and the viability of the rural non-agricultural sector in the future.

TVEs and the State Sector

Reforms have transformed TVEs into the largest (probably also the strongest) component of an expanding collective sector in a mixed economy composed of enterprises of different ownership forms (state, collective, private, and joint) that coexist but also compete with one another for resources that are partly allocated by market forces and partly under bureaucratic management.[22] However, TVEs and SEs do not compete on a level playing field. In terms of access to credit, foreign exchange, energy, centrally allocated raw materials, and state subsidies, the priority of Chinese bureaucrats is to put the state sector (SEs and large urban collective enterprises) ahead of TVEs and private enterprises. However, the same ranking also holds when it comes to the degree and intensity of government regulation and supervision. Thus, even though all enterprises are dependent to some degree on the market and all are affected by government regulations, the lower an enterprise is in terms of the state's priority, the less it is regulated and the more it is dependent on the market. Given the priority ranking of China's bureaucracy, TVEs are often crowded out in the allocation of resources, but, because TVEs are more aggressive and not as closely supervised as SEs, they sometimes succeed in circumventing government regulations and win in the competition for resources and markets.

As long as this mix of market forces and bureaucratic control continues, there will be conflicts between TVEs and SEs, and periodically the state sector will attempt to restrain TVEs. Given the growing importance of rural enterprises to rural prosperity, an outright suppression of TVEs is unlikely, but attempts to restrict and regulate their activities can be expected. Therefore, the future development of TVEs will depend in part on the ability of the government to confine and to manage the conflicts between TVEs and SEs.

To minimize the friction between TVEs and SEs, some have called for a division of labour between the two sectors by restricting TVEs to suitable activities.[23] If by suitable it is meant industries that are suited for dispersed location, then the historical experience of rural industry in other countries would suggest the following types of activities for TVEs: industries that produce highly perishable goods, industries that process a dispersed raw material, industries that produce final products that are bulky or difficult and costly to transport, industries that use production processes in which scale economies are unimportant, service industries that are location specific, labour-intensive industries producing standardized products or components using simple technology, etc. While these categories are suggestive, they are not adequate or workable as guidelines to decision-makers.

Whether or not a TVE should engage in a particular line of activity must be judged on the enterprise's individual merits and based on economic cost-benefit analysis. Furthermore, because of serious defects in China's planning system and because site selection is a highly politicized process, industrial investment and the locations of industrial enterprises (state and collective) are often irrational.[24] In other words, there are SEs as well as TVEs that, for economic and environmental reasons, should be closed or relocated.[25] Thus, if the division of labour between the rural TVE sector and the urban state sector is to be based on economic and location suitability, then all activities, and not just TVE activities, need to be scrutinized.

Because of the underlying tensions in a partially reformed economy, the current mix of market forces and bureaucratic control is not a stable one. Either China will adopt more reforms and move towards some form of market socialism, or there will be a retreat to a restrengthened system of bureaucratic control. Economic reforms have stalled and restarted several times since the early 1980s. The direction of change at present is towards market socialism, and given the strength of current political commitment to reform, it is unlikely that China will return to a Stalinist-type command economy. But the pace and the direction of reform may change. If the government is unable to manage the economy so that there is both growth and stability and if corruption and rent-seeking, already widespread, become destructive, the pressure to slow reform and to reinstate bureaucratic control may be difficult to resist.

If China retreats to greater bureaucratic control, what is the prospect for TVEs? Because TVEs have become sufficiently important in many provinces and to influential groups in the bureaucracy and the Party, they are not without advocates on the national stage.[26] Consequently, the sector is not likely to disappear. But greater bureaucratic control inevitably reduces the role of the market, so activities of individual and team enterprises will undoubtedly be curtailed. However, the larger TVEs will probably be protected. What might happen is that, like SEs, TVEs will also be caged, i.e. integrated into the planned sector. They will produce and receive their raw materials according to the state plan, and of course since they are part of the plan they will no longer compete with SEs and the tension between them will disappear.[27] Of course, once they are caged, TVEs will no longer be a source of competitive stimulus and an impetus for change and economic reform. In time, TVEs under such circumstances would lose their competitiveness like so many SEs already have and become an obstacle to efficient growth.

If reform continues, administrative allocation of resources will be replaced by market allocation. In fact, with the changes implemented in the late 1980s and the early 1990s, the market is now used to allocate most material resources. When allocation is done bureaucratically, efficient enterprises cannot bid resources away from the inefficient, and in a system where the state sector is favoured, rural collective and private enterprises are always at a disadvantage. The increased use of markets to allocate resources will allow TVEs to compete for resources on a more equal basis, but only if the reform package also includes measures that ensure competition in the economy. Otherwise, SEs, given their size and market power, can easily preserve their privileged position in the economy to the detriment of TVEs. Only when all enterprises compete for resources on equal terms (a condition not as yet established) will the allocation of resources be significantly more efficient.

Can TVEs compete with SEs? In other words, are TVEs efficient? Obviously there are no categorical answers to these questions. Of the TVEs we visited, most were producing goods for which there was a market and were making profits. Some TVEs, because they fill a niche or because they serve SEs, do not compete directly with SEs. But we also know that many TVEs made little or no profit. Furthermore, since prices in China do not reflect the real costs and benefits of either inputs or outputs to society, efficiency can not be judged on profitability alone. One suspects that some (perhaps many) TVEs would not survive the scrutiny of a properly conducted cost-benefit analysis. But many would survive such a scrutiny. Furthermore, inefficiency is not a problem confined to TVEs. China's economy is burdened with many inefficient SEs that should be closed down if judged by the same cost-benefit criterion. Even if some TVEs do not survive in a fully reformed economy, reform would nevertheless be good for the rural non-agricultural sector because it would give the efficient rural enterprises an equal chance to compete for scarce resources and because competition would ensure that only efficient rural enterprises will grow.

Ownership and Property Rights

During the 1980s, TVEs developed with little attention paid to the question of ownership. Who are the ultimate owners of TVEs? The local governments, the workers, or the local residents? Because ownership affects TVEs in many ways—e.g. the nature of their relationship with TVGs, their inclination to merge with other enterprises, their willingness to relocate, and their objectives—it has now become an issue of considerable importance.

TVGs and TVEs

During the 1980s, TVGs operated as if they were the sole owners of TVEs. Indeed, the relationship between TVEs and TVGs was similar to that between SEs and the state in pre-reform days, i.e. with enterprise, Party, and government affairs inextricably tangled. In fact, during the 1980s, because of their physical proximity to TVEs and because each TVG had only a handful of TVEs to oversee, TVGs were probably more intimately involved in the affairs of TVEs than the state bureaucracy ever was in the affairs of SEs.

Over the years, TVEs have been accused of inefficiency and poor management, and much of the problem has been attributed to the close ties between TVEs and TVGs. Because TVEs provide TVGs with the means to discharge their duties and achieve their objectives, both economic and non-economic, they are rarely operated as profit-maximizing enterprises.[28] Because TVEs serve rural governments in ways other than as a source of revenue (e.g. they provide jobs for the underemployed), TVGs do not always impose on them a hard budget constraint. It is not unusual for TVGs to use profits from profitable enterprises to prop up others that are barely profitable or unprofitable. Fortunately, the financial resources available to TVGs are limited; otherwise inefficiency would be even more of a problem in the TVE sector. Because TVEs are used to help defray the costs of collective consumption and that of local government, there is also pressure on TVGs, particularly those in poorer regions, to squeeze TVEs for short-term gains to the government at the cost of long-term growth to the enterprise. Personal abuse of power is also a problem. At both the township and village level, because power is concentrated in the hands of the local Party secretary and his or her deputies, an often-heard complaint is that those in authority act like little emperors, contracting out TVEs to their friends and supporters (or appointing them to senior positions) as rewards.

The need to separate Party, government, and enterprise has long been recognized, but past attempts have not been successful.[29] As was discussed in Chapter 7, the dismantling of the commune in 1983 so as to separate the Party, the government, and the economy (*dang, zheng*, and *jing*) in rural areas did not achieve that aim. The contract responsibility system (*chengbao zeren zhi*), introduced in the early and the mid-1980s to give TVEs greater autonomy, was also only partially successful (see Chapter 5). Many now feel that the only way to separate Party, government, and enterprise at the township and village levels is to define and clarify enterprise ownership so that TVEs are answerable not only to TVGs but also to their other legitimate claimants (the community, local residents, and workers).

One reason why efforts to separate Party, government, and enterprise were unsuccessful in the 1980s was because economic reform was incomplete. It was shown in Chapter 7 that because TVGs were underfunded they became increasingly dependent on TVEs for revenue, and this induced local governments to become even more involved in economic activities. This is not likely to change until the tax system is reformed to give local governments greater budgetary decision-making and revenue-raising authority.[30] As long as their financial security is dependent on their control of TVEs, TVGs will want to retain near total control over them. The premature separation of TVGs and TVEs, before a new fiscal arrangement is in place, will only diminish local governments' ability to meet their obligations in their assigned areas of responsibilities, i.e. social services, agriculture, and rural infrastructure.

There is another reason why it has been difficult to separate TVGs and TVEs. In the partially reformed economy that existed in China in the 1980s, without TVGs' connections and support, TVEs would have had great difficulties obtaining inputs and distributing their outputs. TVEs developed as fast as they did in the 1980s only because they were an integral part of TVGs. One suspects, in the 1980s, few TVEs could have afforded or wanted to be totally independent of TVGs. With the deepening of economic reform, TVEs have become less dependent on TVGs for assistance and support, so future attempts to separate TVGs and TVEs may have a better chance to succeed. But, in the final analysis, the real issue is whether TVGs are willing to share in the ownership of TVEs, i.e. whether they are willing to exchange some of the discretionary power they now have as sole owners of TVEs for improved economic efficiency. This issue, like so many others, can be resolved only at the national level. The ownership question needs to be addressed as soon as possible because the longer TVGs believe they are the sole owners of TVEs, the less likely they will be willing to relinquish control in the future. In which case, TVGs could become a serious obstacle to future market-orientated reforms.

Ownership and Location

In rural China, where a TVE is located is determined by its level of ownership. Village and team enterprises are located on village land, usually close to where homes are clustered, and township enterprises are usually situated in the township seats.[31] Because of the linkage between ownership and location, rural industry has developed in a spatially dispersed and fragmented manner; in south Jiangsu there are industries in nearly every village. Furthermore, because each level of ownership has its own set of enterprises, fragmentation has created excess capacity and duplication. The close link between ownership and location is also the reason why rural enterprises remain small, since they rely heavily on local capital. Does it make economic sense for every village to try to develop its own non-agricultural enterprises and the supporting infrastructure?

In the long run, it is questionable whether some of the existing TVEs can survive without substantial protection. TVGs already do what they can to protect TVEs, e.g. through tax relief and subsidies and by buying locally, but because their jurisdictions are limited in size the protection they can provide is fortunately restricted.[32] Otherwise the economy would be even more fragmented than it is currently. Some TVEs are also protected by high transport costs, but with economic development comes lower transport costs, and it is unclear whether all of them can survive the keener competition that is coming. To remain competitive, particularly if reform continues and deepens, many TVEs will need to become larger so they can take advantage of newer technology and capture the available scale economies, many will need to congregate in market towns and cities so as to share infrastructure and diversified technical services, and many will need to be closer to large urban enterprises so as to increase the lateral linkages so important to their long-term viability and success. It is not an accident that the most successful TVEs are also those in close proximity to urban areas.

To remain viable and to achieve a higher level of technological competence and competitiveness, some TVEs will need to be combined and relocated to market towns, county towns, or regional cities.[33] For this to occur, location and ownership need to be separated, or at least the bond between them needs to be made more elastic. Convincing localities that TVEs should be less dispersed will not be easy. Proximity and direct control are a locality's guarantee that its TVEs are operated in the local interest and that profits are used in support of the local economy and for the benefit of local residents. Unless localities can be assured that they will continue to receive the full benefits of ownership when TVEs are moved outside of local jurisdiction, resistance to the separation of location and ownership will be strong.

That private and collective properties were confiscated in the past makes the job of convincing TVGs and villagers to merge their enterprises with those from other villages or townships and to relocate some of them to another jurisdiction a difficult one. It is therefore necessary to make a concerted effort to develop property rights and to redefine and clarify ownership so that both local interests can be protected and rural enterprises that have economic or technological needs for size are willing to merge with other enterprises and/or move to other jurisdictions. The development of property and ownership rights will also make it easier for rural (and urban) capital to move across administrative boundaries and will make such movement more likely to occur.

There is another reason why the ownership question needs to be resolved soon. Because economic reform is gaining momentum and private ownership is now a possibility, an increasing number of rural cadres, particularly in the more prosperous rural areas where TVEs are most developed, have declared their interests in buying the TVEs they now manage, and some are already behaving as if they are the owners. No one disputes the fact that the skills and entrepreneurship of these rural cadres have contributed immensely to the success of TVEs, but they are not the only ones with a legitimate claim to the right of ownership. If the question of ownership is not clarified soon, there is a chance that as reform proceeds these rural cadres will become, by default, the *de facto* if not the *de jure* owners of TVEs. If the question of ownership is resolved without recognizing in a meaningful way the contributions made by these cadres, there is a risk that they will lose interest in helping TVEs to develop in the future. What is needed is a resolution of the ownership question that will on the one hand protect the interests of the community and on the other hand keep the current crop of cadre-entrepreneurs committed to TVE development.

To settle the ownership question, the government is considering the adoption of a stock-share system.[34] Under this scheme, TVEs will be transformed into stock companies, and shares will be held by three groups: the community, the enterprise, and individuals (e.g. staff, workers, and local residents). The number of shares that an individual may purchase will depend on his or her contribution to the enterprise since its founding. The management of each TVE will be responsible to a board of directors representing the stockholders. Details, for example, of how the shares will be divided among the three groups, how stock prices will be determined, how many shares may be purchased by individuals, are still to be announced. Enterprise managers have responded coolly to the proposed scheme partly because they are suspicious of the government's motive and partly because they worry about losing control to the board of directors. Not surprisingly, enterprise managers prefer a system that limits ownership to enterprise members (i.e. management, staff, and workers), that, in effect, allows enterprise members to buy the enterprise from the community.[35] The government will probably experiment with several schemes before selecting one for widespread adoption, so it is still unclear how the ownership question will be resolved. What is certain is that how the issue is finally settled will have a major impact on the direction of TVE development in the long term.

Industrialization Strategy and Rural Industry

For three decades, from 1959 to 1978, China pursued a strategy of forced industrialization that was characterized by brute capital formation and a heavy industry bias. When China discarded the Ten-Year Plan in December 1978, it marked a change in its industrialization strategy. The government announced a readjustment policy to correct basic imbalances in the national economy, specifically the overemphasis on accumulation and the practice of allocating too much investment to heavy industry and productive capital (plant and equipment) and too little to agriculture, light industry, and unproductive capital (infrastructure and housing). Thus, beginning in 1979, the government began to curb capital construction and to reallocate resources from the capital goods sector to the consumer goods sector. The TVE sector has benefited from this shift in policy. Indeed, much of its growth during the 1980s was the result of TVEs filling niches in the local, regional, and national markets and producing consumer goods not adequately supplied by the state sector.

Conditions have changed considerably since the early 1980s. There are now fewer gaps in the markets, the pent-up demand for everyday consumer goods has been substantially reduced, and new TVEs will need to compete not only with a larger number of existing TVEs but also a more responsive state sector. Furthermore, Chinese consumers have become more discriminating and less easy to satisfy. Thus, even if the government continues to support the rapid expansion of the consumer goods industry, it is not likely that the TVE sector can long maintain its double-digit growth by just filling niches and producing simple consumer goods for the domestic market.

But, with the state sector still dominated by heavy industry, the pressure on the government to return to a development strategy that favours heavy industry is strong. Indeed, the evidence suggests that after an initial readjustment towards agriculture and light industry, heavy industry is again expanding rapidly.[36] If China reverts to a Stalinist-type industrialization strategy, it will make continued rapid growth of the TVE sector more difficult. This is because a strategy that favours the significantly more capital-intensive heavy industry will require increased government control of consumption and material allocation to ensure that resources are allocated to priority sectors. Both of these tendencies are unfavourable to the TVE sector, the first because it would reduce domestic demand for consumer goods, what most TVEs produce, and the second because it will be more difficult for TVEs to acquire machinery and equipment, raw materials and energy. A return to a strategy that favours heavy industry will inevitably increase the friction between SEs and TVEs, and thus enhance the likelihood that the government will impose new controls on TVEs. Furthermore, a Stalinist-type industrialization strategy will also reduce farm investment and thus make it more difficult for agriculture to release resources, particularly labour, to rural industry.

Probably the best chance for the continued rapid growth of rural industry is for TVEs to turn outward and make a breakthrough in export markets. Is this likely? Linking rural enterprises to the export sector as a way to maximize foreign exchange earnings to finance the government's priority import programmes was of course the central theme of the coastal development strategy advocated by Zhao Ziyang. After Tiananmen and Zhao's downfall in 1989, the coastal development strategy was attacked but the tactic of putting both ends on the outside (i.e. importing raw materials and using foreign capital to produce goods for export) survived and was the driving force behind China's economic recovery in the early 1990s, particularly after Deng Xiaoping visited Guangdong in January 1992 and called for bolder reforms and the continued opening of the Chinese economy.

China extended its open policy in the early 1990s. In response to complaints from interior provinces that their problems have been ignored and that the playing field is not level in foreign trade because regions faced different rules, the government has muted the coastal emphasis of its export strategy.[37] The playing field was made more level in 1991 when the government announced new foreign trade rules.[38] Then, in May 1992, the government announced plans to extend the open policy to cities in the middle and upper reaches of the Yangzi River.[39] Twenty-three cities in the Yangzi River Valley were given permission to offer preferential policies to attract investment from abroad and other parts of China, and the Yangzi River Joint Development Shareholding Company was formed by Shanghai, Nanjing, Wuhan, Chongqing, and the Bank of Communications to help regions along the river to become more export orientated. The extension of the open door to cities along the Yangzi River is part of a larger plan to integrate 'provinces and regions in central and west China with coastal areas and places along the Yangtze [Yangzi] River' and thus open the interior to the outside world.[40]

The tactic of putting both ends on the outside is attractive to TVEs because it gives them not only improved access to foreign investment and technology but also helps solve their marketing and supply problems. Many TVEs in the coastal provinces have already benefited from this strategy, and many more will benefit in the future. However, it is unlikely that the strategy will have the same impact on TVEs in the coastal provinces as it did earlier on those in Guangdong, and this is not only because of the vastly greater number of TVEs in the coastal provinces.

Putting both ends on the outside combines foreign capital (equipment and machinery), technology (including management, production, and marketing skills), and frequently also raw materials and components with Chinese unskilled labour and land to produce goods primarily for export. The arrangement suits TVEs because land and unskilled labour are their only resources. Foreign companies find the arrangement attractive because it requires the least amount of risk and is also the least complicated to implement since it demands little initial investment, involves the transfer of very simple technology, and requires minimum contact with China's planned economy.

However, because it is suited to only certain types of manufacturing activities (e.g. simple assembly and processing), most foreign companies have only a limited interest in investing in this type of export-orientated arrangement.[41] In the case of China, the interest has come primarily from firms in Hong Kong or Taiwan. Indeed, putting both ends on the outside has succeeded as well as it has in south China only because its two closest neighbours, Hong Kong and Taiwan, have been engaged fortuitously in a major restructuring of their economies. Rapidly rising wages and land prices in Hong Kong and Taiwan have forced many of their industries to relocate their low value-added, labour-intensive activities, and since the 1970s many of them have moved to China and South-East Asia. With the process now close to completion and with the government in Taiwan (worried that its economy is becoming too closely linked with the mainland) encouraging its businesses to consider alternative investment sites (e.g. the Philippines and Vietnam), future growth of this type of investment in China will likely be slower than in the 1980s and the early 1990s. There is another reason why this type of export-orientated investment activity can not continue to grow for long at double-digit rates. China's increasingly prominent position as an exporter of simple labour-intensive consumer goods is bound to invite political pressures on China to control future expansion.

Recognizing the limitation of an open policy that relies too heavily on export-orientated foreign investment, the Chinese government has tried to attract other types of foreign investment. Since 1991–2, China's open policy has focused on the development of the Pudong New Zone (*xinqu*) in Shanghai and the opening of cities in the Yangzi River Valley. The government is investing heavily in this mega-project and is counting on it to transform the Yangzi River Valley into a soaring dragon, with Shanghai as the dragon head and the seven provinces along the Yangzi River as the body.[42] Pudong is called a new zone (*xinqu*) rather than a special zone (*tequ*) because it has been granted even more preferential treatments than the special zones and because its purpose is to attract high-tech, capital-intensive investment rather than the low-tech, labour-intensive, and export-orientated investments that have gone to the special zones and other parts of south China. Given the high-tech emphasis, the development of Pudong is not likely to affect TVEs directly. However, there may be indirect spin-offs. To shift their resources to more sophisticated activities, some SEs in the Yangzi River Valley will want to transfer some of their mature activities to other producers, including TVEs. TVEs that are most likely to benefit from this type of transfer are those in the most developed regions (e.g. suburbs of municipalities and some of the urbanized townships in south Jiangsu) partly because of their propinquity to Shanghai and the other open cities on the Yangzi River and partly because in the past decade they have invested heavily in human capital so are in a strong position to take advantage of new opportunities.

Even if the government does not revert to a Stalinist-type industrialization strategy, the growth of TVEs will probably be slower in the 1990s than in the 1980s. New TVEs will need to compete not only with the large number of existing TVEs but also with a more responsive state sector. There are already many TVEs producing simple consumer goods, and most of the niches in the market that can be filled by TVEs have already been filled. In the coastal provinces, TVEs are increasingly placing their hope on the open policy to attract export-orientated foreign investment and to open new markets. While many TVEs in the coastal provinces will benefit from the open policy, most will continue to serve only the domestic market. In other words, for the majority of TVEs, a breakthrough in export markets is not likely. As yet, the open policy has not been extended to many rural areas in the interior. But, even if this were to change in the near future, its impact on TVEs in the interior will be limited. Putting both ends on the outside is not a realistic possibility for TVEs in China's hinterland. Even if the interior is opened, investors from Hong Kong, Taiwan, and South Korea interested in forming joint ventures with TVEs are likely to continue to focus on the coastal provinces (particularly Guangdong, Fujian, Jiangsu, Zhejiang, and Shandong). The reality is that, compared to the coastal regions, China's interior is significantly less developed, with inferior infrastructure, little history of interaction with foreign countries, weak ties with Hong Kong and Taiwan (the two most important sources of foreign investment), and much less developed rural industry. Thus, much of the direct benefits of an export-orientated industrialization strategy will go to rural enterprises in the coastal provinces, and they will continue to be the primary link between rural industry and the export sector. As for the interior, it is doubtful that output growth in rural industry in the next decade or two can be fast enough to make a significant impact on its rural employment structure.

Rural Industry and Rural Development

Since the mid-1980s, the government has relied on rural industry to play two roles: (1) to absorb redundant labour from agriculture so that peasants can leave the land but not the countryside, and (2) to subsidize and build agriculture (*yi gong bu nong* and *yi gong jian nong*). This strategy is most likely to succeed in those parts of China where the share of industry in the rural economy is relatively large and where industrial growth is potentially vigorous. For example, in the more industrialized townships in *sunan*, as rural industries expand in numbers and in scale, they will not move (nor will they feel a need to move) to urban centres. On the contrary, urbanization will come to them. In fact, the counties between Shanghai and Wuxi will soon be part of an urban corridor. Because TVEs are prosperous and exist in large numbers in these areas, they can provide sufficient support to agriculture to help it achieve sustained growth. Thus, in the more developed areas, by allowing peasants to leave the land but not the countryside, it is possible for the two branches of the rural sector to develop in a co-ordinated and mutually supporting manner.

But given China's size, diversity, population distribution, and regional differences in the level of economic development, it would be naïve to believe that all regions have equal prospects of rural non-agricultural development. International experience as well as recent experience in China suggest that proximity to urban industries is one, if not the most, powerful stimulus to rural non-agricultural activities, particularly high-quality (meaning high-productivity and high-income) ones. Thus, the rural regions with the brightest prospects for rural non-agricultural development are those near urban industrial areas, where urban–rural linkages are strongest and where footloose industries producing standardized products are more likely to develop because such locations offer them the advantages of cheap land and labour and closeness to major markets and sources of technology. Thus many of China's rural counties will never attain the high incidence of rural non-agricultural activity that now exists in some of Jiangsu's more densely populated counties in the south. In other words, a significant portion of China's rural population will not soon, if ever, benefit from the high income associated with the more productive non-agricultural activities, and in fact many will remain untouched by rural non-agricultural development altogether. In these areas, policies such as leave the land but not the countryside and using industry to subsidize and to build agriculture have little meaning. To promote rural development and to help rural people in the less developed areas to achieve a reasonably comfortable standard of living (*xiaokang*), the government cannot rely on rural non-agricultural development alone but must also consider ways to raise agricultural income directly and to increase rural labour (population) mobility.

Agricultural Income

For the foreseeable future, most of China's rural population will continue to earn their living from the land, which means that their income will rise only with an increase in the return on farming. It is beyond the scope of this study to discuss the many factors that have influenced agricultural productivity and income in China.[43] Nevertheless, it is worth while to make the following two observations.

The first is that farm income is closely linked to agricultural output and input prices. In the second half of the 1980s, low procurement prices for agricultural products (and delays by authorities in redeeming promissory notes given to farmers in exchange for grain) and sharp increases in the cost of inputs (particularly agricultural chemicals, diesel fuel, and agricultural machinery) substantially lowered the returns to agriculture and fuelled an exodus from farming. Clearly, to improve farm income and to avoid the chaos of a massive exodus of peasants from the countryside, the government will need to pay closer attention to the relationship between output and input prices. In the past, the government's reluctance to raise urban retail prices of grain significantly limited its ability to influence farm income and farmers' incentives to produce. This apparently has changed. The government increased the urban prices of grain and edible vegetable oils substantially in 1991 and again in 1992. That the government is willing to be more flexible in its food price policy is an encouraging development.

The second observation is that in the long run farm income and agricultural productivity will be determined by the amount of resources invested in agriculture. During the 1980s, with state and collective investments in decline and individual farmers reluctant to invest in agriculture, deterioration in agricultural infrastructure became a serious problem.[44] If rural development is to continue, this trend must be reversed. This is particularly important in the least developed regions and provinces where 'there is a clear need for more investment in both water and fertilizer'.[45]

There are many reasons why farm households have been reluctant to make long-term investments in agriculture—the fear that the government's land policy might change, uncertainty over land-use rights, pent-up demand for consumption and housing, and better returns from other types of investments.[46] This may change now that the government appears willing to make adjustments in its food price policy to improve the return on farming. However, the single most important obstacle to increased private investment in agriculture is the unresolved issue of land ownership. Since the mid-1980s, peasants have been able to inherit and to transfer responsibility land, but apparently they need even more assurance from the government before they are willing to make long-term investments on their farms. Perhaps with more time, as they gain confidence in the permanence of the government's land policy, peasants may begin to invest more heavily in agriculture but it is possible that private investment in farming will not increase significantly until the government consents to the private ownership of land.

There is also a need to increase collective and state investment in agriculture. Many projects related to water conservancy and rural infrastructure are beyond the limited capacity of small farm households to finance and to organize. Even if farm households have the financial capacity, the amount invested is likely to be less than optimal since individual investors can not always capture all the benefits associated with these projects. That many activities (irrigation, water control, seed preparation, afforestation, etc.) are beyond the capacity of individual households, the reluctance of farm households to invest and to maintain rural infrastructure, investment indivisibility, and other problems that emerged in the 1980s under the household responsibility system have convinced Chinese policy-makers that collectives must become more involved in agricultural management. In the early 1990s, the government announced its intention to establish gradually a dual management system in agriculture to integrate household farming with collectively organized investment and support activities.[47] In other words, the daily tasks of farming will be done under the present household responsibility system, but agricultural investment and certain pre- and post-production tasks will be managed collectively. TVEs are expected to help by providing the resources needed to undertake some of the collectively managed activities.

It is too early to know how hard the government will push the implementation of the dual management system, how it will work, or how well it will work. The government has emphasized that the purpose of the dual management system is not to undermine the household responsibility system but rather to undertake those agricultural activities beyond the capacity of individual farm households. However, in view of past excesses, there is always the danger that, in implementing the new system, collective activities will be expanded by curtailing individual initiatives, and thus undo the positive effects that reform has had on agricultural production. It is also worth noting that, except in the most developed townships, TVEs do not have the resources to finance many of these collective agricultural activities. Most of the burden will have to fall on the shoulders of individual farm households. It is interesting that the government is once again placing emphasis on labour mobilization as the basis for agricultural investment.[48] The state must also play a larger role in capital construction. It is difficult to see how agricultural growth in the least developed regions can accelerate without substantial state investment and budgetary support.

Labour Mobility

For decades, Chinese leaders have maintained an extremely restrictive migration policy, particularly when the movement involves the permanent migration of rural residents to urban areas. Among the consequences of this policy are a huge and underemployed rural labour force and a wide disparity between urban and rural incomes. The main reason for the government's promotion of rural non-agricultural development in the 1980s was to relieve some of the pressure that had built up in the countryside for peasants to migrate to cities in search of jobs and higher income. To facilitate the development of rural non-agricultural activities and to provide additional ways of coping with the large number of underemployed rural workers, Chinese leaders have relaxed their restrictive migration policy and adopted what may be described as a balanced and decentralized approach to urbanization.[49] In the early 1980s, China announced that future urbanization would be guided by the following principles: 'control the size of large cities, rationally develop medium-size cities, and actively develop small cities (*kongzhi da chengshi, heli fazhan zhongdeng chengshi, jiji fazhan xiao chengshi)'*.[50] Accordingly, China's migration policy since the early 1980s has been 'to strictly limit migration to big cities; carefully control movement to medium cities; allow freer movement to small cities and towns; and encourage migration from larger to smaller places'.[51]

Chinese leaders believed that this decentralized approach to urbanization and the accompanying relaxation of the restrictive migration policy would prove beneficial to both the urban and the rural sectors, while avoiding at the same time some of the costs of migration and urbanization.[52] For example, a decentralized pattern of urbanization would spread urban–rural linkages, so important and necessary to rural non-agricultural development, to more areas in the countryside. Furthermore, the economic and social costs of moving rural residents to small towns and medium-sized cities probably would be considerably lower than those associated with moving them to the very large cities (say, those with a population over 1 million). And, of course, a decentralized pattern of urbanization would avoid increasing the population and adding to the burden of the already congested metropolitan centres.

In the past, the government relied on the household registration system to enforce its restrictive migration policy. Before a person can migrate, he or she needs to change his or her place of permanent residence (*hukou*), and this can be done only with the permission of the authorities in both the place of origin and the destination. For example, without an urban *hukou*, it is impossible to gain access to most urban jobs, housing, schools, and grain ration. To date, the *hukou* system is still in place and the government has not changed its policy of restricting permanent migration.[53]

What has changed since the early 1980s is that the government has become more tolerant of temporary migrants, i.e. those who have not changed their legal residence but nevertheless reside elsewhere for a period of a few days to several years, and has made it easier for migrants to survive away from their permanent residence. Temporary migrants such as itinerant craftsmen and traders have existed throughout the history of the People's Republic, although their number declined after the formation of communes. In the 1960s and the 1970s, communes also made arrangements with urban collectives and SEs under which peasants were contracted to these urban enterprises as temporary contract workers (*linshi hetong gong*).[54] However, these forms of rural–urban mobility were under strict government control, since rural workers could not work outside their legal residence without the permission of their communes, and those who did faced the risk of fines. This changed dramatically after 1978. The combined effect of the implementation of the household responsibility system, which freed peasants from collective management, the easing of internal travel restrictions, and the implementation of policies that encouraged rural workers to engage in non-agricultural activities was to make it easier and more attractive for peasants to become temporary migrants. Since the early 1980s, large numbers of villagers have moved temporarily to market towns and township cities to work in TVEs, and many have also worked periodically in large cities as contract or temporary workers, on urban construction projects, or as self-employed itinerant pedlars and craftsmen.[55]

To bring prosperity to more of China's rural population, it is not sufficient that peasants be permitted to leave the land but not the countryside (the township); some must be allowed to leave their native villages and townships on a permanent basis.[56] Migration is important because it gives those living in backward and impoverished areas an alternative and is one of the main ways to reduce income inequities. With regional inequities growing under the current economic policy, improving labour mobility is all the more important. Moving people off the land in backward regions to jobs in more developed areas benefits both the migrants and those who remain behind, the former because they are moving to more productive employment, and the latter because they now face a more favourable man–land ratio. Increased labour mobility will also help postpone the premature substitution of capital for labour in those parts of rural China, e.g. south Jiangsu, where TVEs have already experienced labour shortages. Furthermore, increased mobility is needed to facilitate the relocation of some TVEs to towns and secondary cities so that they can capture scale economies, share technical infrastructure, and more easily develop the lateral linkages with other enterprises that are crucial to their future development.

Although the government is now more tolerant of temporary migrants and the *hukou* system is no longer a serious obstacle to labour mobility,[57] there is still much that the government can do to make migration easier and to develop the country's labour-market. One change that might be helpful would be to remove the distinction between temporary residents and permanent residents. Temporary residents can be required to return to their place of permanent residence, so they have less incentive to make a firm commitment to the work unit or to the temporary residence. Evidence from Jiangsu suggests a high turnover rate among temporary migrants employed by TVEs, and this of course makes it more difficult for TVEs to build up their stock of human capital. This was not a problem in the 1980s when TVEs employed mostly unskilled workers, but it will become a problem as TVEs adopt more complex technology and become more dependent on semi-skilled and skilled workers.

Given its vast rural population and the unabated pressure on much of that population to leave the land, China is not likely to remove all restrictions on rural–urban migration, particularly the movement of people into the major cities. But there are many measures that can be adopted that would increase labour mobility and facilitate a redistribution of China's population without violating the spirit of its current policy. In particular, the government should implement its announced policy of actively developing small and medium-sized cities more aggressively and do more to prepare them, particularly those in the more developed and densely populated regions, for the rapid population growth that is coming inevitably. For example, the government should encourage communities in the more developed rural regions that are experiencing labour shortages to accept permanent migrants from less developed rural areas. Rural workers with employment in township and county towns should be permitted to move their families and their legal place of residence to the urban address. The need for such changes is particularly urgent in the more developed parts of China such as south Jiangsu.[58]

To date, relatively few resources have been allocated to the development of small and medium-sized towns, and this has inhibited the movement of rural population to these places. Needless to say, if a larger share of the rural population and more industries are to be concentrated in these towns in the near future, this will have to change. The maintenance and construction of urban infrastructure are financed by state grants, revenue from the urban maintenance and construction tax, and in some cases a share of the profits earned by industrial and commercial enterprises. But many of these resources are not available to rural towns.[59] The elevation of many rural places to urban status after 1984 has helped, but the resources they have to invest in roads and urban infrastructure are still very limited.[60] Aside from the occasional provincial government grants, the only secure support for small town development has come from the TVGs. However, TVGs have no real tax power and only those with access to profitable TVEs have any discretionary spending flexibility. Clearly, to avoid future bottlenecks in development, the central government either has to allocate more of its resources directly to small town development or has to allow local governments new revenue-raising powers so they can be more aggressive in developing small towns and medium-sized cities.

China cannot resist much longer the strong economic forces that are pushing and pulling workers from rural areas. Past restrictions on rural–urban migration have left a legacy of social tensions and economic distortions. If future economic development is to proceed in a relatively efficient manner and if economic progress is to reach a larger share of China's rural population, a significant number of its rural population will need to leave the land and the countryside. For reasons both of equity and of efficiency, rural labour needs to be more mobile. To delay further a redistribution of China's population will only exacerbate rural–urban tensions and make it more difficult to solve the problem in the future.

Because rural non-agricultural development involves more than just the rural sector and is affected by policies in many areas, it must be viewed as an integral part of a much larger development process that involves industrialization, urban development, the movement of resources across both administrative and occupational boundaries, and systemic reform. Thus the future of rural non-agricultural development and whether rural industry will continue to enliven the Chinese economy and provide it with much needed flexibility and competitive stimulus, or become instead an impediment to efficient economic growth, depend critically on whether the government has the will to confront the larger issues discussed above and whether it is able to devise imaginative and workable solutions.

NOTES

1. This may be changing. In 1992, after Deng Xiaoping called for bolder reforms, TVEs in some of the more industrialized counties in *sunan* were given greater operational autonomy, including the power to make investment decisions. However, TVGs are still consulted before important decisions are made.
2. In the early 1990s, TVEs turned to trust companies and the curb market for capital. In 1993 it was almost impossible to get bank credits, partly because the government tightened bank credit in an attempt to cool an overheated economy and partly because government policy kept bank deposit rates below the rate of inflation and thus diverted savings from banks to non-bank financial institutions, e.g. trust companies and the curb market. Unable to borrow from the banking system, TVEs (particularly those in the rapidly developing coastal regions) turned to trust companies for loans at high interest rates. Trust companies were willing to lend to TVEs because they were among the most profitable enterprises in the country and thus willing to pay the high interest rates demanded.

3. Yuan Mingfa, 'Xiangzhen gongye shichang xiaoshou piruan de yuanyin ji duice (The causes and countermeasures for the market weakness faced by rural industry)', *Wuxi jingji (The Economy of Wuxi)*, 4 (1990), 20.
4. A summary of the argument can be found in Zhang Yi, 'Zai xin xingshi xia dui xiangzhen qiye de zai renshi (Re-understanding rural enterprises under new circumstances)', *NJW* 1 (1990), 42.
5. Supporters of rural enterprises have argued that the restraints imposed on the rural non-agricultural sector in 1989–90 went beyond the need to control inflation. For example, in 1989 the state did not increase the target for bank lending to TVEs but did increase it for lending to SEs by tens of billions of *yuan* (see Zhang Yi, 'Zai xin xingshi xia dui xiangzhen qiye de zai renshi', 43).
6. Yu Guoyao and Li Yandong, 'Dangqian xiangzhen qiye mianlin de zhuyao wenti (Major problems currently confronting rural enterprises)', *NJW* 10 (1989), 22–3.
7. Huang Shouhong, 'Xiangzhen qiye shi guomin jingji fazhan de tuidong liliang (Rural enterprises are the driving force in the development of the national economy)', *JY* 5 (1990), 44.
8. e.g. see ibid. 39–46, and Zhang Yi, 'Zai xin xingshi xia dui xiangzhen qiye de zai renshi', 41–5. This and the following paragraph draw heavily on these two articles.
9. e.g. in 1987 the state sector, through the state commercial channel, purchased 96 per cent of the cotton, 96 per cent of the wool, 99 per cent of the silk cocoons, and 96 per cent of the tea produced in China.
10. In 1988, the TVE sector was a net supplier of coal but a net consumer of electric power. But, in terms of coal equivalent, the TVE sector was a net supplier of energy. In addition, many of the locally owned state thermal power stations were constructed with capital contributed by TVEs (see Ch. 5).
11. A comparison of collective enterprises and state enterprises shows that by 1990 'gross value per 100 *yuan* of capital at state-operated enterprises had fallen to half that at collective enterprises, while gross value per worker and employee was only 30 per cent higher' (Robert M. Field, 'China's Industrial Performance since 1978', *CQ* 131 (Sept. 1992), 594). However, because Chinese estimates of industrial output and capital are subject to upward biases (and the bias in output may be more serious for collective enterprises than for state enterprises), the performance of state enterprises may not be as unfavourable as the official data suggest. For a discussion of the upward bias in output see ibid. 582–8, and for a discussion of the upward bias in the capital estimate, see Chen Kuan *et al.*, 'New Estimates of Fixed Investment and Capital Stock for Chinese State Industry', *CQ* 114 (June 1988), 243–66. Correcting these biases would modify but not invalidate the conclusion that during the 1980s collective enterprises in general and TVEs in particular performed substantially better than did SEs.
12. In 1988, the total employment of rural enterprises at all levels (township, village, team, and individual) was 95.5 million (*ZTN* (1990), 400), and if each worker supported two people then rural enterprises provided sustenance to

nearly 200 million. Of the 95.5 million, TVEs employed 49 million (*ZTN* (1990), 394).

13. Huang Shouhong, 'Xiangzhen qiye shi guomin jingji fazhan de tuidong liliang', 39. See Ch. 7 for evidence on the importance of TVEs to local public finance.

14. *BR* (13–19 May 1991), 9.

15. Zhang Yi, 'Zai xin xingshi xia dui xiangzhen qiye de zai renshi', 41.

16. The Chinese estimate that there were approximately 100 million surplus labourers in the countryside in the late 1980s and that another 100 million can be expected by the year 2000. See Mei Fangquan, 'Ben shi jimo Zhongguo nongcun reng jiang shengyu yiyi laodongli (One hundred million surplus labourers in rural China at the end of this century)', *NJW* 12 (1986), 42–4.

17. See Li Peng, 'Report on the Outline of the Ten-Year Programme and of the Eighth Five-Year Plan for National Economic and Social Development', delivered at the Fourth Session of the Seventh National People's Congress on 25 Mar. 1991, *BR* (15–21 Apr. 1991).

18. Han Baocheng, 'Readjustment Improves Rural Enterprises', *BR* (27 Aug.–2 Sept. 1990), 17.

19. Zou Jiahua, 'Report on the Implementation of the 1990 Plan for National Economic and Social Development and the Draft 1991 Plan', delivered at the Fourth Session of the Seventh National People's Congress on 26 Mar. 1991, *BR* (22–8 Apr. 1991), 29.

20. These figures were pieced together from data provided in interviews with the Ministry of Agriculture, Animal Husbandry, and Fishery and the Ministry of Urban and Rural Construction and Environmental Protection in early 1984 and in an article in *Renmin ribao* (*People's Daily*) (18 Mar. 1984), 2. For the implicit assumptions behind the projections, see Samuel P. S. Ho, *The Asian Experience in Rural Nonagricultural Development and its Relevance for China* (Washington, DC: World Bank, 1986), 71–3. More recent projections show similar trends, e.g. see Mei Fangquan, 'Ben shi jimo Zhongguo nongcun reng jiang shengyu yiyi laodongli', 43.

21. See e.g. Zhongguo nongcun fazhan wenti yanjiuzu, 'Guomin jingji xin chengzhang jieduan he nongcun fazhan (The national economy in a new mature phase and rural development)', *JY* 7 (1985), 12–18; Zou Fengling, 'Xiangzhen qiye chengbao hou de jige wenti (Several problems after the contracting out of rural enterprises)', *JQ* 3 (1986), 56–8; Gu Songnian and Yan Yinglong, 'Xiangzhen qiye jingji lilun yanjiu zai shijian zhong kaituo qianjin (Research on the economic theory of rural enterprise as its development begins to move forward in practice)', *JY* 5 (1985), 55–61; Jing Yang, 'Guanyu xiangzhen qiye de "quanshu" wenti (Concerning the problem of "rights and ownership" in rural enterprises)', *JQ* 8 (1985), 17–18; Zhou Qiren *et al.*, 'Xiangzhen qiye de zhengdun yu fazhan (The consolidation and development of rural enterprises)', *Renmin ribao* (*People's Daily*) (4 Oct. 1985); Wu Rong and Wu Zhikang, 'Tan fazhan xiangzhen qiye zhong jige you zhengyi de wenti (Concerning several controversial problems in the development of rural enterprises)', *Xiangzhen jingji shouce* (*Handbook of Xiangzhen Economy*), 1 (1984), and Jiangsu sheng shelian and Jiangsu sheng jingji xuehui, 'Wei xiangzhen qiye jinyibu jiankang fazhan kaituo daolu (Open a

path for the further healthy development of rural enterprises)', *NYW* 4 (1986), 17–21.

22. Rural reforms have been more radical and far reaching than urban industrial reforms. For a discussion of the reform measures introduced so far, see Dwight H. Perkins, 'Reforming China's Economic System', *Journal of Economic Literature*, 26/2 (1988), 601–45. An extremely useful discussion of some of the problems encountered in reforming a socialist economic system is Janos Kornai, 'The Hungarian Reform Process: Visions, Hopes, and Reality', *Journal of Economic Literature*, 24/4 (1986), 1687–737.

23. e.g. it has been suggested that TVEs should concentrate on agricultural processing, the processing of raw materials and the production of components needed by large urban enterprises, and the production of crafts and other labour-intensive goods for export.

24. A much discussed example is the construction of the Baoshan Steel Mill near Shanghai. See Martin Weil, 'The Baoshan Steel Mill: A Symbol of Change in China's Industrial Development Strategy', in US Joint Economic Committee, *China Under the Four Modernizations*, part 1 (Washington, DC: US Government Printing Office, 1982), 365–93.

25. In 1979, the government urged urban factories to transfer to TVEs, in a planned way, 'the production of those products or components that can be produced in rural areas' (see Ch. 2). Some relocation of urban activities to rural areas did occur in the 1980s through subcontracting arrangements between urban enterprises and TVEs, but there has not been any substantial redistribution.

26. Beside Jiangsu, TVEs have become important in Guangdong, Zhejiang, Shandong, Hebei, Liaoning, and the three provincial-level municipalities. The most forceful advocate for TVEs at the national level is the Ministry of Agriculture, Animal Husbandry, and Fishery.

27. Throughout the 1980s, supporters of TVEs have called for the gradual inclusion of TVEs in the state plan. See, for example, Yan Yinglong and Li Zongjin, 'Jiaqiang dui nongcun gongye de jihua guanli (Strengthen the planned management of rural industry)', *NJW* 1 (1984), 55–8.

28. For an interesting discussion of the difficulties faced by TVGs in balancing their many roles, see Song Lina and Du He, 'The Role of Township Governments in Rural Industrialization', in William A. Byrd and Lin Qingsong (eds.), *China's Rural Industry: Structure, Development, and Reform* (Oxford: Oxford University Press, 1990), 348–52.

29. For discussion of why TVEs need to be separated from TVGs, see, for example, Wang Jianhua, 'Xiangzhen qiye de suoyouzi wenti (On the problem of ownership of rural enterprises)', *NJW* 8 (1985), 14–16; and Zou Fengling, 'Xiangzhen qiye chengbao hou de jige wenti'. To reduce the influence of TVGs and to give more direct control to the masses, some provinces have tried to give rural households shares in TVEs. See Liu Huazhen, 'Tantan xiangcun jiti qiye de gufenzhi wenti (A discussion of the stock-share system for rural collective enterprises)', *NJW* 1 (1987), 29–31.

30. For a discussion of possible directions of reform, see World Bank, *China, Revenue Mobilization and Tax Policy* (Washington, DC: World Bank, 1990), 101–8.

31. Some TEs are located on village land, but many of these are either jointly owned by the township and the village or were at one time village enterprises that have since been taken over by the township.
32. Although local governments cannot vary the nominal tax rates or redefine the tax base, since these are set centrally, they are in control of tax assessment and collection and have considerable discretion in granting tax relief without the need to obtain approval from the central government. The evidence suggests that local governments are not reluctant to use this power. See, for example, Xu Shanda, 'Guanyu xiangzhen qiye shuishou zhengce he shuifu wenti de sikao (Some thoughts on tax collection policy and tax burden of rural enterprises)', *NJW* 9 (1987), 49.
33. To eliminate duplication and to capture scale economies, the government has been encouraging TVEs producing similar goods to merge. We saw evidence of this in the surveyed townships.
34. Jiangsu is to implement this system on an experimental basis in the autumn of 1993. The information in this paragraph is based on a discussion with members of the JPASS and several enterprise managers during a visit to *sunan* in May–June 1993.
35. If ownership or part-ownership is not possible, a second choice of many enterprise managers, particularly those in charge of successful TVEs, is to rent the enterprise.
36. Y. Y. Kueh, 'The Maoist Legacy and China's New Industrialization Strategy', *CQ* 119 (Sept. 1989), 436–9.
37. One frequent complaint was that the percentage of foreign exchange earnings from exports that could be kept locally was higher in special economic zones in coastal regions and in Guangdong and Fujian than in other parts of China. Another is that the rate used to convert foreign exchange earnings to RMB varied by regions (e.g. in 1988 it was higher in Shanghai and Zhejiang than in Jiangsu). See *JJN* (1989), 2. 34, and David Zweig, 'Internationalizing China's Countryside: The Political Economy of Exports from Rural Industry', *CQ* 128 (Dec. 1991), 733–5.
38. In 1991, the government introduced new regulations that allowed all regions to keep the same percentage of foreign exchange earnings from exports in a given product category. See 'Learning the Rules of Foreign Trade', *China News Analysis*, 1464 (15 July 1992), 3.
39. *BR* (15–21 June 1992), 10.
40. *BR* (3–9 Aug. 1992), 18–19.
41. For a review of foreign investment in China, see Y. Y. Kueh, 'Foreign Investment and Economic Change in China', *CQ* 131 (Sept. 1992), 637–90.
42. *BR* (15–21 Feb. 1993), 14–19.
43. For an analysis of China's agriculture during the post-Mao period, see *CQ* 116, Special Issue (Dec. 1988), 'Food and Agriculture in China during the Post-Mao Era'.
44. For details on the decline in state and collective support of agriculture, see Robert F. Ash, 'The Peasant and the State', *CQ* 127 (Sept. 1991), 496–503.
45. Kenneth R. Walker, '40 Years On: Provincial Contrasts in China's Rural Economic Development', *CQ* 119 (Sept. 1989), 458.

46. Ash, 'The Peasant and the State', 508–9.
47. See *CQ* 128 (Dec. 1991), 873–4, and 129 (Mar. 1992), 273–4.
48. Robert F. Ash, 'The Agricultural Sector in China: Performance and Policy Dilemmas during the 1990s', *CQ* 131 (Sept. 1992), 553.
49. Balanced and decentralized in contrast to the polarized urbanization of a small number of primate cities.
50. Roughly, large cities are those with a population in excess of 500,000 (particularly those with 1 million or more), medium cities are those with a population between 200,000 and 500,000, and small cities are those with a population below 200,000. For an official explanation of the policy, see Li Mengbai, 'Zhengque renshi he guanche woguo chengshi fazhan de jiben fangzhen (Correctly understand and implement our country's basic policy towards urban development)', presented at the National Conference on Urban Development (4 Dec. 1982), mimeo.
51. Sidney Goldstein, 'Urbanization in China, 1982–87: Effects of Migration and Reclassification', *Population and Development Review*, 16/4 (1990), 674.
52. Some Western economists have also argued in favour of a more balanced and decentralized approach to urbanization. For example, see Eliezer Brutzkus, 'Centralized Versus Decentralized Pattern of Urbanization in Developing Countries: An Attempt to Elucidate a Guideline Principle', *Economic Development and Cultural Change*, 23/4 (1975), 633–52.
53. In the south Jiangsu villages we visited in the late 1980s we were told that only a few temporary migrants from the north who had married local residents were permitted to transfer their permanent residence (see Ch. 5). During most of the 1980s, it was nearly impossible to move permanently from a rural to an urban area or from a small town to a large city. For a discussion of the criteria used to assess requests for transfer of residence to a medium-sized city, see Sidney Goldstein and Alice Goldstein, *Population Mobility in the People's Republic of China* (Honolulu: Papers of the East–West Population Institute, No. 95, 1985), 12–14.
54. The contract was formally between the urban work unit and the commune, and most of the pay went directly to the production team. For more on this practice, see Marc Blecher, 'Peasant Labour for Urban Industry: Temporary Contract Labour, Urban–Rural Balance and Class Relations in a Chinese County', *World Development*, 2/8 (1983), 731–45.
55. While there is no question that the rural population is now more mobile than ever before, it is nevertheless worth noting that much of the movement, both temporary and permanent, is within rural areas (strictly speaking, market towns are not urban places so that population movements from villages to market towns are not considered rural–urban migration), and that inter-regional rural-to-rural movement is still relatively limited. (On this point, see Ch. 5. But there are exceptions. For example, apparently a large number of temporary migrants from other provinces have moved to rural areas in Guangdong. However, Guangdong, and Fujian, have a much more liberal economic environment than do the other provinces.) It is also unclear whether many rural residents have moved permanently, either *de facto* or *de jure*, to urban areas since the early 1980s. The problem is that there are no reliable

statistics on temporary migrants. In theory the law requires visitors to an urban area to register as temporary migrants if the visit is to last longer than three days. In practice, this rule is not enforced, and the collected data are not processed or aggregated. In 1900, Xinhua News Agency reported that temporary migrants may number as high as between 60 and 80 million. But it is unknown how many of these millions were from rural areas and what proportion of them were transients. The best information on rural–urban migration is provided by the 1987 National Survey, which defined migrants to include both permanent migrants and temporary migrants who had lived at the place of destination for at least six months at the time of the survey. The survey found that between 1982 and 1987 there was a net movement of 13 million people from rural to urban areas—5.8 million to cities and 7.2 million to towns—accounting for about 3 per cent of the urban population in 1987 as reported by the survey. But these numbers may be biased upwards since China relaxed substantially the criteria used to define urban areas in 1984 so that 'some of those identified as rural-to-city migrants in the 1987 National Survey may in fact have moved from one rural place to another but came to qualify as rural-to-urban migrants because the destination was reclassified as an urban place after the move was made' (Goldstein, 'Urbanization in China', 685). The above discussion is based on ibid. and Huang Fungyea, *Urban Development and the Migration of Rural Labor in Mainland China*, Economic Papers No. 127 (Taipei: Chung-Hua Institution for Economic Research, 1989).

56. That some time in the future rural residents must be allowed to leave the land and the countryside is recognized by many in China, and the issue has received considerable attention in the Chinese literature. For example, see Cai Long, 'Shi lun li tu ye li xiang (Leave the land and the countryside, an exploratory discussion)', *NJW* 7 (1985), 4–8; Song Guoqing, 'Cheng xiang kafang yu nongmin zhuanyi (Opening cities to the countryside and the shifting of peasants)', *NJW* 7 (1985), 9–12; Chen Sheng, 'Cheng xiang jian laodongli heli liudong yu kaifang nongcun laodongli shichang (The rational movement of labour between cities and the countryside and the opening of rural labour-markets)', *NJW* 10 (1986), 12–15; Gu Xiulin, 'Jingji chengzhang zhong de nongye laodongli zhuanyi (The transfer of agricultural labour as the economy matures)', *NJYS* 5 (1988), 50–3; and Chen Jiyuan, 'Nongcun gaige shi nian de lilun qidi (Theoretical enlightenment from ten years of rural reform)', *NJYS* 6 (1988), 1–9.

57. Most provinces have abandoned the use of grain coupons. Although some jobs are opened only to those with proof of permanent residence, many do not require this, and in most regions (including urban areas) outside workers only need to go through the formality of registering as temporary residents. Temporary migrants can also buy access to housing and the school system. Indeed it is now possible for individuals to buy permanent residence (*hukou*) in rural areas and most urban areas. One reason why there is increased interest in acquiring rural *hukou* is to have better access to rural land.

58. Being the most prosperous and developed of China's provinces and also one where the pressure of population on land is intense, Jiangsu is ripe for rapid urban growth. According to census data, Jiangsu's urbanization rate in 1982

was well below the national average (15.8 per cent as compared to the national rate of 20.6 per cent). With a population of over 60 million, it had only one city over 1 million (Nanjing) and five cities in the range of 500,000–1,000,000. Most of its regional cities and towns were in the small-medium size range (below 500,000). (See Du Wenzhen and Gu Jirui (eds.), *Zhongguo renkou* (*Jiangsu fence*) (*China's Population—Jiangsu*) (Beijing: Zhongguo caizheng jingji chubanshe, 1987), 189–91.)

59. For an explanation of the economic advantages of having an urban designation, see Yok-shiu F. Lee, 'Small Towns and China's Urbanization Level', *CQ* 120 (Dec. 1989), 780–2.

60. In 1984, China relaxed substantially the criteria it used to designate urban areas, and, between 1982 and 1987, the number of rural localities classified as towns rose from 2,660 to 11,103. In fact the significant rise in the level of urbanization (from 19 per cent of total population in 1980 to 52 per cent in 1989) was largely the result of reclassification of rural areas as urban. (See Goldstein, 'Urbanization in China', 676–82, or Huang Fungyea, *Urban Development*, 12–17.)

Appendix
Research Design and Data

THIS appendix describes how the field research was organized and discusses the quality of the information collected in the field. The first section discusses the research design. This is followed by a description of the regions selected for in-depth fieldwork and a discussion of why they were chosen. How the field research was organized and how data were collected are then discussed. Finally, in the fourth section, the data sets are described and their quality assessed.

RESEARCH DESIGN

The research for this study began in 1986, and the fieldwork was developed and conducted in collaboration with the Institute of Economics, Jiangsu Provincial Academy of Social Sciences. Our objectives were: (1) to document the growth, structural changes, and regional variations in rural non-agricultural activities in Jiangsu since the late 1970s; (2) to determine the extent to which rural non-agricultural development has altered the organization and the structure of Jiangsu's rural economy; (3) to study the operations of rural enterprises; and (4) to examine the role of local governments in the development and the management of the rural non-agricultural sector.

To fulfil these objectives, we needed statistics on rural non-agricultural development, disaggregated by counties whenever possible. As well, we needed information on the operation of local governments, the operations of rural enterprises, the relationships between local governments and rural enterprises, and the impact of rural non-agricultural development on rural households. Even though by 1986 China had published a great deal of statistical data, and Chinese economists had written extensively on China's rural reform and her rural non-agricultural development, much of the data needed to analyse rural non-agricultural development in Jiangsu either did not exist or were not available in the required disaggregated form. To obtain the unavailable county data, we turned to the provincial government for assistance, and our efforts to get data released were on the whole successful.[1] The information on the operation of local governments, on enterprises, and on households, however, did not exist and had to be gathered through interviews and surveys.

To collect systematically the needed information required a considerable amount of in-depth field research. Furthermore, much of the fieldwork, in particular the interviews, had to be compressed into two short periods of five to six weeks each. The small size of the research team (five members—four from Jiangsu and one from Canada), a tight budget, and the difficulty of travel in Jiangsu's countryside were other major considerations that influenced the design of the field research.

With these constraints in mind, we decided to limit in-depth field research to three townships (*xiang/zhen*), the administrative unit immediately below the county (*xian*), and more specifically to six villages, two in each of the three townships. Restricting the fieldwork to a limited number of localities allowed us to collect data more carefully and more thoroughly than would have been possible otherwise. The project required the collection of first-hand information from three main sources: township-village governments (TVGs), rural enterprises, and rural households.

From the TVGs, we wanted historical and background information about the local economy and about rural non-agricultural development in the region. Besides statistics, each government was asked to provide a short written description of its experience with rural non-agricultural development. We also requested interviews with village/township leaders, and during field surveys in 1987 and 1988, at least two, and in some instances three, separate in-depth interviews were conducted with cadres from each of the villages/townships.

Given our limited resources, a large-scale survey of rural enterprises was clearly impossible. However, we felt strongly that if we were to gain a clear understanding of the inner workings of the rural non-agricultural sector it would be essential to survey at least a handful of rural enterprises. The primary objective of the survey was not to collect enterprise statistics, although some were collected, but rather to gather qualitative information on how rural enterprises were managed and on the institutional structure of the rural non-agricultural sector. In addition, we hoped the survey would give us a sense of the economic and political environment within which rural enterprises operated and the nature of the relationship between TVEs and TVGs.

Sixteen enterprises, selected from the three chosen townships, were surveyed. Each enterprise was asked to prepare a brief company history, and the TVEs in the sample, with the exception of one that was engaged in tea cultivation and processing, were also asked to complete two lengthy questionnaires.[2] In addition, the research team conducted at least one in-depth interview with each of the enterprises.

When properly administered, questionnaires and interviews can be expected to provide accurate information about enterprises and villages. However, questionnaires and interviews do not necessarily yield reliable information about individual households. This is because, as a rule, peasants do not keep records of their daily activities, and their recollection of receipts, expenses, and labour utilization is usually vague and incomplete. In view of this, we decided to collect information from rural households by conducting a highly labour- and supervision-intensive household record-keeping survey. With the assistance and under the supervision of statistical workers, a sample of 161 rural households, selected from the six surveyed villages, kept daily records of all their economic activities (e.g. production, receipts, expenditures, and labour deployment) for one calendar year (1987).

THE SURVEYED SITES

This section discusses the criteria used in selecting the survey sites, and describes the economic characteristics of the regions selected.

The Three Townships

Fig. A.1 presents Jiangsu's administrative structure in 1986. The province was divided into eleven *diji shi* (regional-level administrative cities), which were composed of forty-one *shixia qu* (urban or suburban districts under the direct jurisdiction of the city proper or municipality)[3] and sixty-four *xian* (county) and *xianji shi* (county-level administrative cities).[4] The sixty-four counties were in turn organized into 1,905 townships (*xiang/zhen*). While townships contain some urban settlements (*zhen/jizhen* or market towns), they are mainly rural areas, composed of villages (*cun*). Our research design called for the in-depth survey of three of these townships—more specifically six villages, two in each of the selected townships.

Since our research focus was on rural non-agricultural development, we wanted the three townships to be from areas that had experienced rapid rural changes since 1978. In addition, we wanted the three townships to be from areas in Jiangsu with different geographic characteristics and with different patterns of rural non-agricultural development. After considering the alternatives, we selected as survey sites Sigang and Hufu in southern Jiangsu and Haitou in northern Jiangsu near its border with Shandong Province (see Fig. 3.1). Once we had identified the three townships, we then selected two villages—one above and one below average (as measured by their 1985 per capita household income)—for intensive field survey. The six villages selected were Jingxiang and Zhashang in Sigang, Yangquan and Xiaojian in Hufu, and Haiqian and Haiqi in Haitou.

Selected economic indicators for Sigang, Hufu, and Haitou are presented in Table A.1. As can be seen, all three have experienced rapid rural changes since 1978, the magnitude of which is suggested by the dramatic shift in labour allocation from agriculture (defined broadly to include cultivation, animal husbandry, forestry, fishery, and sidelines other than industrial activities) to non-agricultural

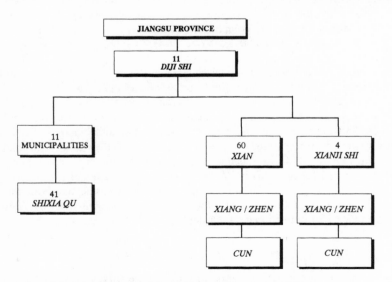

FIG. A.1. Administrative Divisions, Jiangsu Province, 1986

TABLE A.1. *Selected Economic Indicators, 1987, Sigang, Hufu, and Haitou*

	Sigang	Hufu	Haitou
Number of administrative villages	16	22	18
Population	n.a.	24,607	34,623
Agricultural (rural) population	28,497	22,528	33,652
Rural labour	17,998	13,018	15,609
Cultivated area (mu)[c]	26,200	9,959[a]	16,932
Paddy	19,664	8,420	9,940
Rice yield $(jin$[d]$/mu)$	1,006	942	1,138
GVAO (RMB 000)	37,851	15,387	55,849
% from			
Farming	24.56	47.90	22.29
Forestry		25.37	0.45
Animal husbandry		14.66	13.40
Fishery		0.12	63.32
Sidelines[e]	75.44	11.94	0.53
RGVIO (RMB 000)	292,992	106,368	39,937
% from			
Township industry	64.35	34.97	28.54
Village industry	33.70	60.96	38.96
Below village	1.95	4.07	32.49
Rural industry labour	10,245	7,032	1,786
% in			
Township industry	55.61	37.39	24.30
Village industry	43.90	61.82	55.99
Below village	0.49	0.80	19.71
Cultivated area/rural population			
$(mu$/person)	0.92	0.40	0.49
Per capita distributed rural income (RMB)	1,058	834	702
Per capita rural GVIO (RMB)	10,282	4,323	1,153
Per capita gross output from farming			
(RMB)	326	300	360
% rural labour in agriculture			
1978	65	72	95[b]
1987	15	23	69

[a] Area under tea not included.
[b] 1975.
[c] 1 mu = 0.0667 ha.
[d] 1 jin = 0.5 kg.
[e] The GVIO produced by units other than TVEs has been removed and placed under RGVIO (in the category 'below village').

Source: Township-Village Background Data.

activities. In 1978 agricultural workers accounted for 65 per cent of the total rural labour force in Sigang, 72 per cent in Hufu, and 95 per cent in Haitou. By 1987 the share of agricultural labour in the total rural labour force had declined to 15 per cent in Sigang, 23 per cent in Hufu, and 69 per cent in Haitou. However, the data in Table A.1 also suggest that the three townships are at different stages of rural non-agricultural development and have experienced different patterns of rural development.[5] Sigang is among the most industrialized townships in Jiangsu, Hufu is a moderately industrialized township, and Haitou is just beginning to experience rural non-agricultural development. A brief discussion of each township is provided below.

Sigang Township was selected because of its highly developed rural industrial sector and because it is an example of the so-called *sunan* (south Jiangsu)[6] model of rural non-agricultural development—i.e. one that is based on the development of industrial enterprises that are closely linked to the urban sector for key material inputs, markets, and technology and are owned and tightly controlled by township and village governments. Located in the fertile Changjiang (Yangzi) River Delta, Sigang is a part of Zhangjiagang,[7] a *xianji shi* and one of six counties in Suzhou *shi*. When it was first designated a county in 1962, the only industries in Zhangjiagang were traditional mills and handicraft workshops. But, because rural industry developed relatively early in Zhangjiagang, it is now one of the most developed and prosperous counties in China. In 1986 Zhangjiagang boasted the second highest (after Wuxi County) per capita RGVIO among Jiangsu's sixty-four counties, and its rural industry accounted for 82 per cent of the total GVIO produced in the county.

Among Zhangjiagang's twenty-seven townships, Sigang is ranked above average economically. Its per capita rural household income in 1985, at RMB 765, was 47 *yuan* higher than the county average. In 1986, its RGVIO was nearly eight times that of its GVAO, and the structure of its rural non-agricultural sector was typical of *sunan*—i.e. dominated by industrial enterprises that were owned collectively by the township and village governments. In 1987 64 per cent of Sigang's RGVIO was produced by township enterprises, 34 per cent by village enterprises, and only 2 per cent by enterprises with ownership below the village level. Although its per capita RGVIO (RMB 10,282 in 1987) was more than twice the county average, Sigang's rural industrial sector was still not among the most developed in Jiangsu. Nevertheless, by rural standards the industrial enterprises in Sigang are relatively large, use relatively advanced industrial technology, and being urban orientated find themselves frequently in competition with state enterprises.

Hufu was selected in part because its level of rural non-agricultural development was close to being average in Jiangsu. Hufu is in Yixing, one of three counties in Wuxi *shi*. Compared to Zhangjiagang, Yixing is substantially less industrialized. Its per capita RGVIO in 1986, at RMB 2,433, was only half that of Zhangjiagang. Of the forty-seven townships in Yixing, Hufu is the second most prosperous. Its 1987 per capita rural household income and its 1987 per capita RGVIO, at RMB 834 and RMB 4,323 respectively, were both substantially higher than the county average. But, compared to Sigang, Hufu was much less industrialized and much less prosperous, for example its per capita RGVIO in 1987 was less than half that in Sigang.

Another reason for selecting Hufu was that it was less urbanized and farther away from Jiangsu's urban centres than Sigang. Located in a mountainous region adjacent to Zhejiang Province, Hufu, unlike Sigang and Haitou, is a grain deficit area, and its agriculture is heavily dependent on tree crops (e.g. tea) and forestry. Because of its location and geography, Hufu's rural non-agricultural sector, unlike that in Sigang, is closely tied to its natural resources and depends more on traditional skills and technology. Perhaps because of this, village enterprises are much more important in Hufu than in the more developed Sigang. In 1987, 61 per cent of Hufu's RGVIO was produced by village enterprises and only 35 per cent by township enterprises, almost precisely the reverse of Sigang where 64 per cent of its RGVIO was produced by township enterprises and 34 per cent by village enterprises.

Haitou, one of thirty townships in Ganyu County, is located on the coast near the border between Jiangsu and Shandong. Given its geographic position, Ganyu's economy and its agriculture are more like Shandong's than the southern half of Jiangsu. Although Haitou is the most prosperous and probably also the most industrialized township in Ganyu, it is significantly poorer and less industrialized than most townships in the south. Its per capita rural household income is 132 *yuan* lower than Hufu's and 365 *yuan* lower than Sigang's, and its per capita RGVIO is one-fourth that of Hufu and only one-tenth that of Sigang. Reflecting its coastal location, the bulk of Haitou's GVAO is from fishery rather than cultivation. Not surprisingly many of its non-agricultural activities are also related to the sea. Unlike in Sigang and in Hufu, where individual and joint industrial activities account for a negligible share of the rural non-agricultural sector, they are very important in Haitou, accounting for about one-third of its RGVIO in 1987.

In other words, Haitou was selected because of its location in northern Jiangsu, because it was significantly poorer and less developed than townships in southern Jiangsu, and because its rural non-agricultural sector was to a much greater extent driven by individual activities than by collective enterprises.

The Six Villages

The six villages selected for in-depth fieldwork were chosen in the summer of 1986. Among the factors considered were accessibility and the level of development relative to other villages in the same township. In consultation with township officials and on the basis of the 1985 per capita household income statistics, one above average village and one below average village were selected from each of the three townships. In Table A.2 the major characteristics of the selected villages are compared with the average for the township as a whole. It is worth while to point out that because we selected villages in Haitou that were close to the highway that runs parallel to the sea coast, they are more densely populated and are larger and with less cultivated land per person than the average for the township. Also noteworthy is that Jingxiang, the below average village selected in Sigang, developed extremely rapidly between 1985 and 1987 so that its per capita household income climbed from 10th (out of 16) in 1985 to 2nd in 1987.

The six villages differed significantly in size, in endowment, in per capita income, in economic structure and organization, and in their level of rural non-

TABLE A.2. *Characteristics of Surveyed Villages, 1985*

Village	Population	Cultivated area per person (*mu*/person)	Per capita household income	
			RMB	Rank
Average in Sigang	1,745	0.94	765	—
Jingxiang	1,329	1.05	714	10 (16)[a]
Zhashang	2,404	0.84	1,135	1 (16)
Average in Hufu	969	0.47	764	—
Yangquan	1,142	0.56	1,001	2 (22)
Xiaojian	925	0.30	634	18 (22)
Average in Haitou	1,791	0.51	701	—
Haiqian	3,544	0.38	892	1 (18)
Haiqi	2,415	0.38	567	15 (18)

[a] The number in the bracket is the number of administrative villages in the township.

Source: Township-Village Background Data.

agricultural development (see Tables A.2 and A.3). Haiqian, the largest of the six villages, had four times the population of the smallest (Xiaojian), and Zhashang, the most prosperous, had a per capita household income that was more than twice that of the poorest (Haiqi). Jingxiang in Sigang township had two to three times more land per person than the four villages selected for survey in Hufu and Haitou. Not surprisingly, only in the two coastal villages (Haiqian and Haiqi) was fishery (coastal fishing and aqua-culture) an important economic activity. And the importance of fishery also helps to explain why individuals in Haiqian and Haiqi owned so much more in productive fixed assets than those in the other villages. While the vast majority of the workers in Haiqian and Haiqi was engaged primarily in agriculture (broadly defined), very few in Zhashang and Jingxiang considered agriculture their principal occupation. Between these two extremes were Yangquan and Xiaojian, in which about 25–30 per cent of the workers were engaged in agriculture. Township enterprises employed a large number of workers in Jingxiang and Zhashang, somewhat fewer in Yangquan and Xiaojian, and almost none in Haiqian and Haiqi. Village enterprises were also substantially more important in the four southern villages. In fact, in 1987, Haiqi had no village enterprise, and although Haiqian had a few they were not operated by the village but contracted out to individuals. Industrial activities, to the extent they existed in Haiqian and Haiqi, were small scale and family owned and operated.

PREPARATION AND ORGANIZATION OF FIELDWORK

The research team made three extensive visits to the survey sites—in June–July 1986, in May–June 1987, and in April–May 1988. The primary objectives of the

TABLE A.3. *Jingxiang, Zhashang, Yangquan, Xiaojian, Haiqian, Haiqi, Selected Economic Indicators, 1987*

	Sigang		Hufu		Haitou	
	Jingxiang	Zhashang	Yangquan	Xiaojian	Haiqian	Haiqi
Labour (person)	892	1,418	735	939	1,500	1,066
% labour in agriculture[a]	3.7	4.0	29.0	25.0	61.0	81.0
Cultivated area (*mu*)	1,363	2,031	636[d]	282[d]	1,359	914
Paddy (*mu*)	1,301	1,620	264	244	1,250	640
Rice yield (*jin/mu*)	1,056	1,063	1,061	946	1,178	1,080
GVAO (RMB 000)	1,254	1,208	885	360	3,649	2,111
% from						
Farming	44.5	56.5	93.0	65.3	28.2	16.5
Forestry	0.7	1.3	7.0	0.7	0.1	0.7
Animal husbandry	20.6	34.9	n.a.	14.1	11.9	13.2
Fishery	7.3	11.7	n.a.	—	30.2	69.5
Sidelines[b]	—	22.4	n.a.	19.9	29.4	0.1
Village GVIO (RMB 000)	10,146	50,763	4,850	5,004	10,872	2,630
% from						
Village enterprises	100.0	99.0	100.0	100.0	23.2	0.0
Other[c]	0.0	1.0	0.0	0.0	76.8	100.0
Collectively owned fixed assets (RMB 000)	1,592	9,347	1,284	836	790	115
Privately owned productive fixed assets (RMB 000)	96	n.a.	185	207	625	600
Per capita income (RMB)	1,243	1,313	1,210	618	876	569

[a] Agriculture includes farming, forestry, animal husbandry, sidelines, and fishery.

[b] The GVIO produced by units other than TVEs has been removed and placed under village GVIO (in the category other).

[c] Production by team enterprises, partnerships, and individuals. In Zhashang, team enterprises were elevated to village ownership in 1987.

[d] Area under tea not included.

Source: Township-Village Background Data.

initial visit were to make contact with the local governments, to select enterprises and households for in-depth survey, and to select local workers to assist in the fieldwork. The in-depth interviews and the survey work were carried out in 1987 and 1988. In addition to the visits by the entire research team, individual researchers from JPASS periodically visited the survey sites in 1986 and 1987 to ensure that survey preparation and data collection (in particular the household record-keeping survey) were proceeding according to schedule.

Because of limited resources we knew it would be unrealistic to attempt a large household record-keeping survey, and decided to aim for a sample survey of between fifty and fifty-five households in each of the three townships. Sample households were selected in the summer of 1986. Because some of the randomly drawn households either had no adult who was literate and therefore able to keep a daily journal or refused to participate in the survey, we had to approach 252 randomly selected households before we completed the sample of 161 households. Since poor households are more likely to lack the ability to keep records, they are underrepresented in the sample. The population and the sample size in each of the six surveyed villages are given in Table A.4.

Although, in the past, rural residents had been willing to participate in interviews and answer questionnaires without compensation, we decided to provide a small honorarium to those households who agreed to participate in the household record-keeping survey. Unlike interviews or questionnaires, which would take at most an hour or two, the record-keeping survey was for an entire year. Because rural reform has greatly increased the opportunity to earn income, peasants have become more conscious of the opportunity cost of time and few would have agreed to participate in a time-consuming survey unless they were paid.[8] It is a reflection of how much rural China had changed in the past ten years that some of the households we approached considered the honorarium we offered (RMB 5 per month plus a bonus at the end of the survey) inadequate, and agreed to participate in the survey only as a favour to us and to the village leaders.

TABLE A.4. *Household Record-Keeping Survey, Sample Size by Village*

Village	Households in village	Number of households approached	Sample size	Number of statistical workers
Sigang Township				
Jingxiang	362	30	20	2
Zhashang	705	56	33	4
Hufu Township				
Yangquan	323	47	32	3
Xiaojian	262	44	26	3
Haitou Township				
Haiqian	873	45	30	3
Haiqi	635	30	20	2
Total		252	161	17

Enterprises to be surveyed in 1987–8 were also selected in the summer of 1986. Enterprises were chosen from different levels of ownership and from different industries. Since we wanted enterprises with at least a few years of history, enterprises established in 1986 were not considered.

Table A.5 lists the names and the main characteristics of the sixteen enterprises selected for in-depth survey. The sample included seven township enterprises (including one that was a joint equity venture with a Japanese investor), five village enterprises, and four enterprises with ownership at or below the production team level[9] (however during 1987 two of these were changed to village enterprises). Among them were a fishing partnership, a tea cultivation and processing enterprise, an aqua-culture (prawn-breeding) enterprise, a fishnet putting-out system operated as an individual enterprise, a limestone quarry, and eleven manufacturing enterprises in such key rural industries as building materials, textiles, metal and metal products, and food processing. With the exception of the steel cable factory, all were small- or medium-scale enterprises.

Forty-two villagers were selected in the summer of 1986 to help with the fieldwork—seventeen to help supervise the household record-keeping survey, sixteen to collect enterprise statistics, and nine to compile historical and background statistics on the villages and townships.[10] All those selected had at least a junior high school education and most also had some experience in statistical work. For example, the seventeen workers selected to supervise the household record-keeping survey had all previously served as team accountants and were therefore not only familiar with record-keeping but also had considerable knowledge of village affairs. Book-keepers at the sixteen enterprises were selected to compile enterprise statistics, and nine local cadres involved in economic or statistical work at the village or township level were selected to collect village and township statistics. In addition to the forty-two statistical workers, several cadres at each administrative level (the village, the township, and the county) helped coordinate the field research and provided invaluable assistance.

Between 15 September 1986 and 9 October 1986, training classes for statistical workers and members of households selected to participate in the household record-keeping survey were held in the six villages. In total 207 people attended these classes. All the statistical workers attended, as did many of the village and township cadres. Of the 161 households, 137 were represented at the regular training classes, and the remaining twenty-four sent representatives to special make-up classes that were held in late October. The training classes for statistical workers were used to explain the objectives of the questionnaires and of the record-keeping survey, to go over each question in the questionnaires, to explain the economic and technical terms used in the questionnaires, and to explain the procedures to be used at the end of each month to classify and aggregate the daily journal entries. The classes for household members were used to explain the purposes of the household record-keeping survey, to identify those activities and transactions that were to be recorded, to stress the importance of keeping a daily record, and to give assurance that the data collected would be kept confidential and would not be used for tax purposes.

A one-month trial record-keeping survey was conducted in November 1986. In mid-December, all the statistical workers were invited to Nanjing to share ex-

TABLE A.5. *Characteristics of Enterprises Selected for In-depth Survey, 1986*

Name of enterprise	Principal products	Level of ownership	Year founded	Gross receipt or GOV (RMB million)	Number of workers
Zhangjiagang Steel Cable	steel wire and cable	township	1974	40.27	1,826
Zhangjiagang No. 2 Chemical Fibre	polyester fibre	township	1980	24.36	550
Zhangjiagang Radio Material	CP wire	township	1980	2.52	208
Zhashang Textile	woollen textiles	team[a]	1984	5.75	208
Zhangjiagang Hemp Textile	hemp yarn and bags	village	1984	1.36	147
Yixing Copper Material	electrolytic copper, copper rod	township	1978	15.69	441
Yixing Telecommunication Material	circuit boards	township	1981	3.56	230
Hufu Zhanggong Quarry	limestone	village	1966	2.02	577
Yangquan Tea Farm	tea	village	1968	0.54	515[b]
Hufu Bamboo Craft	brooms	partnership[c]	1984	0.09	86
Yuhe Aqua-Culture and Refrigeration	frozen prawns	township[d]	1985	4.75	57
Haitou No. 3 Brick	brick	village	1985	0.37	70
Haitou Agricultural Implements	iron and wooden agricultural implements	township	1959	0.21	63
Haitou Prawn Culture	prawns	township	1984	0.39	—[e]
Zhang Deli Fishnet	fishnet	individual	1981	1.50	8
Chen Zhoulu *lianheti*	coastal fishing	partnership[c]	1983	0.06	16[f]

[a] Became a village enterprise in 1987 through merger.

[b] Of which 250 were seasonal workers.

[c] What the Chinese called *xin jingji lianheti* (new economic associations). Hufu Bamboo Craft was converted to a jointly owned (private–village) enterprise in 1987.

[d] A Sino-Japanese joint equity venture with 25% of the investment from Japan, 25% from the county, and 50% from the township.

[e] Township-owned ponds were contracted out to 51 contractors.

[f] Each of the 8 households in the association contributed two workers.

Sources: Enterprise Background Data and Enterprise Interview Notes.

periences, to resolve recording and classification problems encountered during the trial survey, and to finalize plans for the actual survey that was to begin on 1 January 1987.

During the survey year (1987), each statistical worker received a monthly honorarium of RMB 10. To encourage further the prompt and accurate collection and processing of data, statistical workers were also informed that a bonus fund of RMB 10 per statistical worker would be made available to the villages to award year-end bonuses to those who had performed particularly well during the year. Although we recommended that bonuses be awarded to only 70 per cent of the statistical workers, the final decision was left to the villages. Ultimately, the villages decided to award bonuses to all statistical workers but to link the size of the bonus to performance. The best workers received year-end bonuses of RMB 15 and the less conscientious RMB 5. However, given its relatively small size, the bonus probably did not significantly improve the quality of the fieldwork. Two other factors were more important: (1) local cadres were surprisingly committed to the survey work, in part because it was supported by the provincial government but also in part because they believed the research would benefit their work,[11] and (2) many statistical workers developed a genuine interest in the survey and took pride in their work.

THE DATA SETS

The fieldwork produced eight sets of data: Township-Village Background Papers, Enterprise Background Papers, Township-Village Background Data, Enterprise Background Data, Survey of Migrant Workers, Household Record-Keeping Survey, Township-Village Interview Notes, and Enterprise Interview Notes. Below we briefly describe the data sets, how they were collected in the field, the major difficulties encountered, and their perceived weaknesses.

Township-Village Background Papers

In 1986, each township or village government was asked to prepare a short (8–10 pages) paper on rural non-agricultural development in the region, the problems encountered, and prospects for future development. These papers were completed and forwarded to the Institute of Economics, JPASS, in Nanjing in early 1987.

Enterprise Background Papers

Each of the sixteen enterprises selected for the in-depth survey was asked to provide a written company history, describing how the enterprise started, the difficulties encountered and how they were overcome, and their future plans. These papers were also completed in early 1987.

These two sets of papers were read by members of the research team for background before formal survey work began in the summer of 1987. While they contained some of the basic facts we were after, the stories they told were usually incomplete and sometimes misleading. Nevertheless, they served as a check-list of events and issues that needed to be covered in the interviews. In fact, questions about several of the background papers raised during interviews led to some very useful and stimulating discussions.

Township-Village Background Data

This large body of data was collected by the nine township-village statistical workers in response to a Township-Village Questionnaire that we distributed in late 1986. The questionnaire requested a wide variety of data in such areas as: population, labour force, agriculture, rural non-agricultural activities, local public finance, labour mobility, and per capita household income. In most cases we requested data for all years since 1962, but in some cases, e.g. local public finance, age structure, and per capita household income, we only requested data for the survey year (1987) or for selected benchmark years (1975, 1978, 1985, and 1987). After the questionnaires were completed, they were checked by a member of the research team for omissions and obvious recording errors, and were collected after corrections were made in the field in 1988.

As a rule the data provided by the townships were more complete than those provided by the villages. For example, only two villages were able to provide pre-1978 data. Record-keeping at the village level appeared to be surprisingly patchy, particularly in recent years. Apparently, after the introduction of the household responsibility system, some records were discarded or turned over to the township and others have not been maintained. The quality of village data is also uneven. Many statistics series have unexplained gaps in them and some data are internally inconsistent. We were able to trace some of the data problems to inconsistencies and poor bookkeeping in the original records.

Enterprise Background Data

This body of data was collected with the help of two questionnaires—Enterprise Background Questionnaire (A and B)—that were administered to the eleven township-village industrial enterprises in our sample. Questionnaire A requested a wide range of detailed statistics about the enterprise as well as information about enterprise decision-making power and the background of key personnel. For the more common statistics, e.g. output, wage payment, and employment, we requested data for all years since 1980 (only four of the sample enterprises were established before 1980). For the others, e.g. wage structure, structure of work-force by education and age, key raw materials by origin, and sales by destination and by marketing channel, we requested data for only 1987 since they had to be assembled or collected especially for us. Questionnaire B requested qualitative information on such matters as the enterprise's incentive system, the content of its lateral economic relations with other units, the reasons for modifications in its products or production process, and its future plans. Both questionnaires were distributed to the statistical workers at the end of 1986 and were collected in the field in the spring of 1988.

We encountered considerable difficulties in collecting the enterprise data. Some of the problems were caused by poor design and ambiguous wording in some of the questions used in the questionnaire. However, most of these were discovered fairly early and corrected. What we failed to appreciate sufficiently was the relatively low skills of many of the bookkeepers who had the responsibility of assembling and compiling the data requested. Another serious problem was that many of the enterprises were small and did not keep complete records. Once we recognized that some of the statistical workers needed additional assistance, a member of the research team was assigned to work with them to make certain that the data were compiled correctly. However, we did not always succeed in eliminating all the internal inconsistencies in the collected data, in part because the original records were themselves often inconsistent.

Enterprise Background Questionnaire B produced relatively little useful information. The main reason for this was that the questionnaire dealt with issues (e.g. new products and production process, lateral relationships with other enterprises, etc.) that were not as yet important in many of the enterprises we surveyed. Another reason was that several of the statistical workers, contrary to our instruction, completed the questionnaire on their own without consulting with the enterprise manager. But, as bookkeepers, they were ill equipped to answer questions in areas that were on the whole outside their expertise, and consequently gave answers that were incomplete or incorrect. However, this problem was not serious since we were able to obtain much of the information we wanted from the in-depth interviews we conducted with the enterprise managers.

Survey of Migrant Workers

During our fieldwork in Hufu in the summer of 1987, we discovered that one of the sixteen enterprises, Zhanggong Quarry, employed a large number of workers from outside the township. Because we knew so little about the rural labour-market, we seized the opportunity and asked the quarry to administer a short, hastily designed questionnaire to fifteen of its migrant workers.[12] The survey asked for information on the background of the migrant workers, their monthly income in Hufu, their monthly income if they had remained at home, how much income they remitted home in the previous year, etc.

Household Record-Keeping Survey

This survey was conducted with two objectives in mind: (1) to measure household income by source, and (2) to measure the allocation of family labour between agricultural and non-agricultural activities. To collect the information, the surveyed households kept a daily record of their production, sales, purchases, and how they utilized their family labour. In addition, at the beginning of the record-keeping year, each household was surveyed to determine its size and composition, the level of education achieved by its working members, and the condition of its house. Of the 161 households that began the survey, 160 completed the one-year project. One household was dropped after four months when its only literate member died.

Each household kept three journals—one for cash transactions, one for transactions in kind, and one for labour. At the end of each month, the statistical workers collected the journals, classified the entries recorded that month, and transcribed them to pre-prepared worksheets.[13] The worksheets were mailed to the Institute of Economics in Nanjing every quarter and the data entered into an IBM PC and processed. To aggregate the data, transactions in kind were converted to values using average annual prices deduced from the records of cash transactions. (Households were required to record both the quantities and the values of their sales and purchases.) Since the two surveyed villages selected in each township were close to each other, they were considered part of the same market and only one price was calculated for each township. The price calculations were based on the records of thirty households, ten from each township. The average 1987 prices of the main commodities transacted in kind in each township are presented in Table A.6.

The record-keeping survey was conducted between 1 January 1987 and 31 December 1987 under the supervision of seventeen statistical workers (i.e. each worker supervised eight to twelve households). During the first month of the survey, each statistical worker visited the households under his or her supervision twice weekly to assist with journal entries and to make sure that daily records were properly kept. Subsequently there were fewer regular visits, but, as neighbours, the statistical workers were in frequent contact with the participating households, kept a close watch on them, and made sure that records were kept. Most households were quite conscientious about recording their activities.[14] However, to make sure, statistical workers also examined the journals every month for omissions or incorrect entries.[15]

While the participating households were on the whole co-operative and quite willing to provide information, there was one notable exception. Even after repeated assurances that the data would be kept confidential, many households refused to reveal the size of their bank balances.[16] Fortunately this failure is not

TABLE A.6. *Prices of Selected Commodities, Sigang, Hufu, and Haitou, 1987 (Annual Average in RMB)*

Commodity	Sigang	Haitou	Hufu
Grain (*jin*)[a]	0.24	0.25	0.21
Vegetable (*jin*)	0.22	0.14	0.35
Pork (*jin*)	1.18	0.80	2.00
Egg (*jin*)	2.13	1.88	1.88
Poultry (*jin*)	1.98	7.67[b]	2.25
Fish[c] (*jin*)	2.59	0.42[d]	—
Fruit (*jin*)	0.26	—	0.20
Rabbit fur (*jin*)	33.63	—	—
Chestnut (*jin*)	—	—	1.50
Timber (cubic metre)	—	—	400.00
Bamboo (piece)	—	—	2.08

[a] 1 *jin* = 0.5 kg.
[b] RMB per head.
[c] Includes prawn and shellfish.
[d] A large proportion of the catch in Haitou was the less expensive shellfish.

Source: Household Record-Keeping Survey.

critical to our research as information on bank balances is not needed to estimate either household income or the allocation of household labour.

When using the survey results, we must remember that our samples exclude households that lacked the ability to keep daily records. In other words, poor households are underrepresented in all our samples. However, because illiteracy is closely correlated with poverty, poor households are likely to be more under-represented in the samples selected from the poorer villages. This is confirmed when we compared village estimates of average per capita household income of all households in the village with that of households included in the sample.[17] In all cases, the average per capita income of households included in the sample is higher than the village average, and the upward bias becomes increasingly more serious as we move from the richer to the poorer villages.[18] For example, in Zhashang (the most developed of the six villages), the average per capita income of households included in the sample was 12 per cent higher than the village average, but in Haiqi (the poorest of the six villages) the average per capita income of households in the sample was 89 per cent higher than the village average.

The quality of the information collected varied. In general households were more candid about their expenditures than their receipts and kept better records of their cash transactions than their in kind transactions. For example, only the most important in kind transactions (e.g. the production of grain, vegetable, poultry, and pork) were recorded, and even these were not always carefully kept. Thus, in a number of households, we found sales of some of these goods to be greater than production.[19] Because of this, we suspect the estimates of both income in kind and consumption in kind to be downward biased. However, it should be noted that the surveyed villages were highly commercialized so that non-market activities were relatively unimportant. We also suspect that some of the households in the survey that were involved in self-employed non-agricultural activities outside the village (e.g. itinerant traders and craftsmen) were less than candid about their earnings.

The data collected by the record-keeping survey contain useful information, but because poor households are underrepresented in the samples and because of the other shortcomings discussed above, the data set must be used with considerable care. We need to remember that the average per capita income is biased upward, and that the bias may be quite serious for the poorer villages (e.g. Haiqi). In addition, since poor households are underrepresented in the sample, income inequality measures derived from the survey findings (see Chapter 8) are also biased. Specifically, incomes are more unequally distributed at the village level, particularly in the poor villages, than suggested by the survey results.

Interviews

In-depth interviews were conducted at least once with each of the sixteen enter-prises and at least twice (once in 1987 and again in 1988) with township and village officials. The interviews were unstructured. Since we already had back-ground papers, we did not request a formal briefing. Nevertheless, following the usual practice in China, most interviews began with a short briefing which was then followed by questions. (When there was a second interview, it began immediately with questions.) While we generally had in mind a list of questions or issues that we wanted to discuss, we were quite willing to allow the discussion to move in different directions or to topics that came up during the interview that were of interest to us. Most interviews lasted two hours.

All the interviews involved three or four members of the research team and several people from the other side (the enterprise, the village, or the township). (All participating members from the research team took notes.) The interviewees were co-operative and responsive, and discussions were open and frank.[20] Even though the interviews were in-depth and the interviewees co-operative, it is still unclear that we obtained an accurate history of the enterprise (or village/township). The problem is not that the interviewee intentionally distorted the facts but rather that the facts were presented from a particular point of view, and in some cases the interviewee no longer had a clear grasp of the facts because the events occurred long ago. We tried to minimize the problem by interviewing several people from each unit and also by talking with outsiders who were familiar with or had extensive dealings with the unit. However, this was not always possible since some of the key cadres had been promoted and transferred to other regions. Despite these shortcomings, we believe the interviews provided useful and reasonably complete information.

While in the field, members of the research team met in the evenings to compare interview notes, and one member of the team was assigned to write up the notes of each interview, which then became the official Enterprise Interview Notes or Township-Village Interview Notes.

NOTES

1. Within a year or two of our request, most of the county data we asked for were published in the *Jangsu jingji nianjian* (*Economic Yearbook of Jiangsu*). I would like to believe that our efforts to get the county data released facilitated the publication of these data.
2. The tea enterprise was not asked because the questionnaires were designed with industrial enterprises in mind. Altogether, eleven enterprises completed the two questionnaires.
3. The region takes the name of the principal city in the region and this causes some confusion since, in Chinese, both the *diji shi* (the region) and the city proper or municipality are called 'city' (*shi*). For example Nanjing *shi* may refer either to the municipality of Nanjing or to the much larger region of Nanjing. To avoid confusion, we shall hereafter use 'city' to refer to the city proper and shi to refer to the region. Thus, Nanjing, one of the eleven *diji shi*, is organized into ten *shixia qu* (directly administered by Nanjing City) and five *xian*, each with its own *xian* government.
4. Hereafter, we shall refer to both *xian* and *xianji shi* as 'county'.
5. Another indication that the three townships were at very different levels of development is the fact that the percentage of surveyed households (selected randomly for the record-keeping survey) with three or more children was less than 10 per cent in Sigang, 21 per cent in Hufu, and 58 per cent in Haitou (Household Record-Keeping Survey).

6. Traditionally, *sunan* (south Jiangsu) encompasses three regions: Suzhou, Wuxi, and Changzhou.
7. Zhangjiagang, until late 1986, was Shazhou County, which was formed in 1962 from parts of Jiangyin and Changshu counties.
8. The State Statistical Bureau also provides a small honorarium to those farm households that participate in its annual record-keeping surveys.
9. Enterprises that are owned either by an individual, jointly by several individuals (partnership), or collectively by one or more production teams.
10. In late 1986, JPASS formally appointed all of them as statistical workers.
11. e.g. those responsible for rural work in Ganyu County found the results of the household record-keeping survey so interesting that they have decided to continue the survey in Haitou Township indefinitely.
12. The quarry was instructed to select the fifteen workers randomly, but it is uncertain that it followed our instruction.
13. Two sets of journals were used, and households always had one set for record-keeping. At the end of each month, the two sets were exchanged.
14. There were a few delinquent households who had to be reminded repeatedly to record their activities on a daily basis.
15. These monthly checks caught several surprisingly large omissions. Perhaps the most glaring one was the failure of the head of one household to record the expenses incurred for his daughter's wedding. (We do not know why he excluded these expenditures. Perhaps, not certain how he should handle these unusual expenditures, he decided to ignore them altogether.) Once we caught the omissions, he immediately agreed to provide the information, and shortly thereafter produced two pages of itemized expenses.
16. We made several attempts at estimating the bank balances of these households using such information as the per capita bank balance in the township, the size of the household, and what we knew about recent economic activities in the households. However, we decided that none of the estimates was sufficiently reliable to use.
17. In all the surveyed villages, the village accountant made annual estimates of per capita household income for each household in the village. The estimates of income received from the collective are probably fairly good since there are records but the estimates of income from individual activities and of remittances are only guesses, albeit educated guesses.
18. Village estimates of per capita household income (RMB) for the entire village and for the households included in the village sample are shown in the table.

Village	Entire village	Households in sample
Sigang Township		
Jingxiang	1,243	n.a.
Zhashang	1,313	1,467
Hufu Township		
Yangquan	1,210	1,662
Xiaojian	618	920

Village	Entire village	Households in sample
Haitou Township		
Haiqian	876	1,183
Haiqi	569	1,078

19. In these cases, rather than accepting a negative in kind consumption, we assumed zero consumption.
20. This was helped by the fact that the research was conducted over a period of three years so that we had time to develop a close and friendly relationship with those we interviewed.

Bibliography

JOURNALS AND REFERENCE WORKS CITED

Beijing Review.

China Quarterly.

Guomin shouru tongji ziliao huibian, 1949–1985 (Compilation of National Income Statistical Material, 1949–1985), compiled by SSB (Beijing: Zhongguo tongji chubanshe, 1987).

Jiangsu jingji nianjian (Economic Yearbook of Jiangsu), compiled by Jiangsu jingji nianjian bianji weiyuanhui (Nanjing: Nanjing daxue chubanshe).

Jiangsu nianjian (Jiangsu Almanac), compiled by Jiangsu nianjian bianzuan weiyuanhui (Nanjing: Nanjing daxue chubanshe).

Jiangsu sheng guomin jingji tongji ziliao (Jiangsu Provincial National Economic Statistical Material), compiled by Jiangsu sheng geming weiyuanhui jihua weiyuanhui (1976).

Jiangsu sheng nongcun jingji shouru fenpei tongji ziliao, 1986 (Statistical Material on the Distribution of Rural Income in Jiangsu Province, 1986), compiled by JPSB (Beijing: Zhongguo tongji chubanshe, 1987).

Jiangsu sishi nian (Forty Years of Jiangsu), compiled by JPSB (Beijing: Zhongguo tongji chubanshe, 1989).

Jiangsu tongji nianjian (Statistical Yearbook of Jiangsu), compiled by JPSB (Beijing: Zhongguo tongji chubanshe).

Jiangsu sheng 1986 nian nongcun zhuhu diaocha ziliao huibian (Compilation of Material on Survey of Rural Household, Jiangsu Province, 1986), compiled by Jiangsu sheng nongcun chouyang diaocha dui (1987).

Jingji tizhi gaige shouce (Manual of Economic System Reform), compiled by Wang Jiye and Zhu Yuanzhen (Beijing: Jingji ribao chubanshe, 1987).

1985 nian Jiangsu sheng nongcun zhuanyehu, xin jingji lianheti tongji ziliao (Statistical Material on Specialized Households and New Economic Associations in Rural Jiangsu, 1985), compiled by JPSB (Beijing: Zhongguo tongji chubanshe, 1986).

1978 nian Jiangsu sheng nongye tongji ziliao (Statistical Material on Agriculture, Jiangsu Province, 1978), compiled by JPSB (Beijing: Zhongguo tongji chubanshe, 1979).

1986 nian Jiangsu sheng xiangzhen qiye tongji ziliao (Statistical Material on Rural Enterprises, Jiangsu Province, 1986), compiled by Jiangsu sheng xiangzhen qiye guanliju (1987).

Quanguo beicun laodongli qingkuang diaocha ziliao ji (Collection of Material from the National Survey of Labour Conditions in 'One Hundred' Villages), ed. Yu Dechang (Beijing: Zhongguo tongji chubanshe, 1989).

Renmin ribao (People's Daily).

Statistical Yearbook of China (English Edition), compiled by SSB (Hong Kong: Economic Information and Agency).

Statistical Yearbook of the Republic of China, compiled by Republic of China, Directorate-General of Budget, Accounting, and Statistics.

The Globe and Mail.

The People's Republic of China, 1949–1979: A Documentary Survey, ed. Harold C. Hinton, 4 vols. (Wilmington, NC: Scholarly Resources Inc., 1980).

Xiangzhen qiye zhengce fagui xuanbian (Selected Laws and Regulations concerning Rural Enterprises), compiled by the Nong mu yu ye bu xiangzhen qiye ju (Beijing: Xinhua chubanshe, 1987).

Zhongguo jingji nianjian (Economic Yearbook of China), compiled by Zhongguo jingji nianjian bianji weiyuanhui (Beijing: Jingji guanli chubanshe).

Zhongguo gongye jingji tongji ziliao, 1949–1984 (China Industrial Economic Statistical Material, 1949–1984), compiled by SSB (Beijing: Zhongguo tongji chubanshe, 1985).

Zhongguo nongcun tongji nianjian (Statistical Yearbook of Rural China), compiled by SSB (Beijing: Zhongguo tongji chubanshe).

Zhongguo nongye de guanghui chengjiu, 1949–1984 (The Glorious Achievements of China's Agriculture, 1949–1984), compiled by SSB (Beijing: Zhongguo tongji chubanshe, 1984).

Zhongguo nongye nianjian (Agricultural Yearbook of China), compiled by Zhongguo nongye nianjian bianji weiyuanhui (Beijing: Nongye chubanshe).

Zhongguo tongji nianjian (Statistical Yearbook of China), compiled by SSB (Beijing: Zhongguo tongji chubanshe).

Zhongguo tongji zhaiyao (Statistical Survey of China), compiled by SSB (Beijing: Zhongguo tongji chubanshe).

Zhonghua renmin gongheguo xingzhenqu huatuce (Map of People's Republic of China by Administrative Areas) (Beijing: Zhongguo ditu chubanshe, 1987).

BOOKS AND ARTICLES CITED

ADELMAN, IRMA, and SUNDING, DAVID, 'Economic Policy and Income Distribution in China', in Bruce L. Reynolds (ed.), *Chinese Economic Reform How Far, How Fast?* (Boston: Academic Press, 1988).

American Rural Small-Scale Industry Delegation, *Rural Small-Scale Industry in the People's Republic of China* (Berkeley, Calif.: University of California Press, 1977).

ANDERSON, DENNIS, 'Small Industry in Developing Countries: A Discussion of Issues', *World Development*, 10/11 (1982), 913–48.

—— and LEISERSON, MARK, *Rural Enterprise and Nonfarm Employment* (Washington, DC: World Bank, 1978).

—— and —— 'Rural Nonfarm Employment in Developing Countries', *Economic Development and Cultural Change*, 28/2 (Jan. 1980), 227–48.

ASH, ROBERT F., 'The Peasant and the State', *CQ* 127 (Sept. 1991), 493–526.

—— 'The Agricultural Sector in China: Performance and Policy Dilemmas during the 1990s', *CQ* 131 (Sept. 1992), 545–76.

BLECHER, MARC, 'Peasant Labour for Urban Industry: Temporary Contract Labour, Urban–Rural Balance and Class Relations in a Chinese County', *World Development*, 2/8 (1983), 731–45.

BRUTZKUS, ELIEZER, 'Centralized Versus Decentralized Pattern of Urbanization in Developing Countries: An Attempt to Elucidate a Guideline Principle', *Economic Development and Cultural Change*, 23/4 (1975), 633–52.

BU RUIZHI, ZHANG ZAIMING, and DING HUA, 'Jiangsu nongmin shouru ji zhichu de shuliang fenxi (A quantitative analysis of peasant income and expenditure in Jiangsu)', *NJYS* 4 (1988), 13–19.

BUCK, JOHN L., *Land Utilization in China* (Chicago: University of Chicago Press, 1937).

BYRD, WILLIAM A., 'Entrepreneurship, Capital, and Ownership', in William A. Byrd and Lin Qingsong (eds.), *China's Rural Industry: Structure, Development, and Reform* (Oxford: Oxford University Press (for the World Bank), 1990).

—— and LIN QINGSONG (eds.), *China's Rural Industry: Structure, Development, and Reform* (Oxford: Oxford University Press (for the World Bank), 1990).

CAI LONG, 'Shi lun li tu ye li xiang (Leave the land and the countryside, an exploratory discussion)', *NJW* 7 (1985), 4–8.

CHAN, ANITA, and UNGER, JONATHAN, 1982. 'Grey and Black: The Hidden Economy of Rural China', *PA* 55/3 (1982), 452–71.

CHEN JIYUAN, 'Nongcun gaige shi nian de lilun qidi (Theoretical enlightenment from ten years of rural reform)', *NJYS* 6 (1988), 1–9.

CHEN KUAN, JEFFERSON, GARY, RAWSKI, THOMAS, WANG HONGCHANG, and ZHENG YUXIN, 'New Estimates of Fixed Investment and Capital Stock for Chinese State Industry', *CQ* 114 (June 1988), 243–66.

CHEN LIANGBIAO, 'Xiangzhen qiye yi gong bu nong de lilun jichu ji qi duice yanjiu (A study of the theoretical foundation for and the policy of using industry to subsidize agriculture)', *NJW* 3 (1987), 42–6.

CHEN SHENG, 'Cheng xiang jian laodongli heli liudong yu kaifang nongcun laodongli shichang (The rational movement of labour between cities and the countryside and the opening of rural labour-markets)', *NJW* 10 (1986), 12–15.

CHEN WAN (trans. J. A. Williams), 'Rural Reform in Southern Jiangsu Province', *China Spring Digest* (July/Aug. 1987), 14–16.

CHEN WENHONG, 'Zhongguo xiangzhen gongye de wenti (The problems of rural industry in China)', *Guangjiaojing*, 160 (Jan. 1986), 63–73.

CHENG XIANGQING, LI BAIJUN, and XU HUAFEI, 'Siying qiye fazhan mianlin de zhuyao wenti (Main problems confronting the development of private enterprises)', *NJW* 2 (1989), 24–6.

CCPCC, 'Zhonggong zhongyang guanyu 1984 nian nongcun gongzuo de tongzhi, zhongfa yihao (Circular of the Central Committee of the Chinese Communist Party on Rural Work during 1984, Document No. 1)', in *Nongcun gongzuo tungxun (Rural Work Communications)*, 7 (1984), 3–8. An English translation of this document is available in *CQ* 101 (Mar. 1985), 132–42.

—— and SC, 'Guanyu shixing zheng she fenkai jianli xiang zhengfu de tongzhi (Notice concerning the separation of government administration from the commune and the establishment of township governments)' (12 Oct. 1983), in *ZJN* (1984), 9. 9.

—— and —— 'Dangqian nongcun jingji zhengce de ruogan wenti (Some questions concerning current rural economic policies)' (2 Jan. 1983), in *XQZFX* 81–90.

—— and —— 'Guanyu kaichuang shedui qiye xin jumian de baogao (Report concerning the start of a new phase in commune–brigade enterprises)', No. 4 Document (1 Mar. 1984), in *XQZFX* 111–28.

—— and —— 'Guanyu 1986 nian nongcun gongzuo de bushu (Concerning the deployment of rural work in 1986)', Document No. 1 (1 Jan. 1986), in *XQZFX* 163–70.

CHUTA, ENYINNA, and LIEDHOLM, CARL, 'Rural Small-Scale Industry: Empirical Evidence and Policy Issues', in Carl K. Eicher and John M. Staatz (eds.), *Agricultural Development in the Third World* (Baltimore: Johns Hopkins University Press, 1984).

CONNER, ALISON E. W., 'China's Provisional Regulations Governing Private Enterprises', *East Asian Executive Reports* (Oct. 1988), 9–11.

—— 'Private Sector Shrinking under Intense Criticism and Increasing Controls', *East Asian Executive Reports* (Dec. 1989), 10–13.

DU WENZHEN and GU JIRUI (eds.), *Zhongguo renkou (Jiangsu fence)* (*China's Population—Jiangsu*) (Beijing: Zhongguo caizheng jingji chubanshe, 1987).

FEI XIAOTONG, 'Xiao chengzhen da wenti (Small towns, big issues)', in Jiangsu sheng xiao chengzhen yanjiu ketizu (ed.), *Xiao chengzhen da wenti* (*Small Towns, Big Issues*) (Jiangsu renmin chubanshe, 1984).

—— 'Xiao chengzhen zaitansuo (A further inquiry into small towns)', in Jiangsu sheng xiao chengzhen yanjiu ketizu (ed.), *Xiao chengzhen da wenti* (*Small Towns, Big Issues*) (Jiangsu renmin chubanshe, 1984).

—— *Rural Development in China: Prospect and Retrospect* (Chicago: University of Chicago Press, 1989).

FEWSMITH, JOSEPH, 'Rural Reform in China: Stage Two', *Problems of Communism* (July–Aug. 1985), 48–55.

FIELD, GARY S., *Poverty, Inequality, and Development* (New York: Cambridge University Press, 1980).

FIELD, ROBERT M., 'China's Industrial Performance since 1978', *CQ* 131 (Sept. 1992), 577–607.

GAMBLE, SIDNEY D., *Ting Hsien: A North China Rural Community* (New York: Institute of Pacific Relations, 1954).

GELB, ALAN, 'Workers' Incomes, Incentives and Attitudes', in William A. Byrd and Lin Qingsong (eds.), *China's Rural Industry: Structure, Development, and Reform* (Oxford: Oxford University Press (for the World Bank), 1990).

GOLDSTEIN, SIDNEY, 'Urbanization in China, 1982–87: Effects of Migration and Reclassification', *Population and Development Review*, 16/4 (1990), 673–702.

—— and GOLDSTEIN, ALICE, *Population Mobility in the People's Republic of China* (Honolulu: Papers of the East–West Population Institute, No. 95, 1985).

GRIFFIN, KEITH, 'Income Differentials in Rural China', *CQ* 92 (Dec. 1982), 706–13.

—— and SAITH, ASHWANI, *Growth and Equality in Rural China* (Bangkok: ILO-ARTEP, 1981).

GU JIRUI (ed.), *Xiangzhen qiye shouce* (*Handbook for Rural Enterprises*) (Beijing: Zhongguo qingnian chubanshe, 1985).

GU SONGNIAN and YAN YINGLONG, 'Xiangzhen qiye jingji lilun yanjiu zai shijian zhong kaituo qianjin (Research on the economic theory of rural enterprise as its development begins to move forward in practice)', *JY* 5 (1985), 55–61.

GU XIULIN, 'Jingji chengzhang zhong de nongye laodongli zhuanyi (The transfer of agricultural labour as the economy matures)', *NJYS* 5 (1988), 50–3.

GU YIKANG, 'Nongcun gongyehua bixu yu nongye jiyuehua tongbu fazhan (Rural industrialization and intensive agriculture must be developed simultaneously)', *NJW* 8 (1985), 7–10.

HAN BAOCHENG, 'Readjustment Improves Rural Enterprises', *BR* (27 Aug.–2 Sept. 1990), 17–25.

HARE, DENISE, 'Rural Non-agricultural Activities and their Impact on the Distribution on Income: Evidence from Farm Households in Southern China', unpublished paper (Nov. 1991).

HO, SAMUEL P. S., 'Decentralized Industrialization and Rural Development: Evidence from Taiwan', *Economic Development and Cultural Change*, 28/1 (1979), 77–96.

—— *Small-Scale Enterprises in Korea and Taiwan*, World Bank Staff Working Papers No. 384 (Washington, DC: World Bank, 1980).

—— 'Economic Development and Rural Industry in South Korea and Taiwan', *World Development*, 10/11 (1982), 973–90.

—— 'Rural Nonagricultural Development in Asia: Experiences and Issues', in Yang-Boo Choe and Fu-Chen Lo (eds.), *Rural Industrialization and Non-Farm Activities of Asian Farmers* (Seoul: Korea Rural Economics Institute/Asian and Pacific Development Centre, 1986).

—— *The Asian Experience in Rural Nonagricultural Development and its Relevance for China*, World Bank Staff Working Paper No. 757 (Washington, DC: World Bank, 1986).

—— 'Off-Farm Employment and Farm Households in Taiwan', in R. T. Shand (ed.), *Off-Farm Employment in the Development of Rural Asia* (Canberra: National Centre for Development Studies, ANU, 1986).

—— and HUENEMANN, RALPH, *China's Open Door Policy* (Vancouver: University of British Columbia Press, 1984).

—— GU JIRUI, YAN YINGLONG, and BAO ZONGSHUN, *Jiangsu nongcun fei nonghua fazhan yanjiu (Research on Rural Non-agricultural Development in Jiangsu)* (Shanghai: Shanghai renmin chubanshe, 1991).

HUANG FUNGYEA, *Urban Development and the Migration of Rural Labor in Mainland China*, Economic Papers No. 127 (Taipei: Chung-Hua Institution for Economic Research, 1989).

HUANG, PHILIP C. C., *The Peasant Family and Rural Development in the Yangzi Delta, 1350–1988* (Stanford, Calif.: Stanford University Press, 1990).

HUANG SHOUHONG, 'Xiangzhen qiye shi quomin jingji fazhan de tuidong liliang (Rural enterprises are the driving force in the development of the national economy)', *JY* 5 (1990), 39–46.

ISLAM, RIZWANUL, 'Non-Farm Employment in Rural Asia: Issues and Evidence', in R. T. Shand (ed.), *Off-Farm Employment in the Development of Rural Asia* (Canberra: National Centre for Development Studies, ANU, 1986).

—— (ed.), *Rural Industrialisation and Employment in Asia* (New Delhi: International Labour Organisation, Asian Employment Programme, 1987).

Jiangsu sheng zhexue shehui kexue lianhehui (ed.), *Sunan moshi xin tansuo (A New Exploration of the Sunan Model)* (Shanghai: Shanghai renmin chubanshe, 1987).

Jiangsu sheng shelian and Jiangsu sheng jingji xuehui, 'Wei xiangzhen qiye jinyibu jiankang fazhan kaituo daolu (Open a path for the further healthy development of rural enterprises)', *NYW* 4 (1986), 17–21.

JING WEI, 'Economic Newsletter', a 3-part series on economic development in Jiangsu, *BR* (14 Nov. 1983), 17–22 (part 1); (28 Nov. 1983), 17–22 (part 2); (12 Dec. 1983), 20–3 (part 3).

JING YANG, 'Guanyu xiangzhen qiye de "quanshu" wenti (Concerning the problem of "rights and ownership" in rural enterprises)', *JQ* 8 (1985), 17–18.

KADA RYOHEI, 'Off-Farm Employment and the Rural–Urban Interface in Japanese Economic Development', in R. T. Shand (ed.), *Off-Farm Employment in the Development of Rural Asia* (Canberra: National Centre for Development Studies, ANU, 1986).

Kai chuang xiangzhen qiye xin jumian (*Start A New Phase For Rural Enterprises*) (Beijing: Zhongguo shedui qiye baoshe and Nong mu yu ye bu shedui qiye guanliju, 1984).

KORNAI, JANOS, 'The Hungarian Reform Process: Visions, Hopes, and Reality', *Journal of Economic Literature*, 24/4 (1986), 1687–737.

KUEH, Y. Y., 'Economic Reform in China at the *"xian"* Level', *CQ* 96 (Dec. 1983), 665–88.

—— *Economic Planning and Local Mobilization in Post-Mao China*, Contemporary China Institute Research Notes and Studies No. 7 (London: School of Oriental and African Studies, University of London, 1985).

—— 'The Maoist Legacy and China's New Industrialization Strategy', *CQ* 119 (Sept. 1989), 420–47.

—— 'Foreign Investment and Economic Change in China', *CQ* 131 (Sept. 1992), 637–90.

LARDY, NICHOLAS R., *Economic Growth and Distribution in China* (London: Cambridge University Press, 1978).

—— 'Overview: Agricultural Reform and the Rural Economy', in *China's Economy Looks Toward The Year 2000*, i (Washington, DC: US Government Printing Office, 1986).

—— *Foreign Trade and Economic Reform in China, 1978–1990* (Cambridge: Cambridge University Press, 1992).

LATHAM, RICHARD J., 'The Implications of Rural Reforms for Grass-Roots Cadres', in Elizabeth J. Perry and Christine Wong (eds.), *The Political Economy of Reform in Post-Mao China* (Cambridge, Mass.: The Council on East Asian Studies, Harvard University, 1985).

'Learning the Rules of Foreign Trade', *China News Analysis*, 1464 (15 July 1992), 1–9.

LEE, YOK-SHIU F., 'Small Towns and China's Urbanization Level', *CQ* 120 (Dec. 1989), 771–86.

LI HONGRU and ZHANG NING, 'Jiangyou xian nongcun xin jingji lianheti diaocha (An investigation of new economic associations in the countryside of Jiangyou County)', *JD* 3 (1984), 5–9.

LI MENGBAI, 'Zhengque renshi he guanche woguo chengshi fazhan de jiben fangzhen (Correctly understand and implement our country's basic policy towards urban development)', presented at the National Conference on Urban Development (4 Dec. 1982), mimeo.

LI PENG, 'Report on the Outline of the Ten-Year Programme and of the Eighth Five-Year Plan for National Economic and Social Development', delivered at the Fourth Session of the Seventh National People's Congress on 25 Mar. 1991, *BR* (15–21 Apr. 1991), 1–24.

LI QINGZENG, 'Lun woguo nongcun laodongli de guosheng wenti (A discussion of the problem of surplus labour in rural China)', *NJW* 10 (1986), 8–11.

LIN JUHONG, 'Xin jiu shengchan guanxi jiaoti zhi ji de nongcun fuye (Rural sidelines at the transition between new and old production relations)', *NJYS* 3 (1988), 51–62.

LIN, JUSTIN YIFU, 'Rural Factor Markets in China After the Household Responsibility System Reform', in Bruce L. Reynolds, *Chinese Economic Policy Economic Reform at Midstream* (New York: Paragon House, 1989).

LIN ZHENG, 'The Discussion on Chinese Economic Development Strategy', *Chinese Economic Studies*, 25/1 (1991), 80–9.

LIU HUAZHEN, 'Tantan xiangcun jiti qiye de gufenzhi wenti (A discussion of the stock-share system for rural collective enterprises)', *NJW* 1 (1987), 29–31.

LIU SUINIAN and WU QUNGAN (eds.), *China's Socialist Economy: An Outline History, 1949–1984* (Beijing: Beijing Review, 1986).

LIU, YIA-LING, 'Reform From Below: The Private Economy and Local Politics in the Rural Industrialization of Wenzhou', *CQ* 130 (June 1992), 293–316.

LO, CHOR-PANG, 'The Geography of Rural Regional Inequality in Mainland China', *Transactions, Institute of British Geographers*, NS 15/4 (1990), 446–86.

LU GAOXIN, 'Nongcun geti gong shang hu weihe zai zheli da fudu xiajiang? (Why the large decline in individual industrial and commercial enterprises?)', *NJW* 5 (1987), 58–9.

Lun shedui qiye (On Commune-Brigade Enterprises) (Beijing: Zhongguo shedui qiye baoshe and Nongye bu renmin gongshe qiye guanli zongju, 1982).

LUO XIAOPENG, 'Ownership and Status Stratification', in William A. Byrd and Lin Qingsong (eds.), *China's Rural Industry: Structure, Development, and Reform* (Oxford: Oxford University Press (for the World Bank), 1990).

LUO HANXIAN, 'Wenzhou moshi yu shichang jingji (The Wenzhou model and market economy)', *NJW* 9 (1986), 10–12.

MA HONG (ed.), *Xiandai Zhongguo jingji shidian (A Record of Contemporary Chinese Economic Affairs)* (Beijing: Zhongguo shehui kexue chubanshe, 1982).

—— and SUN SHANGQING (eds.), *Zhongguo jingji jiegou wenti yanjiu (Research on the Problems of China's Economic Structure)*, 2 vols. (Beijing: Renmin chubanshe, 1981).

MARTIN, MICHAEL F., 'Defining China's Rural Population', *CQ* 130 (June 1992), 392–401.

MEI FANGQUAN, 'Ben shi jimo Zhongguo nongcun reng jiang shengyu yiyi laodongli (One hundred million surplus labourers in rural China at the end of this century)', *NJW* 12 (1986), 42–4.

MENG XIN, 'The Rural Labor Market', in William A. Byrd and Lin Qingsong (eds.), *China's Rural Industry: Structure, Development, and Reform* (Oxford: Oxford University Press (for the World Bank), 1990).

MO YUANREN, 'Jingying chengbao zeren zhi zengtian le shengji (Management contract responsibility system has increased in vitality)', *JD* 4 (1985), 68–74.

332   Bibliography

Mo Yuanren (ed.), *Jiangsu xiangzhen gongye fazhan shi* (*History of the Development of Rural Industry in Jiangsu*) (Nanjing: Nanjing gongxueyuan chubanshe, 1987).

Mukhopadhyay, Swapna, and Chee Peng Lim (eds.), Development and Diversification of Rural Industries in Asia (Kuala Lumpur: Asian and Pacific Development Centre, 1985).

Naughton, Barry, 'False Starts and Second Wind: Financial Reforms in China's Industrial System', in Elizabeth J. Perry and Christine Wong (eds.), *Political Economy of Reform in Post-Mao China* (Cambridge, Mass.: Council on East Asian Studies, Harvard University, 1985).

——'The Third Front: Defence Industrialization in the Chinese Interior', *CQ* 115 (Sept. 1988), 351–86.

Niu Renliang, Song Guangmao, and Ding Baoshan, '1988 nian yilai jinsuo de zongti xiaoying fenxi (An analysis of the overall effects of the retrenchment since 1988)', *JY* 5 (1990), 13–22.

Nong mu yu ye bu shedui qiye ju erchu, '1982 nian renjun fenpei shouru sanbai yuan yishan de fu xian he fu dui qingkuang (Circumstances of wealthy counties and brigades with per capita income above 300 yuan in 1982)', in Jingji diaocha bianjizu, *Jingji Diaocha* (*Economic Investigation*) (irregular serial), 2 (1983), 77–80.

——'Nongye shouru fanfan xian de tongji ziliao (Statistic material on counties whose agricultural income doubled)', in Jingji diaocha bianjizu, *Jingji Diaocha* (*Economic Investigation*) (irregular serial), 2 (1983), 72–6.

Nongcun diaocha bangongshi, 'Dui bai jia nongcun siying qiye diaocha de chubu fenxi (A preliminary analysis of a survey of 100 private enterprises)', *NJW* 2 (1989), 18–23.

——'Nongcun gaige yu fazhan zhong de ruogan xin qingkuang (Several new situations in the midst of rural reform and development)', *NJW* 3 (1989), 52–7; 4 (1989), 40–5.

Nongcun diaocha lingdao xiaozu, 'Quanguo nongcun shehui jingji dianxing diaocha qingkuang zonghe baogao (Summary report of the survey of typical social and economic areas in rural China)', *NJW* 6 (1986), 4–13.

'Nongcun gongfuye dui nongye de buchang ji qi dui zeren zhi de yingxiang (The use of rural sideline industry to subsidize agriculture and its effects on the responsibility system)', *JD* 2 (1983), 53–7.

Nurkse, Ragnar, *Problems of Capital Formation in Underdeveloped Countries* (Oxford: Basil Blackwell & Mott, Ltd., 1953).

Oi, Jean C., 'Peasant Grain Marketing and State Procurement: China's Grain Contracting System', *CQ* 106 (June 1986), 272–90.

——'Commercializing China's Rural Cadres', *Problems of Communism* (Sept.–Oct. 1986), 1–15.

Pan Shui, Zhou Yaozhong, and Zhu Huaming, ' "Yi gong bu nong" bixu jianchi ("To use industry to subsidize agriculture" must be upheld)', *NJW* 8 (1985), 17–19.

People's Government of Hufu Township, 'Hufu xiang guanyu xiangcun qiye renqi mubiao zeren zhi de yijian (Suggestions concerning the target responsibility system for TVEs during the [contractor's] period of tenure, Hufu Township)' (1988), Document No. 3.

——'Hufu zhen 1988 niandu baifen kaohe xize (Detailed standards, 100 point assessment, Hufu Township, 1988)' (1988).

PERKINS, DWIGHT H., 'Reforming China's Economic System', *Journal of Economic Literature*, 26/2 (1988), 601–45.

——and YUSUF, SHAHID, *Rural Development in China* (Baltimore: Johns Hopkins University Press (for the World Bank), 1984).

PERRY, ELIZABETH J., and WONG, CHRISTINE (eds.), *The Political Economy of Reform in Post-Mao China* (Cambridge, Mass.: Council on East Asian Studies, Harvard University, 1985).

POMFRET, RICHARD, 'Jiangsu's New Wave in Foreign Investment', *China Business Review* (Nov.–Dec. 1989), 10–15.

PRIME, PENELOPE B., 'The Impact of Self-Sufficiency on Regional Industrial Growth and Productivity in Post-1949 China: The Case of Jiangsu Province', Ph.D. diss., University of Michigan, 1987.

RAWSKI, THOMAS G., *Economic Growth and Employment in China* (Oxford: Oxford University Press (for the World Bank), 1979).

RISKIN, CARL, 'Small Industry and the Chinese Model of Development', *CQ* 46 (Apr./June 1971), 245–73.

——'China's Rural Industries: Self-reliant Systems or Independent Kingdoms', *CQ* 73 (Mar. 1978), 77–98.

——*China's Political Economy* (Oxford: Oxford University Press, 1987).

ROCCA, JEAN-LOUIS, 'Corruption and Its Shadow: An Anthropological View of Corruption in China', *CQ* 130 (June 1992), 402–16.

ROLL, CHARLES R., and YEH, K. C., 'Balance in Coastal and Inland Industrial Development', in US Congress, Joint Economic Committee, *China: A Reassessment of the Economy* (Washington, DC: US Government Printing Office, 1975), 81–93.

SCHUMPETER, JOSEPH, *The Theory of Economic Development* (Cambridge, Mass.: Harvard University Press, 1934).

SELDEN, MARK, *The People's Republic of China: A Documentary History of Revolutionary Change* (New York: Monthly Review Press, 1979).

SHAND, R. T., *Off-Farm Employment in the Development of Rural Asia*, 2 vols. (Canberra: National Centre for Development Studies, ANU, 1986).

SHEN LIREN, XU YUANMING, YAN YINGLONG, WU XIANMAN, CHEN SHENG, and XU BINGTAO, *Xiangzhen qiye yu guoying qiye bijiao yanjiu (Comparative Studies of Rural Enterprises and State Enterprises)* (Beijing: Zhongguo jingji chubanshe, 1991).

SHIH, J. T., 'Decentralized Industrialization and Rural Nonfarm Employment in Taiwan', *Industry of Free China* (Aug. 1983), 11–20.

SICULAR, TERRY, 'Agricultural Planning and Pricing in the Post-Mao Period', *CQ* 116 (Dec. 1988), 671–705.

Sigang Township, 'Sigang zhen changzhang renqi mubiao zeren zhi hetong shu (Contract of targets during the factory manager's tenure under the responsibility system, Sigang Township)' (Jan. 1988).

Sigang Township, 'Sigang zhen cun ganbu zeren zhi hetong (Contract under the responsibility system for village cadres in Sigang Township)' (1 Jan. 1988).

SIGURDSON, JON, *Rural Industrialization in China* (Cambridge, Mass.: Council on East Asian Studies, Harvard University, 1977).

SIMON, DENIS F., 'China's Drive to Close the Technological Gap: S&T Reform and the Imperative to Catch Up', *CQ* 119 (Sept. 1989), 598–630.

SINHA, RADHA, 'Rural Industrialisation in China', in E. Chuta and S. V. Sethuraman (eds.), *Rural Small Scale Industries and Employment in Africa and Asia* (Geneva: ILO, 1984).

SMIL, VACLAV, *The Bad Earth* (Armonk, NY: M. E. Sharpe, Inc., 1984).

SONG DAHAI, 'Xiangzhen qiye shixing zichan gufen zhi de diaocha (An investigation of the implementation of the capital stock-share system in rural enterprises)', *NJW* 1 (1987), 35–8.

SONG GUOQING, 'Cheng xiang kafang yu nongmin zhuanyi (Opening cities to the countryside and the shifting of peasants)', *NJW* 7 (1985), 9–12.

SONG LINA and DU HE, 'The Role of Township Governments in Rural Industrialization', in William A. Byrd and Lin Qingsong (eds.), *China's Rural Industry: Structure, Development, and Reform* (Oxford: Oxford University Press (for the World Bank), 1990).

'Sowing in Tears', *China News Analysis*, 1323 (1 Dec. 1986), 1–9.

STALEY, EUGENE, and MORSE, RICHARD, *Modern Small Industry for Developing Countries* (New York: McGraw-Hill, 1965).

State Council, 'Guanyu fazhan shedui qiye ruogan wenti de guiding (shixing caoan) (Regulations on some questions concerning the development of commune–brigade enterprises (draft for trial use)), 3 July 1979', in *ZJN* (1981), 2. 96–100.

——'Guanyu shedui qiye guanche guomin jingji tiaozheng fangzhen de ruogan quiding (Concerning various stipulations to be implemented by commune–brigade enterprises according to the direction of adjustment of the national economy)', in *ZJN* (1982), 3. 13–15.

——'Guowuyuan guanyu tiaozheng nongcun shedui qiye gongshang shuishou fudan de ruogan guiding (State Council regulations regarding the adjustment of the burden of industrial-commercial taxes on CBEs)', in *ZJN* (1982), 3. 55.

——'Guanyu jin jibu huoyue nongcun jingji de shi xiang zhengce (Ten policies to further enliven the rural economy)', in Wang Jiye and Zhu Yuanzhen (eds.), *Jingji tizhi gaige shouce* (*Manual of Economic System Reform*) (Beijing: Jingji ribao chubanshe, 1987).

SSTC, *Zhongguo kexue jishu zhengce zhinan kexue jishu baipishu dierhao* (*A Guide to China's Science and Technology Policies, Science and Technology White Paper No. 2*) (Beijing: Kexue jishu wenxian chubanshe, 1987).

——*Zhongguo kexue jishu zhengce zhinan kexue jishu baipishu disihao* (*A Guide to China's Science and Technology Policies, Science and Technology White Paper No. 4*) (Beijing: Kexue jishu wenxian chubanshe, 1990).

Su Zhe Yue xiangzhen qiye chenggong zhi lu (*The Successful Path of Rural Enterprises in Jiangsu, Zhejiang, and Guangdong*) (Guangzhou: Guangdong renmin chubanshe, 1985).

SURLS, FREDERIC M., 'China's Agriculture in the Eighties', in US Congress, Joint Economic Committee, *China's Economy Looks Toward The Year 2000*, i (Washington, DC: US Government Printing Office, 1986).

TAM, ON-KIT, 'Rural Finance in China', *CQ* 113 (Mar. 1988), 60–76.

TAO XIAOYONG, 'Xiangzhen qiye laodong yonggong moshi de jingjixue kaocha

(An economic investigation of employment patterns in rural enterprises)', *NJYS* 4 (1988), 20–5.

TAYLOR, JEFFREY R., 'Rural Employment Trends and the Legacy of Surplus Labour, 1978–86', *CQ* 116 (Dec. 1988), 736–66.

TONG NONG, 'Woguo nongmin shouru jiegou de xin bianhua (New changes in the income structure of Chinese peasants)', *JD* 4 (1985), 113–16.

TRAVERS, S. LEE, 'Bias in Chinese Economic Statistics: The Case of the Typical Example Investigation', *CQ* 91 (Sept. 1982), 478–85.

——'Getting Rich Through Diligence: Peasant Income after the Reforms', in Elizabeth J. Perry and Christine Wong (eds.), *The Political Economy of Reform in Post-Mao China* (Cambridge, Mass.: Council on East Asian Studies, Harvard University, 1985).

——'Peasant Nonagricultural Production in the People's Republic of China', in US Congress, Joint Economic Committee, *China's Economy Looks Toward The Year 2000* (Washington, DC: US Government Printing Office, 1986), 376–86.

VAN DER GAAG, JACQUES, *Private Household Consumption in China: A Study of People's Livelihood*, World Bank Staff Working Paper No. 701 (Washington, DC: World Bank, 1984).

VEECK, GREGORY, and PANNELL, CLIFTON W., 'Rural Economic Restructuring and Farm Household Income in Jiangsu, People's Republic of China', *Annals of the Association of American Geographers*, 79/2 (1989), 275–92.

VERMEER, E. B., 'Income Differentials in Rural China', *CQ* 89 (Mar. 1982), 1–33.

WALKER, KENNETH R., 'Trends in Crop Production', *CQ* 116 (Dec. 1988), 592–33.

——'40 Years On: Provincial Contrasts in China's Rural Economic Development', *CQ* 119 (Sept. 1989), 448–80.

WANG JIAN, 'The Correct Strategy For Long-term Economic Development— Concept of the Development Strategy of Joining the "Great International Cycle"', *Chinese Economic Studies*, 25/1 (1991), 7–15.

WANG JIANHUA, 'Xiangzhen qiye de suoyouzi wenti (On the problem of ownership of rural enterprises)', *NJW* 8 (1985), 14–16.

WATSON, ANDREW, 'The Reform of Agricultural Marketing in China since 1978', *CQ* 113 (Mar. 1988), 1–28.

——FINDLEY, CHRISTOPHER, and DU YINTANG, 'Who Won the "Wool War"?: A Case Study of Rural Product Marketing in China', *CQ* 118 (June 1989), 213–41.

WEIL, MARTIN, 'The Baoshan Steel Mill: A Symbol of Change in China's Industrial Development Strategy', in US Joint Economic Committee, *China Under the Four Modernizations*, part 1 (Washington, DC: US Government Printing Office, 1982).

WEITZMAN, MARTIN, *The Share Economy: Conquering Stagflation* (Cambridge, Mass.: Harvard University Press, 1986).

'Woguo difang zhongxiao gangtie jingji xiaoyi fenxi (Analysis of the economic effectiveness of small-medium iron and steel works in our country)', in Jingji diaocha bianjizu, *Jingji Diaocha* (*Economic Investigation*) (irregular serial), 2 (1983), 15–21.

Wong, Christine P. W., 'Rural Industrialization in the People's Republic of China: Lessons from the Cultural Revolution Decade', in US Congress, Joint Economic Committee, *China Under the Four Modernizations*, part 1 (Washington, DC: US Government Printing Office, 1982).

——'Central–Local Relations in an Era of Fiscal Decline: The Paradox of Fiscal Decentralization in Post-Mao China', *CQ* 128 (Dec. 1991), 691–715.

World Bank, *China, Rural Finance: A Sector Study* (Washington, DC: World Bank, 1982).

——*China, The Health Sector* (Washington, DC: World Bank, 1984).

——*China, Revenue Mobilization and Tax Policy* (Washington, DC: World Bank, 1990).

——*China, Macroeconomic Stability and Industrial Growth under Decentralized Socialism* (Washington, DC: World Bank, 1990).

Wu Rong and Wu Zhikang, 'Tan fazhan xiangzhen qiye zhong jige you zhengyi de wenti (Concerning several controversial problems in the development of rural enterprises)', *Xiangzhen jingji shouce* (*Handbook of Rural Economy*), 1 (1984).

——and Wu Defu, 'Suzhou shi fazhan nongcun chanqian chanhou fuwu de zuofa (Methods used in Suzhou *shi* to develop activities that provide services to agriculture), *JD* 4 (1985), 54–61.

Xia Zifen (ed.), *Shanghai xiangzhen qiye jingji keji fazhan zhanlue he zhangce wenti yanjiu* (*Research on Problems Concerning the Strategy and Policy for the Economic and Technological Development of Rural Enterprises in Shanghai*) (Shanghai: Shanghai shehui kexue yuan chubanshe, 1988).

Xiao He, 'A Dialogue on the "Great International Cycle"—Interview with Comrades Wang Jian and Pei Xiaolin, Authors of the Theory of the "Great International Cycle"', *Chinese Economic Studies*, 25/1 (1991), 28–33.

Xu Shanda, 'Guanyu xiangzhen qiye shuishou zhengce he shuifu wenti de sikao (Some thoughts on tax collection policy and tax burden of rural enterprises)', *NJW* 9 (1987), 46–51.

Yan Yinglong, 'Lun sunan xiangzhen gongye fazhan zhong de huanjing baohu wenti (On the problems of environmental protection as rural industry develops in *sunan*)', *Jiangsu jingji tantao* (*Jiangsu Economic Inquiry*), 6 (1988), 29–32.

——and Li Zongjin, 'Jiaqiang dui nongcun gongye de jihua guanli (Strengthen the planned management of rural industry)', *NJW* 1 (1984), 55–8.

Yang, Dali, 'China Adjusts to the World Economy: The Political Economy of China's Coastal Development Strategy', *PA* 64/1 (1991), 42–64.

Yang Mingguang, 'Yiyang diqu nongcun xin jingji lianheti de qingkuang (The situation of new economic associations in rural Yiyang)', *JD* 3 (1984), 9–16.

Yangquan Village, 'Jingji zeren hetong shu, Yangquan cun (Economic responsibility contract, Yangquan Village)' (1987).

——'Jingji zeren hetong shu, Yangquan cun (Economic responsibility contract, Yangquan Village)' (1988).

Yotopoutos, Pan A., and Nugent, Jeffrey B., *Economics of Development: Empirical Investigations* (New York: Harper & Row, 1976).

Yu Binghui, 'Zhengzhi, jingji, shehui dui Wenzhou moshi de zai kaocha (A re-examination of the political, economic, and social aspects of the Wenzhou model)', *NJYS* 2 (1988), 9–18.

YU GUOYAO and LI YANDONG, 'Dangqian xiangzhen qiye mianlin de zhuyao wenti (Major problems currently confronting rural enterprises)', *NJW* 10 (1989), 22–7.

YUAN MINGFA, 'Xiangzhen gongye shichang xiaoshou piruan de yuanyin ji duice (The causes and countermeasures for the market weakness faced by rural industry)', *Wuxi jingji* (*The Economy of Wuxi*), 4 (1990), 19–21.

Zenyang ban hao nong gong shang lianhe qiye (*How to Successfully Operate Agricultural–Industrial–Commercial Integrated Enterprises*) (Zhongguo shedui qiye bao, 1982).

ZHANG PINGQUAN, 'Xinjiang nongmin shouru fenpei chayi ji qi pingjia (An appraisal of income differentials among peasants in Xinjiang)', *NJW* 10 (1987), 39–43.

ZHANG RENSHOU, YANG XIAOGUANG, and LIN DAYUE, 'Wenzhou moshi de tese ji qi yiyi (The characteristics and the meaning of the Wenzhou model)', *NJW* 9 (1986), 4–9.

ZHANG YI, 'Zai xin xingshi xia dui xiangzhen qiye de zai renshi (Re-understanding rural enterprises under new circumstances)', *NJW* 1 (1990), 41–5.

ZHANG YIXIN, 'Guanyu nongcun jiating jingying fazhan he zhuanhua de tantao (A probe into the development and transformation of family production in rural areas)', *NJW* 7 (1987), 12–16.

Zhangjiagang Party Committee General Office, 'Guanyu 1988 nian xiangzhen jiguan ganbu zeren zhi kaohe yijian (Some suggestions about the assessment of cadres in township [party and government] organizations under the responsibility system)' (14 Mar. 1988).

Zhangjiagang Steel Cable Factory, *Zhangjiagang shi gang sheng chang lishi* (*History of Zhangjiagang Steel Cable Factory*) (Dec. 1987).

ZHAO YUJIANG, 'Management of Extrabudgetary Funds', in Bruce L. Reynolds (ed.), *Reform in China: Challenges and Choices* (Armonk, NY: M. E. Sharpe, 1987).

ZHAO HONG and CHEN XUN, 'Nonghu jingji xingwei fenxi (An analysis of the economic behavior of peasant households)', *JNW* 11 (1987), 53–7.

Zhashang Village, '1988 nian jingji chengbao hetong (Economic responsibility contract, 1988)' (Jan. 1988).

Zhonggong Foshan diwei bangongshi, 'Foshan diqu zhuanyehu zhongdianhu bai hu diaocha (An investigation of 100 *zhuanyehu* and *zhongdianhu* in the Foshan area)', *JD* 1 (1983), 64–73.

Zhonggong Wuxi shiwei zhengze yanjiushi, 'Wuxi xian fazhan, xiangzhen gongye de jiben jingyan (The basic experience of rural industry in the development of Wuxi County)', *NJW* 11 (1984), 43–7.

Zhongguo nongcun fazhan wenti yanjiuzu, 'Guomin jingji xin chengzhang jieduan he nongcun fazhan (The national economy in a new mature phase and rural development)', *JY* 7 (1985), 3–18.

Zhongguo nongkeyuan fu chuan diaocha baogaozu, 'Guanyu nongmin zhong liang jijixing wenti (On the problem of peasants' enthusiasm for growing grain)', *NJW* 8 (1986), 30–3.

Zhongguo shehui kexue yuan jingji yanjiusuo (ed.), *Zhongguo xiangzhen qiye de jingji fazhan yu jingji tizhe* (*The Economic Development and Economic System of China's Rural Enterprises*) (Beijing: Zhongguo jingji chubanshe, 1987).

ZHOU QIREN, BAI NANSHENG, and CHEN XIWEN, 'Xiangzhen qiye de zhengdun yu fazhan (The consolidation and development of rural enterprises)', *Renmin ribao* (*People's Daily*) (4 Oct. 1985).

ZOU FENGLING, 'Xiangzhen qiye chengbao hou de jige wenti (Several problems after the contracting out of rural enterprise)', *JQ* 3 (1986), 62–3.

ZOU JIAHUA, 'Report on the Implementation of the 1990 Plan for National Economic and Social Development and the Draft 1991 Plan, *BR* (22–8 Apr. 1991), 26–33.

ZWEIG, DAVID, 'Internationalizing China's Countryside: The Political Economy of Exports from Rural Industry', *CQ* 128 (Dec. 1991), 716–41.

Index